Complex-Valued Modeling in Economics and Finance

Sergey Svetunkov

Complex-Valued Modeling in Economics and Finance

Springer

Sergey Svetunkov
National Mineral Resources
University - Mining University
St. Petersburg, Russia

ISBN 978-1-4614-5875-3 ISBN 978-1-4614-5876-0 (eBook)
DOI 10.1007/978-1-4614-5876-0
Springer New York Heidelberg Dordrecht London

Library of Congress Control Number: 2012954285

Printed on acid-free paper

Springer is part of Springer Science+Business Media (www.springer.com)

Preface

As early as the mid-eighteenth century, mathematicians discovered a way to use models involving complex variables. Since then, work with complex variables has been progressing, with the theory of complex variables emerging as a branch of mathematics. Nowadays this theory is widely implemented in all of the natural sciences as work with complex variables makes it possible to describe adequately more complicated processes than do real variables. Economics as an object of scientific research and cognition is no less complicated than the natural sciences, which is why complex variables may be applied in economics to give a more precise description of the processes involved and to build even more complicated models than those that can be built using real variables. Complex variables are sometimes used in certain branches of economic and mathematical simulation, but this study considers application of models of economic and mathematical simulation exclusively in the form of models of complex variables.

Models and mathematical methods of working with complex variables are considered in the study not as some alternative to real-variable models and methods but as an instrument complementing and expanding the existing arsenal of economic and mathematical modeling. It is of principal importance that complex economics provides the economist with a new instrument of research, and the more flexible the instrument, the more diverse the tasks that can be solved by the researcher

The first chapter of this work presents basic principles of complex economics and certain data from the theory of functions of complex variables that are necessary for an understanding of further actions in the formation of complex economic theory.

Very often, understanding a certain meaning of mathematical operations requires graphic methods of describing these operations. Since in the study in question complex economics involves functions of complex variables, it is necessary to know the characteristics of these functions including their graphical representation. This task is solved in the second and third chapters of the study, which consider conformal mappings of basic functions of complex variables. These chapters do not simply state the respective branches of the theory of functions of a complex variable encountered in various textbooks. Textbooks on the theory of functions of a complex variable, in the section devoted to conformal mappings, do not consider, for instance, exponential complex functions with a complex coefficient; the need for such treatment has not existed. Textbooks consider exceptional conformal mappings of this exponential complex function with real exponent. For the purposes of complex economics, there is a need to use complex exponents not only for exponential functions.

The fourth chapter presents an instrument of practical application in complex economics – complex econometrics. The chapter provides only the basic principles of complex econometrics because it is practically impossible for a group of scientists working in this field to develop or adapt to complex econometrics all the branches of real-variable econometrics. Moreover, it is simply impossible to present the entire scientific discipline of econometrics in one chapter. Here we will substantiate and adapt to complex economics the basic sections of correlation and regression analysis of mathematical statistics – calculation of complex coefficients of pair correlation, least-squares method for evaluating coefficients of complex models, method of construction of confidence limits for obtained statistical estimates, and new coefficients showing the adequacy of econometric constructions. The results obtained are sufficient for solving subsequent tasks of complex economics and developing complex econometrics. The ideas of statistical characteristics of complex random variables currently in use in mathematical statistics has led to a deadlock. This can be seen in calculations of complex coefficients of pair correlation – the obtained contradictory results that follow from standard situations testify to their erroneous character. This made it necessary to devise other principles of statistics of complex random variables that underlie new and consistent conclusions and recommendations.

The fifth chapter contains the results of an investigation of one of the simplest types of economic models of complex variables – production functions of complex arguments, where actual production results depend on a complex argument, that is, production resources represented in the form of a complex variable. These functions possess some very important properties applicable in the successful solution of certain economic tasks. Here, we also demonstrate one remarkable feature of complex argument models – the sustainability of their assessments under multicollinearity.

Chapter 6 discusses production complex variable functions, more complicated models than complex argument ones. Here, the complex production result is represented in the form of dependence on a complex resource. Since functional relationships between two complex variables may have various forms, this chapter considers the basic ones.

The seventh chapter uses a case that has been insufficiently explored in the mathematics of complex variables; the existing theory operates with only one complex variable and is therefore called the theory of functions of a complex variable. Chapter 7 involves multifactor complex models, i.e., models of several complex variables. The development of the theory of functions of several complex variables in mathematics finds very little application in economics. This is why in the sixth chapter we state for the first time the principles and approaches of the theory of multifactor functions of complex variables. This was necessitated by our wish to build complex production functions that are more applicable to real economic processes, and that could be done only by increasing the explanation factors used in complex economic models. This chapter presents the properties and characteristics of simple multifactor complex models. The variety of possible applications of the models and methods of complex economics is not limited only to production function models. However, production functions provide a good example for seeing the advantages and disadvantages of complex models. In Chaps. 5 and 6 this is done by comparing such models with basic models of production functions with real variables.

Another good example that demonstrates the advantages of complex variable models is their application in the analysis of stock markets. Chapter 8 shows how to use complex indices of economic conditions and how to use the properties of complex numbers to obtain phase portraits of stock markets that could allow us to reveal laws that remain hidden with the use of real variables. The materials of Chap. 9 aim at showing other ways of developing complex economics than those specified in previous chapters.

In this study we make reference to the literature used in the form of footnotes. At the end of the study we give a complete list of all the publications by those scientists working on the formation of complex economics. If necessary, the reader may refer to these sources. The large volume of scientific results provided by this study could not have been obtained without the support of the Russian Foundation of Fundamental Research. The grants allocated by the foundation on a competitive basis from 2006 till 2010 rendered invaluable financial and moral assistance.

The main ideas, hypotheses, and materials stated in this study belong to the author; however, these hypotheses would never have become a well-balanced theory without the active involvement of a group of scientists, chief among whom was Dr. Ivan S. Svetunkov. This collaboration laid the foundations of complex economics on which basis many other scientific results were obtained. We are extremely grateful to Prof. G.V. Savinov who reviewed my first works, written in cooperation with I.S. Svetunkov, and later formulated a number of interesting proposals published in various articles. A very important contribution in the creation of various sections of complex economics was made by Dr. T.V. Koretskaya, E.V. Sirotina, and A.F. Chanysheva. Their work and the results obtained by them are discussed in the corresponding sections of the book. Some partial conclusions, recommendations, and new scientific results obtained by other young scientists are stated in various parts of the book.

Insofar as the materials presented in this study are new and this is the first time they have been systematized in this way and used for the proposed scientific purposes, the author understands that some points may be subject to debate or contain inaccuracies. It is every scientist's mission, having obtained a new scientific result, to encourage comprehensive scientific discussion and get to the truth. This is why any constructive criticism of the study is welcome. Comments and remarks may be mailed directly via www.sergey.svetunkov.ru.

St. Petersburg, Russia Sergey Svetunkov

Contents

Chapter 1
Theoretical Basis of Complex Economy

In the first chapter of the study we state axiomatic provisions of the theory of economic modeling using methods of the theory of functions of a complex variable (TFCV). Here we state the principles of economic modeling using complex variable functions, introduce the main concepts of the theory of complex economies, and explain the essence of the basic model of the new theory of economic modeling. We show that the methods of the new modeling theory are not alternative but complementary to the existing methods of economic and mathematical modeling due to new models and modeling methods. For the reader to understand the ideas of the study the first chapter provides the data of the complex variable function theory underlying the proposed economic modeling theory. The concept of *complex variable* is briefly explained, and its basic properties and the essence of mathematical operations are specified, as are certain data on the Minkowsky geometry and the Laplace transform.

1.1 Complex Economies as a New Branch of Economics and Mathematical Simulation

Problems of decision making that permanently arise in the course of economic management at any level of a hierarchy – from the workplace and production area to the global economy –require the processing of significant arrays of information to provide information support of the decision-making process. If at the lowest level of the decision-making hierarchy – i.e., in the workplace – it is enough to have intuitive expert assessments, since the decision-making problem is trivial and the quantity and composition of the information being processed with respect to the situation seem to be elementary, every higher level of the hierarchy involves ever more complicated tasks. Here it is not possible to do without mathematical methods that are becoming increasingly more complicated with the increase in the

S. Svetunkov, *Complex-Valued Modeling in Economics and Finance*,
DOI 10.1007/978-1-4614-5876-0_1, © Springer Science+Business Media New York 2012

complexity of problems, which changes for every next higher level of economic management.

Nowadays the arsenal of mathematical methods and models used in economics is quite vast. However, this variety by no means guarantees a successful solution of management problems. On the contrary, there are at least tens of directions of economics and mathematical modeling where models of real variables cannot exceed their capacities and describe economics somewhat incompletely. Under these conditions, economists either fully refuse to use mathematical methods or transfer economics and mathematical models from the sphere of practical solutions to the sphere of the theory of conditional objects having little to do with a real economy. In this case scientists must impose restrictions and assumptions that could transform these models from a set of abstract images into a set of idealized images possessing properties that no real economic object possesses. Let us take a model of "an eternal individual" as being similar to those in the annals of various theoretical branches of economics. The property of "eternal" life cannot be attributed to any real individual; moreover, it completely contradicts reality. The construction of such models and serious discussion thereof in scientific circles demonstrate an inability of modern science to make any progress in the solution of practical tasks of economic modeling.

The limitation of economics and mathematical models of real variables are clear. Attempts to develop them by including in the models new variables or complicating the computational tools using more powerful computational equipment is an important trend in the improvement of economics and mathematical modeling that must not be denied. However, nowadays there is a palpable need for other principles of economics and mathematical modeling, these principles being represented by TFCV.

It should be noted that economists have long faced situations where, during the development and implementation of certain models, they had to calculate imaginary roots. The bravest ones studied the behavior of these complex variable models as one of the most interesting phenomena in economic modeling but never went further. Nobody provided any practical recommendations and suggestions for a wider application of complex variables.

From the economic literature we know about attempts to apply the Laplace transform to economic modeling, where the simulation of complex processes described by real variable models involves their transformation into complex variable models that provide easier operation. A solution of the problem in the area of complex variables is followed by the inverse transformation into the area of real variables.

Laurent Z-transforms applicable to the problem of predicting socioeconomic dynamics were proposed by V.R. Semyonychev [1] as a modification of the discrete Laplace transform. In this case a nonlinear model is transformed by means of a Z-transform into a complex variable model, which provides reparameterization of the original model. This simplifies the evaluation of the coefficients of the original, nonlinear trend. Here a TFCV tool is used as an instrument, to support the application of real variable models in economics.

There are other particular examples of using complex variables for modeling special economic problems. However, they do not refer to the presentation of the economy as an object for modeling by TFCV methods.

It should be noted here that nowadays it is impossible to calculate anything in natural, engineering, and technical sciences without complex variables. Problems of hydro- and gas dynamics, the theory of elasticity, calculation of electric contours and electric transition processes, the physics of the micro and macro world, aircraft building, and many other branches of modern science use complex variables as the basic mathematical instrument of modeling. However, there is no such instrument in economics.

About 100 years ago scientists started using complex variable functions theory to describe nonuniform fields, to model complex flows, and to describe rotating fields, and they began obtaining complex-variable models that much more easily describe complex objects and phenomena than real-variable models. The TFCV providedscientists with a convenient instrument of complex-object modeling. However, economists still ignore the power and variety of the instruments of this theory.

I think that all of this is due primarily to the habit of using a number of general scientific methods and principles like the analogy method and the principle of simplicity.

The principle of simplicity teaches us to use simple models if there is no need to use more complex ones. The analogy method stubbornly gets scientists who think about the very possibility of using the TFCV in economics to search for rotating fields and the economic meaning of the real and imaginary constituents of a complex variable. Since it is convenient to investigate a model of an immortal individual, no results will carry any consequences for economic practice, but these models make it possible to conduct numerous modeling experiments, obtain various trajectories of the calculated variables giving them various names from among concepts in economics, certain scientists believe that the situation with the instruments of economic modeling is satisfactory and it is unreasonable to "multiply entities in excess of those needed" – it is sufficient to apply the available mathematical tools.

And if we try to use the analogy method to find situations in economics where complex variable models describe economic processes, such as, for example, transition processes that take place in AC circuits with rotating electromagnetic fields, then we will not find such situations.

Scientists who study economics and mathematical modeling miss an obvious fact – the complex variable itself may be considered as a model that characterizes the properties of an object in a more complex way as it consists of two real variables, not one as in real-variable models.

When we study an economic index like gross profit, G, we understand that it allows us to evaluate only one aspect of a complex economic phenomenon that represents the results of an economic process. It is no accident that when it comes to decision making, nobody will be satisfied with just maximum gross margin as a way of understanding a situation and making the right decision. It is only in modern economic theory that company behavior is explained on the basis of maximum gross profit. The real economy involves another, no less important, index – production costs C. Then, correlating the gross margin with the production costs we obtain profitability. Since it is profitability that shows both costs and results, i.e., is an index of the economic efficiency of the production process, it is used as another parameter for economic decision making.

In real economic practice, to describe some production process using models of real variables, scientists must simulate both the gross margin and the production costs. Since it is not convenient and rather costly to build two models, scientists build one model adding the gross income and the production costs, which give gross output, the very parameter considered in economic and mathematical modeling as the main production result.

The desire to simultaneously model two economic variables – gross income and production costs – is easily satisfied if we consider the production result as a complex number. In this case this complex number behaves as a model showing the production results. For the case under consideration it may be represented as follows:

$$z = G + iC \tag{1.1}$$

where i is an imaginary unit possessing the following property: $i^2 = -1$.

Considering and modeling the new number z we take into account both the gross income G and the production costs C as they are integral characteristics of the complex number. This means that in dealing with one complex variable, one will be dealing with two real variables. Therefore, the use of a complex variable like (1.1) as a model connecting two economic variables in one unit makes it possible, on the one hand, to obtain a more compact notation and, on the other hand, include in an economic and mathematical model more detailed information on the modeled object, and, in addition, consider them in interrelation.

But if these had been the only innovations introduced in economics and mathematical modeling by application of complex variables, then perhaps it would not have been worth the trouble. Economic parameters and processes simulated by means of complex variables are much more extensive than may appear at first glance. In fact, if we just add up the real and imaginary parts of variable (1.1) we obtain the known value – the gross margin:

$$Q = G + C \tag{1.2}$$

And if we find the ratio of the real to the imaginary part, we get the arctangent of the polar angle of the complex number (1.1) and ... profitability at cost:

$$r = \frac{G}{C} \tag{1.3}$$

Thus, modeling the behavior of only one complex variable makes it possible to study the character of change not only of the two original variables but also of a number of their additional derivative parameters. In the case under consideration we model four important economic parameters simultaneously.

But that's not all! Complex numbers may be presented not only in arithmetic but also in exponential and trigonometric forms. For that, a complex number considered on a complex plane is presented in polar coordinates. It is characterized by the absolute value and a polar angle. The absolute value of a complex number (1.1) determined as

$$R = \sqrt{G^2 + C^2} \tag{1.4}$$

is unique in the system of technical economic analysis and represents a new economic parameter showing the production scale. Its usage in practice may enlarge the diagnostic apparatus of, for example, a branch of economics such as the analysis of economic activity. The ratio of gross margin Q to scale R may also provide an additional production characteristic whose properties may be useful for economic analysis. There can be even more examples. Each application of complex-variable models provides ever new possibilities for more detailed economic modeling.

Thus, even a simple presentation of economic parameters and factors in the form of a complex number (1.1) provides many new possibilities for economic and mathematical modeling. However, mathematical operations with complex numbers provide nontrivial results compared to operations with real numbers. For that very reason there is a branch of mathematics called the theory of functions of a complex variable (TFCV). Use of this new mathematical tool in economics extends the instrumental basis of economic modeling since complex-variable models describe the interrelation between variables in a different way than real-variable models. Often it is easier to describe very complicated relations between real variables using TFCV models and methods than using models of real variables.

As follows from the TFCV, any complex function may ultimately be represented as a system of two functions with real variables, but these functions with real variables are often so complicated that they cannot be applied in practice – simple models of complex variables have very complicated analogs in the world of real variables. Appropriate examples will be given later in this study.

Why do complex-variable models appear to be more preferable to real-variable ones in simulating complex processes? What is the "sacred" meaning of this property?

To answer this question, let us look at a geometric interpretation of each number. A real number represents a point on a numerical axis having a zero point and going to + or – infinity (Fig. 1.1, point Y).

This real number is characterized by the distance from the zero point to this number. If the number is to the left of the zero point, it will be negative; if it is to the right of the zero point, it will be positive.

As follows from mathematical notation (1.1), a complex number represents a point not on an axis but on a complex plane. Therefore, to define it a given point on a complex plane unambiguously, it is not sufficient to have one characteristic.

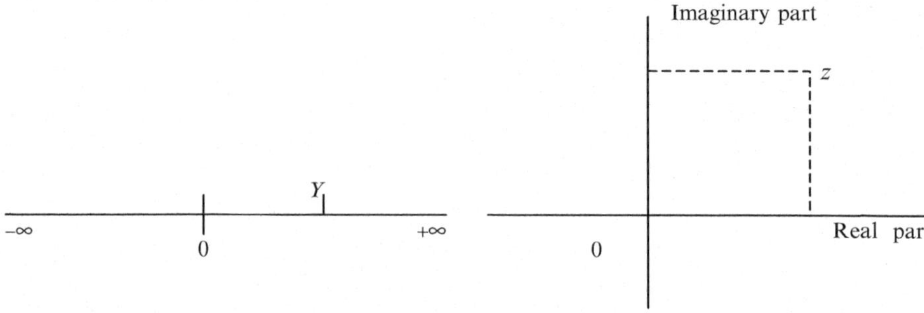

Fig. 1.1 Geometric interpretation of real (y) and complex (z) numbers

We need two coordinates – a segment on the axis of the real part and a segment on the axis of the imaginary part (Fig. 1.1, point z).

When we perform any mathematical operations with two real variables, we do it only with these variables, and when we perform similar operations with two complex numbers, for example, multiply one complex number by another, we perform a mathematical operation with four real numbers simultaneously.

It should be noted that the foregoing discussion does not mean that mathematical operations with complex numbers are better than the same operations with real numbers or that complex-variable models are better than real-variable models. It only means that mathematical operations with complex numbers give *different* results, whereas mathematical models of complex economic variables simulate different economic processes. In some cases models of complex variables will better describe economic processes than models of real variables; in other cases they will do it worse.

But it is evident that the presentation of a pair of economic parameters in the form of a complex number, as was done in the case of production result (1.1), presents economists with a new opportunity to use the theory of complex-variable functions for the purpose of economic simulation. In this theory, functions where variables are complex numbers are called *complex functions*. Since the material developed and proposed in this study demonstrates the application of the TFCV in economics and mathematical modeling, we propose a short but precise name – *complex economics* – as a general definition of this new branch of economics and mathematical modeling.

Thus, complex economics is a branch of economics and mathematical modeling where variables are complex values of economic parameters.

1.2 Basic Concepts of the TFCV

Since in economic and mathematical modeling complex variables have not been used as independent-variable operations that would allow for the formation of original economic and mathematical models, few economists are familiar with their properties and basic rules for dealing with them. This is why the object of this section is to state the basic concepts of the TFCV without which an untrained reader would not understand complex economics. Those who are familiar with the TFCV may skip this section.

In mathematics we often need to solve problems that do not have roots in the area of real numbers. For example, we need to find the root of the equation

$$x^2 + 4 = 0.$$

The solution gives the following roots:

$$x_{1,2} = \frac{\pm\sqrt{-16}}{2} = \pm 2\sqrt{-1}.$$

It is evident that in the area of real numbers this equation does not have solutions because there is no square root of minus one in this field. However, since impossibility of solving such problems leads to considerable limitations on computational capabilities, the so-called "imaginary unit" was introduced in mathematics, i.e., the number $i = \sqrt{-1}$. The square root of this number will definitely be equal to minus one: $i^2 = -1$.

Then the preceding equation has the following solution, which is imaginary:

$$x_{1,2} = \pm 2i$$

We can work further with the obtained imaginary number just as with the solution of the foregoing quadratic equation. But roots of a quadratic equation may not be solely real or imaginary. They may contain both real and imaginary parts. For example, if we solve the equation

$$x^2 + x + 2,5 = 0$$

then its roots, taking into account the introduced concept of "imaginary number," will be represented by two numbers:

$$x_{1,2} = \frac{1 \pm \sqrt{1 - 4 \times 1 \times 2,5}}{2}$$

The first root of the equation under consideration is

$$x_1 = 0,5 + 1,5i,$$

while the second one is

$$x_2 = 0,5 - 1,5i.$$

Because the roots of the equation represent a number consisting of two parts in which there are both a real and an imaginary part, we call it a complex number.

Thus, a complex number may be said to be a numerical pair consisting of two parts – a real and an imaginary one:

$$z = x + iy, \tag{1.5}$$

where x is a real part of the complex number, iy is its imaginary part, x and y are real numbers, and i is the imaginary unit that, as shown above, satisfies the equality

$$i = \sqrt{-1} \text{ or } i^2 = -1. \tag{1.6}$$

Complex numbers allow for the same operations as real ones. However, taking into account specific properties of the imaginary unit, these operations have an original character not inherent in operations with real numbers.

The main problem that an economist faces when representing an economic parameter in the form of a complex number is the complexity of the economic interpretation of the imaginary part. The main question may be stated as follows: where in a real economy would one meet imaginary numbers in general and imaginary units in particular? What is the meaning of an imaginary unit? The answer is that there is no meaning, neither economic nor physical. An imaginary unit is a mathematical rule, nothing more.

Where in real life do economists come across imaginary numbers and imaginary units? Nowhere! Where in real life could an economist encounter logarithms? Again, nowhere! There are no logarithms in the world around us. A logarithm is a mathematical rule that makes it easy to solve practical problems, including in economics. In the same way, an imaginary unit, which, as we have already mentioned, can be considered a mathematical rule, may be used to solve in a very convenient way a whole range of applied tasks in various spheres of human activity.

Using rules specified by conditions (1.5) and (1.6) we can use new mathematical operations, obtain new mathematical results, and form new mathematical models. It should be noted right away that in no sphere of the natural sciences are there processes where one can observe imaginary numbers or units. Complex numbers represent a mathematical model that may or may not describe some actually existing phenomena. If scientists decide to use complex variables for modeling real processes, they set the rules in advance according to which they always associate one component of a complex process to the real part and the other to the imaginary part of the complex variable.

In the same way, there are no phenomena in an economy that could be associated to a real or imaginary part because there are no phenomena where these parts are explicitly delineated. Like scientists in other fields, we will set rules according to which we will be able to represent economic phenomena in the form of complex numbers and complex variables. And wherever this representation of a complex socioeconomic object allows us to obtain a more precise description of its

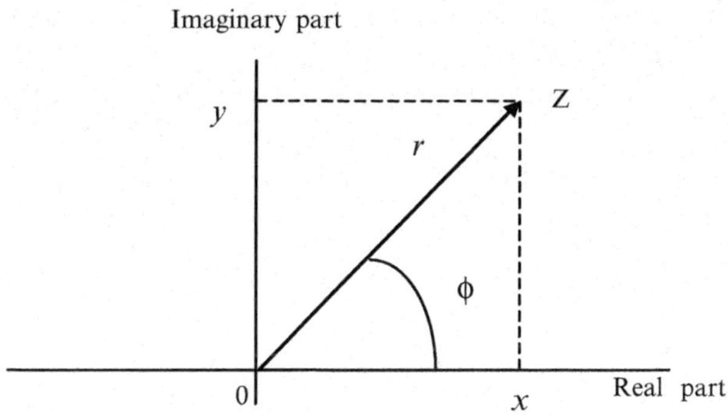

Fig. 1.2 A complex number (1.5) considered as a vector

properties, we will use models of a complex variable or several complex variables instead of models of real variables. Complex numbers may be represented graphically as in Fig. 1.1. The representation of a number on a plane but not on a numerical axis provides a number of new properties of complex numbers that are rather important for their further application in theory and practice, which is why we now turn to its graphical interpretation.

Since, unlike a real variable, a complex variable consists of two parts, it is these two parts that determine a complex plane. In the graph in Fig. 1.2 two axes are plotted that by definition are orthogonal – the axis of the real part and that of the imaginary part of a complex number. It should be pointed out that we have a plane in a Cartesian coordinate system and on its axes we place real numbers x and y. On the horizontal axis we place the real part of the complex variable and on the vertical axis its imaginary part.

Any point lying on a complex plane determined by the aforementioned axes represents a complex number, even if this point is on the axis of real numbers. In this case it represents a complex number with the imaginary part equal to zero.

In a Cartesian plane a complex number (1.5) may be considered a vector (Fig. 1.2) that comes from the point of origin and ends at the point (x, y). Then any complex number may be represented in polar coordinates by means of the vector length and the polar angle:

$$z = x + iy = r(\cos\varphi + i\sin\varphi). \tag{1.7}$$

Here, z is a polar angle (Fig. 1.2) and r a polar radius that in the present case is called the absolute value of a complex number (vector length). Clearly the absolute value of a complex number is equal to

$$r = \sqrt{x^2 + y^2}. \tag{1.8}$$

The polar angle may be found as

$$\varphi = arctg\frac{y}{x} + 2\pi k. \tag{1.9}$$

where k is a whole number. Sometimes a polar angle is called an argument of a complex number.

Two complex numbers are equal to each other only when their real and imaginary parts are equal. This means that, for example, the equation

$$y_r + iy_i = f_r(x_r) + if_i(x_i)$$

may be considered a more compact form of the following set of equations:

$$\begin{cases} y_r = f_r(x_r), \\ y_i = f_i(x_i). \end{cases}$$

The trigonometric form of a complex variable is especially convenient for multiplying complex numbers by each other. In addition to complex number (1.7), let there, be another complex number:

$$w = \rho(\cos\psi + i\sin\psi).$$

By multiplying z and w we will calculate the product of these variables. Skipping the elementary derivation we obtain the following result:

$$zw = r\rho(\cos\varphi + i\sin\varphi)(\cos\psi + i\sin\psi) = r\rho(\cos(\varphi+\psi) + i\sin(\varphi+\psi)).$$

This formula is known as Moivre's formula; it says that the absolute value of the product of complex numbers is equal to the product of modules, and its argument is equal to the sum of the arguments of the factors. Moivre's formula makes it much easier to perform such operations with complex numbers as involution and extraction of the square root of a complex number. In fact, to find the square of a complex number, we need to square its module and multiply the polar angle by 2.

In 1748, in his book "Introduction to the Analysis of Infinitesimals," L. Euler proved the formula bearing his name, that is [2]

$$e^{i\varphi} = \cos\varphi + i\sin\varphi. \tag{1.10}$$

Using Euler's formula any complex number z with module r and argument φ may be written in the following (exponential) form:

$$z = re^{i\varphi}. \tag{1.11}$$

This form is also very convenient for multiplying two complex numbers z_1 and z_2 and performing other operations. Indeed, using (1.11) again multiply complex number z by another complex number w:

$$z_1 \times z_2 = r_1 e^{i\varphi_1} r_2 e^{i\varphi_2} = r_1 r_2 e^{i(\varphi_1+\varphi_2)}. \tag{1.12}$$

Since the absolute value of a complex number may be presented in exponential form

$$r = e^{\ln r},$$

complex number (1.11) may be presented in another form, namely,

$$z = e^{\ln r + i\varphi},$$

which makes it possible to calculate the logarithms of a complex number.

Taking into account the fact that an argument of complex number (1.9) is determined up to a term that is a multiple of 2π, the logarithm of complex number z may easily be calculated as follows:

$$\ln z = \ln(e^{\ln r + i\varphi}) = \ln r + i\varphi + 2\pi k \tag{1.13}$$

That is, the logarithm of a complex number is a periodic function. As a rule, in practice we use the main logarithm value taking $k = 0$. The polar angle of complex number φ is called the argument of a complex number for short and is denoted by

$$\varphi = \mathrm{Arg} z\ . \tag{1.14}$$

The argument of a complex number is not determined uniquely but up to a term multiple of 2π:

$$\mathrm{Arg} z = \arg z + 2\pi k, \tag{1.15}$$

where k is a whole number and arg$z$ is the main value of the argument determined by the condition

$$-\pi < \arg z \leq \pi. \tag{1.16}$$

For brevity's sake and to formalize the language of mathematical problems, the name of the real part of complex number z is abbreviated and referred to as Re(z). The imaginary part is referred to as Im(z). Taking into account these designations it may be easy to specify some range in a complex plane z on which the set of points satisfies the condition

$$\mathrm{Im}\, z^3 > 4.5$$

To find this range, we should find the imaginary part of the complex variable (z^3) and then substitute the value found into the specified inequality. Incidentally, describing a set of points in a plane using, for example, a Cartesian coordinate system and applying real numbers is much more difficult and would involve a system of several nonlinear inequalities. It is evident that complex variables provide a compact representation of this problem.

A complex plane consists of various areas in which a function of a complex variable may be determined. The complex variable function is said to be set to

$$w = f(z) \tag{1.17}$$

if a law is specified according to which each point z from the range of permissible values is associated with a certain point or an aggregate of points w. In the first case, where there is a correlation with one point, the function (1.17) is called single-valued; in the second case, where there is a correlation of each point from z with a set of points from w, the function is called multivalued.

If $z = x + iy$ and $w = u + iv$, then setting the complex variable function $w = f(z)$ is equal to setting two functions of two real variables:

$$\begin{cases} u = u(x, y), \\ v = v(x, y). \end{cases} \tag{1.18}$$

Using complex-number theory one can provide the functional dependence for any pair of real numbers. Situations where one can establish a functional association for a pair of values are quite frequent in economics.

Thus, for example, such parameters as total costs involved in production C and gross income G may serve as the results of any production activity. Then the complex variable of production results may be represented in the following form:

$$G + iC.$$

Production resources used as the real subjects of economic activity are quite diverse. However, all this diversity in the theory of production functions is limited to two resources: capital K and labor L, which also may be represented as one complex variable:

$$K + iL.$$

So far we have not discussed what variables refer to real or imaginary parts of complex numbers. We will do this later on. The important thing is that using an imaginary unit we can to combine two economics parameters into one complex number. Just this presentation of economic parameters of production shows that the model connecting production resources with production results may have the form of a complex-variable function:

$$G + iC = f(K + iL). \tag{1.19}$$

Any model generated by this relationship (1.19) will describe a real production process with a certain degree of accuracy. And it will do so in a different way than models of real variables.

Besides these pairs of values, other pairs of economic parameters may also be identified. Linking them together by an imaginary unit we obtain complex economic variables. Using them to perform mathematical operations will give us other results than those that those obtained by economists today using models of real variables.

More often, in economics several parameters (more than two) depend on several factors (more than two). This is why we would like to simulate this dependence, i.e., to use a mathematical equation to correlate an aggregate of economic parameters with the aggregate of factors influencing them – to use hypercomplex numbers. However, the attempt to introduce a system of numbers containing three units did not yield positive results. We succeeded in building a system of numbers with four imaginary units. In this case we get a so-called system of quaternions, i.e., numbers such as

$$A = a + ib + jc + kd, \tag{1.20}$$

where a, b, c, d are real numbers and i, j, k are imaginary units.

Operations with quaternions are complicated, which does not allow for their application to any practical purposes and is still an area of idealized research. In the field of quaternions, the commutative property of multiplication does not work, which leads to various confusions. Thus, for the equation

$$x^2 + 1 = 0$$

there is an endless number of roots:

$$X = ip + jq + kr, \quad \text{where } p^2 + q^2 + r^2 = 1.$$

That is why the quite evident desire to describe the dependence of a certain complex indicator represented by quaternion (1.20) from another economic indicator represented in the form of another quaternion is not yet feasible.

In the present work we will use only complex numbers like (1.5) or consider functions of complex variables (1.17). If necessary, we will use the conclusions and proposals of the TFCV.

The last important property of complex variables that should be mentioned here is the concept of infinity. For real variables it is evident (Fig. 1.1). When the numerical axis tends to the right, to the area of positive numbers, this means plus infinity. If it tends to the left of the zero point to the area of negative numbers, it indicates a tendency to minus infinity. If we try to determine in the same way infinity for a complex variable, then we will fail as complex numbers are presented not on a numerical axis but in a complex plane, where each of the axes determining the complex plane tends to infinity. The axis of real numbers has both plus infinity and minus infinity, as does the axis of imaginary numbers. Then the question is how to define an infinite complex number.

This question was answered by B. Riemann. Let us consider a sphere S touching a complex plane at zero point [3]. Let P be a point of sphere S opposite the zero point. Each point z of complex sphere S is associated with point M, which is the point of intersection of sphere S with the segment connecting points z and P. The sequence $\{z_n\}$ converging to infinity corresponds to a sequence of points of sphere S converging to point P. This is why point $z = \infty$ is associated with point P on a Riemann sphere.

1.3 Axiomatic Core of the Theory of the Complex Economy

All theories are based on certain assumptions taken without proof. The first group of such assumptions refers to axioms, which are known to be taken without proof because they are obvious. The second group represents postulates, which are conclusions made in other branches of science and therefore taken without proof as the proof has already been done by other researchers.

In the theory of a complex economy let us postulate provisions of the theory of functions of complex variables taking its basic recommendations without proof; anyone who has doubts may always consult the numerous textbooks and scientific works on this branch of mathematics.

We will also postulate the main conclusions of theoretical economics or applied sections of economics. For example, we need not prove the interrelation between the wages of a worker and his productivity; this has already been done in the branch of economics that deals with labor management.

Let us present those provisions that seem evident and therefore do not require proof, i.e., are axiomatic, but without which complex economics as a theory cannot exist.

The first axiomatic provision says that practically all economic indicators used by economists to make judgments about the economy represents some generalized or aggregated values that may easily be presented as the sum of two terms that, with a certain degree of confidence, may be called the “active part” and the “passive part.” This idea is based on numerous classifications adopted in economics.

For example, according to a given classification, the labor resources of any enterprise may be divided into an active part (industrial and production personnel) and a passive part (nonproduction personnel). Or a family’s expenses may be divided into an active part, connected with the immediate satisfaction of current needs, and a passive part, connected with the satisfaction of future needs. The same may be said of the division of the final product of any country, which may be represented in the form of two constituents – consumption (active part) and accumulation (passive part).

Since the active and passive parts of some indicator or factor have different influences on other economic indicators, their total influence can very logically represented in the form of a complex variable, where the active component will be said to be its real part and the passive one will refer to its imaginary part.

In the aforementioned example of electrical engineering, for instance, in alternating current modeling, the active part will be real and the reactive imaginary. Alternating currents occur in a rotating electromagnetic field that creates electric current, voltage, capacity, and power variables in the conductor. Transmission of the electric power along some circuits encounters active and reactive resistance. This could seem like the semantic content of the real and imaginary parts of a complex variable. However, in practice, classification of the active part of electricity parameters as the real part of a complex number, as well as the classification of the reactive components as the imaginary part is conditional; one could just as easily swap them, that is, classify the active power as the imaginary part and the reactive as the real one. Actually, the very concepts of "active part" and "reactive part" represent a rule, a matter of preliminary agreement on the part of scientists. For example, if we classify an active current as the imaginary part of a complex variable and its reactive component as the real part, i.e., do the opposite of what is currently accepted in electric power engineering, the form of the applied mathematical models will change slightly but the computational process and, more importantly, the results will not change at all. It's just that when the theory of functions of complex variables was used for the first time in electrical engineering scientists agreed on how to classify each part. This has become a RULE that is now beyond question. Some scientists even seriously believe that according to its physical properties, the reactive part of electromagnetic power exactly matches the imaginary part of a complex variable. This is certainly not the case.

Similarly, let us, in the theory of the complex economy, agree that in a complex economy there is a RULE according to which the active part of an economic indicator will refer to the real part of a complex number and the passive one to its imaginary part.

Now we should mention several important conditions restricting the scope of complex numbers in economics. To use the tools of the theory of functions of complex variables in economics, a combination of two economic indicators in one complex variable should satisfy the following obvious conditions defined by the features of complex numbers:

1. These indicators should be two characteristics of the same process or phenomenon, i.e., they should show various sides of this phenomenon.
2. In addition, they should have the same dimension or be dimensionless. Also, they should be of the same scale.

Why is it necessary to take into account the first condition if, according to the rules, the real and imaginary parts are independent (orthogonal) of each other? Should they be considered two sides of the same coin? Yes, they should, because any complex variable that is formed from two real ones will further be considered an independent variable unit. Figuratively speaking, it carries within itself information about its two components and demonstrates the influence of each on some result. These values should reflect various aspects of the same phenomenon; otherwise their combination in one variable becomes senseless. These variables may be in a close functional dependence or be completely independent, but the main condition is that they should carry information about a certain common process. Such characteristics of a complex number as its module and argument make sense only when the components of the complex number reflect a common content.

The second condition requiring the same dimension of the components of a complex number stems from the specific properties of complex numbers. In fact, how can the absolute value of complex number (1.8) be calculated if the real and imaginary parts have different dimensions, for example, dollars and pieces? It is impossible to square them as $(\text{dollar})^2$ cannot be added to $(\text{piece})^2$. In the same way, when calculating the polar angle, we should find the ratio of the imaginary to the real part and then the arctangent of the result. If the real and imaginary parts have different dimensions, then nothing can be done as the tangent of an angle is a dimensionless value; it cannot be measured in dollars/pieces.

In economics, a considerable part of parameters may be transformed with monetary units of measurement, for example, labor costs can be determined not in "man-hours" but in the cost of wages – the size of the wage fund at the enterprise or subdivision. This is why for most real economic problems this condition seems quite feasible. However, when each of the variables has its own unit, they should be brought to some relative dimensionless values using the best method for the selected model.

1.4 Basic Model of a Complex Economy

Economic and mathematical models operating with real variables are based on the idea that some economic parameter y is represented as being dependent on another parameter x. This dependence may be described by means of a function when parameter x is correlated with only one parameter y:

$$y = f(x) \,. \tag{1.21}$$

Since each of the variables of the function (1.21) is an aggregate of real numbers that can be shown as an array of points on the numerical axis going through the zero point from minus infinity to plus infinity, and the distance from zero to this point is a number in the selected scale and shows the number itself, model (1.21) demonstrates that for each point on the real variable axis x there is one and only one point on the real variable axis y. These axes may be plotted on one plane in any order, for example, parallel to each other. However, the most informative is the location of these numerical axes when they cross each other at a right angle (perpendicular to each other) and their crossing point is the zero point on each of the axes. In this case the model may be examined in a Cartesian coordinate system.

Model (1.21) may be made more complicated in any number of ways, by adding new variables or by making it multifactorial. Then parameter y will depend on

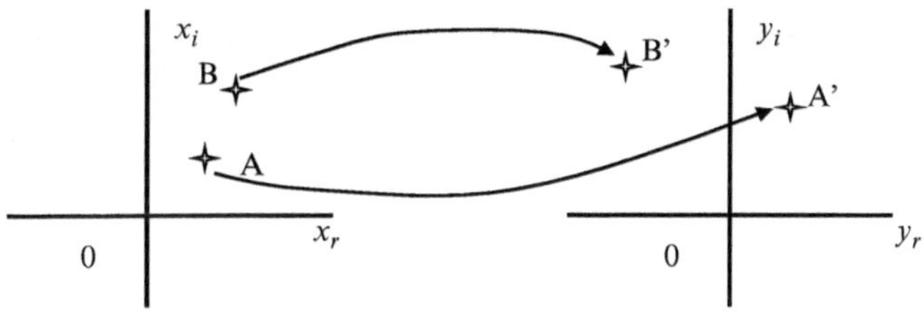

Fig. 1.3 Conformal mapping of points of complex plane x on complex plane y

several variables, and the graph of this dependence will be three-dimensional, four-dimensional, or generally multidimensional subject to the number of variables.

Function (1.21) is basic for constructing models in the area of real variables, its graphical interpretation on the Cartesian plane being its additional characteristic.

In the same way, in a complex economy we consider the basic model of dependence of one complex variable on another. If we know the properties of this dependence, then we can determine which processes it can describe, as well as switch to multifactor complex models and to systems of complex models. However, multifactor complex models and systems of complex equations are not as common as their analogs in the area of real variables.

A very basic model represents a functional dependence of one complex variable $y_r + iy_i$ on another $x_r + ix_i$:

$$y_r + iy_i = f(x_r + ix_i) \ . \tag{1.22}$$

In the TFCV, this function is called complex.

As any complex variable represents a point in a Cartesian coordinate system, equality (1.22) means that one point in a complex plane of variables x is correlated with a point (and in some cases several points) in the complex plane of variables y. This correlation is shown in Fig. 1.3.

In this figure, to the left we see a complex plane of variables x with two points A and B. According to rule (1.22) each of these points is correlated with some point in the plane of complex variables y. Point A is correlated with point A', and point B is correlated with point B'. Thus, using complex function (1.22) any set of points in complex plane x is mapped onto complex plane y. This is why a graphic representation of a complex function is called conformal mapping. This concept is quite suitable for purposes of the present study, though we could use a clearer mathematical definition of conformal mapping, for example, a *mapping of a neighborhood of point x_0 onto a neighborhood of point y_0 provided by function* (1.22) *is called conformal if in point x_0 it (the mapping) possesses a property to preserve the angles between lines and the property of permanent stretching* [4]. This means that not all the mappings of the points of one complex plane onto another plane by means of function (1.22) will be conformal, only those where the curves in the first plane transfer to curves in the other plane, so that the angle between the tangents to these curves in the first plane would correspond to some angle to tangents in the second

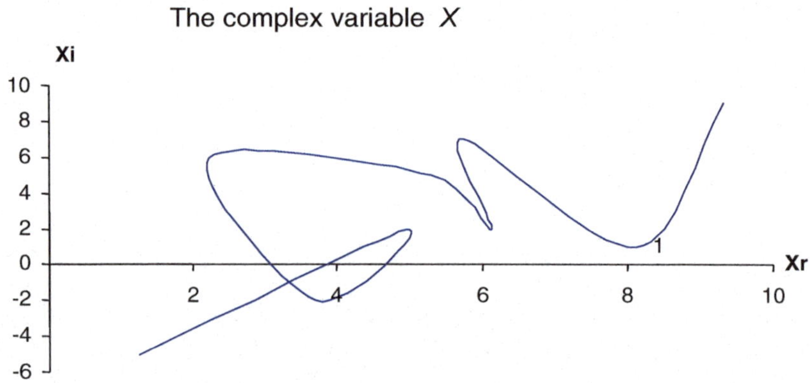

Fig. 1.4 A change in complex variable x

plane, and an infinitesimal circle with the center at point x_0 would correlate with an infinitesimal circle of the second complex plane with the center at point y_0. Complex functions to be considered in the present work possess this property, which is why we will consider conformal mapping as a graphical interpretation of the functional dependence between two complex parameters.

If by means of complex function (1.22) each point in complex plane x is correlated with one and only one point in complex plane y, this conformal mapping is called univalent. But in the TFCV we frequently face situations where, using complex function (1.22), each point in complex plane x is correlated with several points in complex plane y. As we have already mentioned, this function is called multivalued and a graphical representation of the conformal mapping is called multivalent. The phenomenon of multivalence is to be considered more in detail below, in the section where multivalence is revealed as a property of a complex function.

From Fig. 1.3 it is easy to see that a graphical visualization of functions of complex variables is worse than that of functions of real variables. An economist studying empirical data who uses graphical analysis of the conformity of real variables to each other gets an idea of the type and direction of this correlation and knows exactly what form – quadratic, exponential, etc. – the linear relationship has.

If he or she tries to study conformal mapping of one real changing complex variable onto the plane of another complex variable using some function and does this graphically, then most often he or she will not obtain any idea about a correlation between the complex variables.

Thus, for example, the graph in Fig. 1.4 shows a change in complex variable $x = x_r + ix_i$. Figure 1.5 shows a change in another complex variable $y = y_r + iy_i$.

The researcher knows that between these two complex variables there is some correlation whose character she does not know. Her task is to determine what form of complex function should be applied to model this correlation.

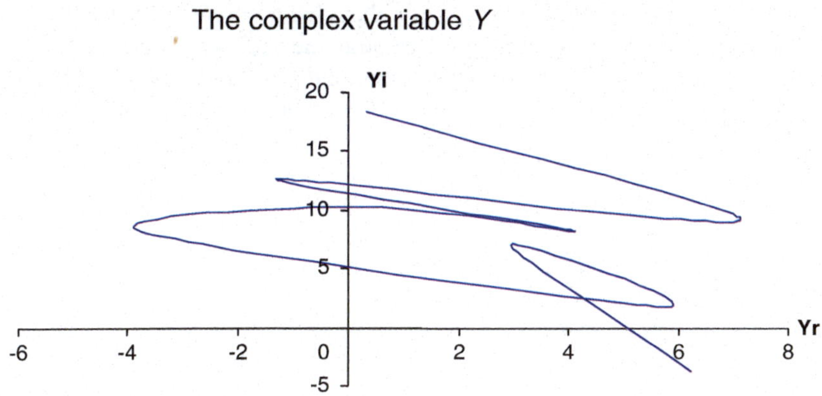

Fig. 1.5 Change in complex parameter y corresponding to change in complex parameter x in Fig. 1.4

A visual comparison of the graphs in Figs. 1.4 and 1.5 creates a stable impression that if there is a relationship between these two complex variables, it has a complex nonlinear character.

However, in fact we have a simple linear complex function like

$$y_r + iy_i = (1 + i)(x_r + ix_i).$$

This example clearly shows how difficult it is to use complex variable functions in practice, at least where a graphical interpretation of ongoing processes is required.

On the other hand, in modeling the economy, we more often deal with smooth trends of indicator changes, for example, a smooth trend of change in one complex indicator x will be transformed into a smooth trend of another complex indicator y if the same linear relationship exists between them that is not revealed visually in the graphs of Figs. 1.4 and 1.5.

According to the properties of complex numbers, basic model (1.21) can be represented as a system of two equalities with real variables:

$$\begin{cases} y_r = f_r(x_r; x_i) \\ y_i = f_i(x_r; x_i) \end{cases}. \tag{1.23}$$

For example, the simplest linear complex-value function

$$y_r + iy_i = (a_0 + ia_1)(x_r + ix_i) \tag{1.24}$$

can be represented as a system of two equalities:

$$\begin{cases} y_r = a_0 x_r - a_1 x_i \\ y_i = a_0 x_i + a_1 x_r \end{cases}. \tag{1.25}$$

It follows from the foregoing discussion that change in one of the indicators of complex variable x leads to a change in both the real and imaginary parts of complex variable y, that is, complex-value functions are multifactor by definition.

In the domain of real numbers we also encounter multifactor models, for example,

$$y = 5 + 3x_1 + 2x_2 \ . \tag{1.26}$$

However, in the domain of real variables, let us set the following problem for model (1.26): find a pair of values of influencing variables (x_1, x_2) for a given value y^*. This problem does not have just one solution, since from equality (1.26), for the given y^*, one can obtain an equation with two unknowns:

$$3x_1 + 2x_2 = y^* - 5 \ . \tag{1.27}$$

For complex-value function (1.22) this problem (in the absence of multivalence) can easily be solved, for example, for linear function (1.24), given known values of y_r^* and y_i^*, one can easily find only one pair of values of complex variable x:

$$x_r + ix_i = \frac{y_r^* + iy_i^*}{a_0 + ia_1} \ . \tag{1.28}$$

For example, if an enterprise wishes to find the best combination of production resources to obtain a certain profit, in the domain of real variables we should solve an optimization problem, and in a complex-value economy we should build an inverse function, as is done in (1.28).

The model of a complex argument is a special case of the basic model (1.22):

$$y_r = f(x_r + ix_i) \ . \tag{1.29}$$

This model can be represented in general form as a complex-value function where the imaginary part is equal to zero:

$$y_r + i0 = f(x_r + ix_i) \ . \tag{1.30}$$

This means that it is a function of real variables but presented in a complex form.

For a whole series of economic tasks the inverse function is of interest:

$$x_r + ix_i = f(y) \ . \tag{1.31}$$

It is a function of complex variables with a real argument. A simple example of this function may be as follows:

$$x_r + ix_i = y^{a_0 + ia_1}. \tag{1.32}$$

Here, a change in one variable entails a simultaneous change in two variables. If we model such a situation using real variables, then we should use a system of two equations. A complex economy provides compact modeling of complicated processes, and this is not its only advantage.

1.5 Some Data on Minkowsky's Geometry

Complex variables "open the door" to a fascinating world of various ideas about the world around us and its models. All the strength of this mathematical tool is demonstrated in theoretical physics, especially in the theory of relativity. Application of the TFCV gives the physicist a graphical interpretation of space curvature and time deceleration or acceleration. This interpretation is connected with the name of Minkowsky – an outstanding mathematician from Konigsberg.

Since this tool may also be used in economics, Minkowsky's geometry should be considered in the first chapter of a study containing the methodological foundations of a complex economy. The first successful attempt to apply the properties of Minkowsky's geometry to a solution of economic problems was taken up by Ivan Svetunkov to verify the adequacy of models in the real world. The main results of this approach are specified in Sect. 4.10.

First, it should be noted that so far the complex number $y_r + iy_i$ has been considered in the Euclidean plane, i.e., in a plane where we place real numbers on x and y axes. On the vertical axis of the Euclidean plane we deliberately wrote "imaginary part" or "iy_i" as in Fig. 1.6. The imaginary number is shown by means of a real number.

That is, a complex number consisting of a real and an imaginary part is represented as a real number in vector form. Then what is the difference between a complex number and a two-dimensional vector? Is there any point in using the theory of complex variable functions where one could use, for example, vector algebra? In this geometrical interpretation these questions are obvious, but there are no simple unambiguous answers.

To understand the power of the TFCV, we should switch from the Euclidean plane to a pseudo-Euclidean plane. To do that, on the horizontal axis we place real numbers (real part of a complex variable) and on the vertical axis an imaginary number (imaginary part of a complex variable). Then we obtain a complex plane (1.7), and points therein will have an absolutely different interpretation than points in a Euclidian plane. This difference means the following. For the Euclidean plane the scale of each of its axes (x- and y-axes) is determined by a one-unit-interval segment. On the horizontal axis, to the right of the zero point we place a "+1" segment, to the left a "–1" segment. In the same way, on the vertical axis, above the zero point we place a "+1", below a "–1." Then the coordinates of the one unit point are (1;1).

However, a complex number consists of a real and an imaginary part and a complex unit should have the coordinates (1;i). This is why along the horizontal

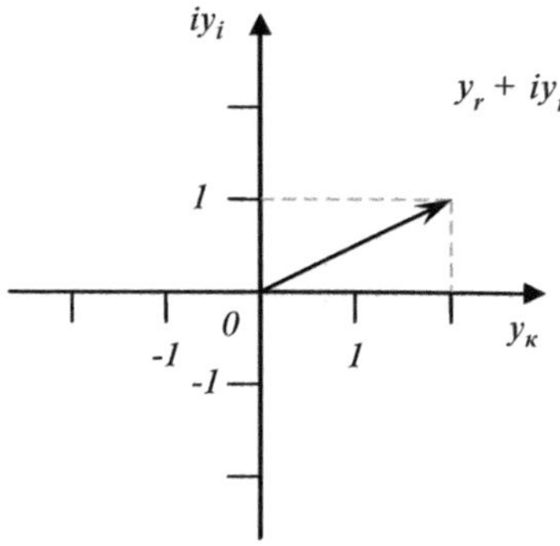

Fig. 1.6 Complex number in Euclidean plane

axis we should place real numbers and on the vertical axis not a "+1" or "–1" but imaginary numbers. Then in the complex plane, on the horizontal axis, we place a "+1" to the right and "–1" to the left of the zero point, but on the vertical axis we place "+*i*" above and "–*i*" below the zero point. Then any imaginary component of a complex number will be placed on this axis in the following way: we should multiply *i* by *y* *y* times. The principal difference in the geometric interpretation is that distances in this pseudo-Euclidean complex plane will be mapped in a completely different way than in the Euclidean plane; therefore, curves will also be mapped differently.

Let us ask one question: what distance does a vector drawn from the zero point to the point corresponding to the complex number $y_r + iy_i$ have?

Or, in other words, what length does it have? The obvious answer turns out to be incorrect. The length of the vector drawn from the zero point to the given one does not characterize the distance, as it does in the Euclidean plane. Since by definition this distance is equal to the square root of the sum of squares of the point coordinates, we obtain the following expression for the complex pseudo-Euclidean plane:

$$|z_t| = \sqrt{y_r^2 + (iy_i)^2} = \sqrt{y_r^2 - y_i^2}\,. \tag{1.33}$$

Thus distances and lengths in the complex pseudo-Euclidean plane have a different interpretation than the same characteristics of Euclidean planes.

Fascinating properties follow from formula (1.33). For example, when the absolute value of the real part of a complex variable is more than the absolute value of the imaginary part, the vector length is a real number. When the absolute value of the real part of a complex variable is less than the absolute value of the imaginary part, the radicand (1.33) becomes negative and the distance then becomes an imaginary number! And when the absolute value of the real part of a complex variable is equal to the absolute value of the imaginary part, if follows from (1.33) that the distance from the point to the origin of coordinates is equal to zero!

It is known that in the Euclidean plane only a zero vector [with coordinates (*0;0*)] can have zero length, and in the pseudo-Euclidean plane (Fig. 1.7) nonzero

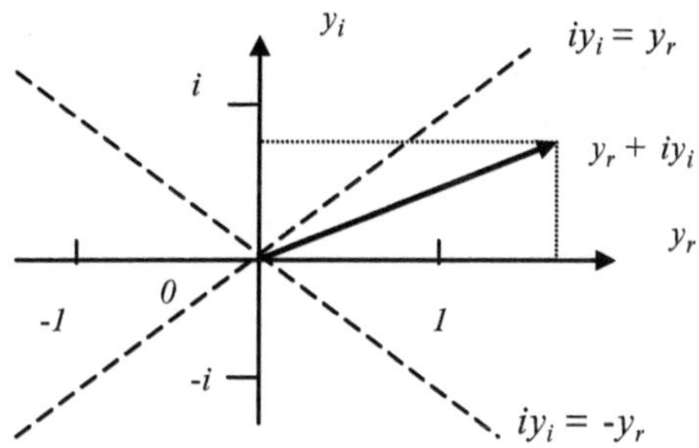

Fig. 1.7 Complex number in pseudo-Euclidean plane

vectors can have zero length. There are many vectors in a plane whose lengths are equal to zero, and all of them, as follows from (1.33), will satisfy the condition

$$|y_r| = |y_i| \tag{1.34}$$

or, as follows from the preceding expression, one of the following conditions:

$$y_r = y_i, \tag{1.35}$$

$$y_r = -y_i \;. \tag{1.36}$$

Vectors whose coordinates satisfy condition (1.35) or (1.36) lie on the corresponding axes in the pseudo-Euclidean plane and have zero lengths. These lines are isotropic. In Fig. 1.7 isotropic lines are shown as dotted lines. They divide the plane into four sectors. In this pseudo-Euclidean plane all the vectors with real lengths will lie either in the right or in the left sector, though vectors with imaginary lengths will lie either in the upper or in the lower sectors. In some economic problems this property may appear rather useful in the identification of various objects.

Now economists can understand how physicists imagine the curvature of space and time as well as the basis for the astonishing hypotheses regarding the instantaneous traversal of a point over huge distances in space because there are numerous points in the pseudo-Euclidean space where the distance between them is equal to zero! We do not insist on extending Minkowsky's geometry to the economic space, however, the mere possibility of using this new mathematical tool for economic modeling should be noted.

1.6 Laplace Transform

As was said at the very beginning of this study, the use of complex variables in economic modeling is negligible. In this smallest branch of economic and mathematical methods and models the section on the Laplace transform is the most popular. Laplace's transforms are actively used to solve systems of differential and integral equations, calculate transfer functions of dynamics systems, calculate transitional electromagnetic processes in electric contours, etc. This transform underlies so-called operational calculus whose tools may also be applicable in several economic problems. Let us show the essence of this transform to determine its applicability in economics.

It is obvious that multiplying a complex number by a real number we can obtain a new complex number. In the same way, multiplying any real variable function *f(x)* by some complex number we can obtain a complex function of a real variable:

$$(p_r + ip_i)f(x) = p_r f(x) + ip_i f(x) \ . \tag{1.37}$$

In general, there are many methods of transferring a real number or its function to the domain of complex numbers. For example, we can transform a real variable function f(x) into a complex function of a real variable by a more complicated method than multiplication, for example:

$$e^{(p_r+ip_i)f(x)} = Re^{i\varphi}, \ \ R = e^{p_r f(x)}, \varphi = p_i f(x). \tag{1.38}$$

However, any such transformation, a transfer of operations from the domain of real numbers to the domain of complex ones, should make sense – it should lead to new results or help solve some problems. That is why from the great diversity of such possible transformations the most popular is the so-called Laplace transform. Its essence is as follows.

Let there be a function *f(t)* in the domain of real variables about which we know that it satisfies the following conditions:

1. It is integrable in any finite interval of the time axis *t*.
2. It is equal to 0 for negative *t*, $f(t) = 0$ for $t < 0$.
3. If *t* increases, the absolute value of the function can also increase but not faster than some exponential function $|f(t)|e^{at}$.

This function is called *the original function.*

Let there also be some complex number *p*. We call Laplace's transform of real variable function *f(t)* a complex variable function *F(p)* determined by the formula

$$F(p) = \int_0^\infty f(t)e^{-pt}dt \ . \tag{1.39}$$

This function, resulting from transform (1.39), is called *the image function*. The correlation between the original function and the image function is written using various mathematical symbols. We will use the following notation:

$$f(t) \doteqdot F(t) \ . \tag{1.40}$$

It should be noted that for any original function image *F(p)* is determined in the semiplane

$$\mathrm{Re}\, p > M\ . \tag{1.41}$$

Laplace's transform turned out to be a very efficient tool for solving numerous complex mathematical problems using differential and integral equations since the main operations with the original functions correspond to mathematical operations with image functions that are easier than those with the originals.

It would not make sense to go into more detail regarding Laplace's transform and the rules and properties of the interrelations of the original functions and the images as there are many specialized publications on operational computation, let us concentrate here on demonstrating the use of Laplace's transform in solving problems.

For example, say we need to solve the differential equation

$$x' - x = 1 \tag{1.42}$$

for initial conditions $x(0) = 1$ (Kochi's problem).

To solve this problem, let us use Laplace's transform. We know that the original unit function

$$f(t) = \{0,\ \text{if}\ t < 01,\ \text{if}\ t \geq 0(1.43)$$

corresponds to the image function

$$F(t) = \frac{1}{p} \tag{1.44}$$

By means of Laplace's transform, for any constant $\omega > 0$

$$f(\omega t) \doteqdot \frac{1}{\omega} F\left(\frac{p}{\omega}\right) \tag{1.45}$$

and the derivative of any image function has the following mapping in a complex plane:

$$f'(t) \doteqdot pF(p) - f(0) \tag{1.46}$$

Taking into account all this, a mapping of the problem of the solution of a differential equation of real variables (1.42) to the domain of complex variables using transforms (1.44)–(1.46) will have the following form:

$$pX(p) - f(0) - X(p) = \frac{1}{p}\ . \tag{1.47}$$

This means that we have an elementary algebraic equation for $x(p)$ instead of a differential equation (1.42). Substituting the initial conditions $x(0) = 1$ into this algebraic equation we get a simple solution:

$$(p-1)X(p) - 1 = \frac{1}{p} \rightarrow X(p) = \frac{p+1}{p(p-1)} = \frac{2p-(p-1)}{p(p-1)} = \frac{2}{p-1} - \frac{1}{p}. \quad (1.48)$$

From the image obtained it is easy to move on to the originals since it is known that an exponential function has the following mapping:

$$e^{\lambda t} \doteqdot \frac{1}{p-\lambda} \quad (1.49)$$

and it is the first term on the right-hand side of (1.47) that has this form.

Then the original of the first term will be written as follows: $2e^t$.

The second term on the right-hand side of solution (1.48) represents just the image of the unit function (1.43).

Then the following will be the solution to the original differential equation (1.42):

$$x(t) = 2e^t - 1. \quad (1.50)$$

This example can easily persuade us that mathematical operations with images of various functions of real variables meeting the aforementioned requirements are much easier than operations with the originals. And since Laplace's transform for a wide variety of functions of real variables that can serve as original functions were calculated long ago and tabulated, the solution to the most diverse and complex problems of operational calculus is considerably simplified.

In economics, use of differential equations, much less systems of differential and integral equations, is quite rare and is found mainly in the construction of idealized mathematical models, but when it does happen, scientists achieve success when they use Laplace's transforms.

References

1. Semyonychev VK, Semyonychev EV (2006) Information systems in economics. Econometric modeling of innovations. Part I. Publishing house of Samara State Aerospace University, Samara
2. Euler L (1961) Introduction in analysis of infinitesimals, vol 1. State Publishing House of Literature in Physics and Mathematics, Moscow
3. Shabunin MI (2002) Theory of functions of complex variable. Unimediastyle, Moscow
4. Krasnov ML, Kiselev AI, Makarenko GI (2003) Complex variable function: problems and examples with detailed solutions. Editorial URSS, Moscow

Chapter 2
Properties of Complex Numbers of a Real Argument and Real Functions of a Complex Argument

Important results can be obtained if we apply simple complex-value models in economic modeling – complex functions of a real argument and real functions of a complex argument. This chapter focuses on the properties of these models and the possibility of using them in economic practice.

Complex models of a real argument represent the dependence of a complex variable on a real argument. This dependence can be obtained only if one uses a function that transforms real variables into complex ones. The Laplace transform is a well-known transformation method; however, this chapter focuses on other methods widely applied in economics. Real models of complex argument solve another problem – the transformation of a complex variable into a real one. The properties of the simplest models of this type are considered in this chapter with respect to economic modeling.

2.1 General Problem of Conformal Mapping in Complex-Valued Economics

Before using a tool of the theory of functions of a complex variable (TFCV) in economics it is necessary to study the properties of this tool [1]. One of the methods for understanding these properties is provided by conformal mapping of points from one complex plane to another. With reference to various cases of the TFCV, conformal mapping provides problems of varying degrees of complexity. We will consider the simplest cases since an understanding of conformal mappings of elementary complex-valued functions will allow researchers to choose the proper complex-valued function for modeling.

Thus, we can say that conformal mapping is a convenient graphical method for understanding how, by means of a given function, one complex variable in a complex plane of an argument is mapped to another complex variable modeling the value of the variable of the complex result.

S. Svetunkov, *Complex-Valued Modeling in Economics and Finance*,
DOI 10.1007/978-1-4614-5876-0_2,

Since we work in the sphere of complex numbers, any real number may be represented as a complex number with zero imaginary part. Then we obtain three types of functions to be used in economic modeling.

The first type represents the relationship between a complex variable and a real argument:

$$y_r + iy_i = F(x_r + i0) = f(x_r) + if(x_r). \tag{2.1}$$

It is a complex function of a real variable.

The second type comes up when a complex argument is associated with a real result:

$$y_r + i0 = F(x_r + ix_i) = f_r(x_r, x_i) + if_i(x_r, x_i) \Leftrightarrow \begin{cases} y_r = f_r(x_r, x_i), \\ f_i(x_r, x_i) = 0. \end{cases} \tag{2.2}$$

It is a function of a complex argument.

The third type is the relationship between a complex variable and a complex result:

$$y_r + iy_i = F(x_r + ix_i) = f_r(x_r, x_i) + if_i(x_r, x_i). \tag{2.3}$$

It is a complex function (a function with complex values).

The TFCV considers mainly conformal mappings of the third type. However, in economics we can use all three as models of a complex-valued economy. This is why it is essential to examine in depth the properties of all three types of functions. This chapter will focus on the properties of the first two types.

2.2 Complex Functions of a Real Argument

The complex functions of a real argument represent a certain "mapping" of a set of real numbers on a numerical axis to the plane of complex variables:

$$y_r + iy_i = F(x + i0) = f(x) + if(x). \tag{2.4}$$

This function transforms real variables and the respective functions to complex variables and the respective functions.

Situations where one variable influences two others are quite frequent in economics. For example, in marketing, consumers are grouped in particular categories – segments where the basic indicator is a similar reaction of all consumers of this segment to a product and its marketing support. This means that consumers with similar levels of income (if we categorize by income) will react similarly to a given price and buy the same quantity of the product at that price. This in turn means that the price y_r and the consumption volume y_i depend on the level of income x. With this knowledge, one can look at the reaction to goods by consumers from various segments as being subject to an increase in income of each segment and model this reaction by a function of a real argument (2.4).

The variety of possible functions of a real argument that may be put forth to model the aforementioned economic processes is limited only by the imagination of the researcher creating the model. This is why in this section we deal only with the simplest functions and their properties.

A linear model of a real argument,

$$y_r + iy_i = (a_0 + ia_1) + (b_0 + ib_1)(x + i0) = (a_0 + b_0x) + i(a_1 + b_1x), \qquad (2.5)$$

is of little interest because any change in the argument entails a directly proportional change in the real and imaginary parts of the complex result. This means that for any change in the real argument – linear or nonlinear – we have a line in a complex plane whose slope and position thereon is completely determined by the values of a complex proportionality coefficient.

Nonlinear transformations of a real variable to a complex plane are of practical interest. The first of these methods is the complex involution of a real argument:

$$y_r + iy_i = (a_0 + ia_1)x^{(b_0+ib_1)}. \qquad (2.6)$$

The proportionality coefficient that can be placed before the argument that is subject to involution can be not only complex, but also real or imaginary. Let us represent the complex function of a real argument in exponential form, then (2.6) will be written as

$$y_r + iy_i = \sqrt{a_0^2 + a_1^2}x^{b_0}e^{i(arctg\frac{a_1}{a_0}+b_1\ln x)}. \qquad (2.7)$$

To simplify the notation of the complex proportionality coefficient, let us write it as $A = \sqrt{a_0^2 + a_1^2}$ and the polar angle as $\alpha = arctg\frac{a_1}{a_0}$.

The equality of the real and imaginary parts of this equation may be represented as the following system:

$$\begin{cases} y_r = Ax^{b_0}\cos(\alpha + b_1\ln x), \\ y_i = Ax^{b_0}\sin(\alpha + b_1\ln x). \end{cases} \qquad (2.8)$$

It is clear that both the real and imaginary parts of this complex real variable function change with an increase in the argument according to the cosine (real part) and sine (imaginary part) law. Taking into account the fact that the real argument in these trigonometric functions is not direct but a logarithm, with a uniform increase in the real argument, periods of oscillation of both the real and imaginary parts of the function under consideration will increase. A logarithm limits the function domain;

since a logarithm of zero does not exist, the zero point is not included in the function domain.

If we consider the result (2.7) in the complex plane, the points of this function will be located as follows. The module of this function

$$r = Ax^{b_0} \tag{2.9}$$

will increase with an increase in the argument $x > 0$ for $b_0 > 0$ and decrease for $b_0 < 0$, and the polar angle will increase,

$$\varphi = \alpha + b_1 \ln x, \tag{2.10}$$

if $b_1 > 0$ and decrease (move in a clockwise direction) if $b_1 < 0$.

Hence, it is easy to see that in the complex plane function (2.6) is mapped subject to the values of the complex exponent in the form of a convergent or divergent spiral.

Let us consider a special case of function (2.6), where time t acts as the argument:

$$y_r + iy_i = (a_0 + ia_1)t^{(b_0 + ib_1)}. \tag{2.11}$$

This function represents a complex trend and may be used in practice in certain economic situations.

As follows from the aforementioned properties of the function under consideration, the character of a complex trend will be fully determined by its coefficients. Here are some interesting types of such trends.

Thus, if we use the trend

$$y_{rt} + iy_{it} = t^{(-0,5+i10)}, \tag{2.12}$$

then each of the components of the complex-valued trend will look like Figs. 2.1 and 2.2.

The same form of the trend but with other coefficients

$$y_{rt} + iy_{it} = t^{(0,25+i0,35)} \tag{2.13}$$

models completely different dynamics (Figs. 2.3 and 2.4).

Trends like those shown in Figs. 2.3. and 2.4 are quite frequent in the domain of real variables; however, models describing the dynamics of trends like 2.1 and 2.2 are quite rare in studies on socioeconomic processes, except for stock markets.

The next model of a real argument may be a complex exponential function of the real argument. It may be presented as follows:

$$y_r + iy_i = (a_0 + ia_1)e^{(b_0 + ib_1)x}. \tag{2.14}$$

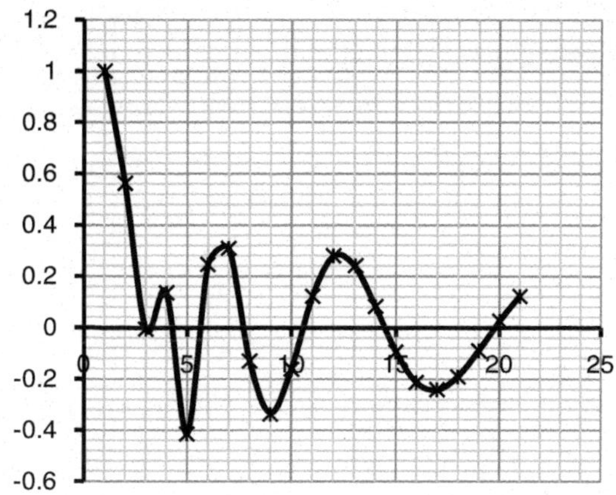

Fig. 2.1 Dynamics of real part of complex trend (2.12)

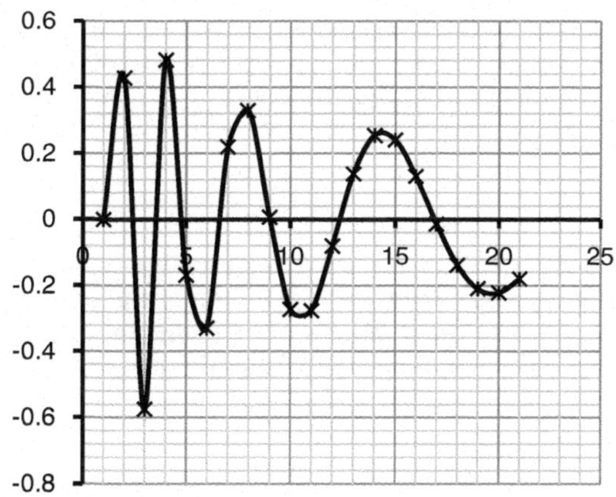

Fig. 2.2 Dynamics of imaginary part of complex trend (2.12)

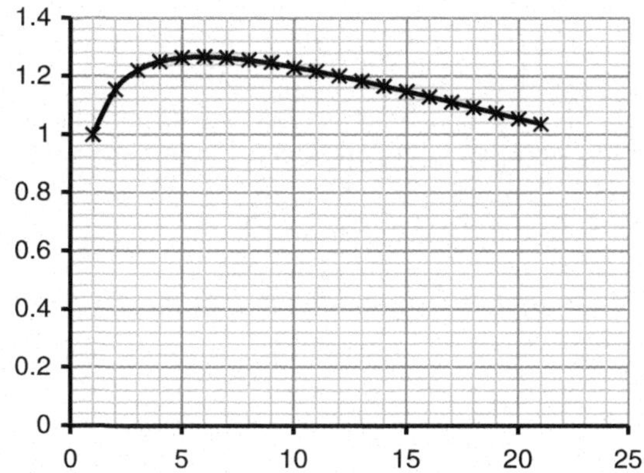

Fig. 2.3 Dynamics of real part of complex trend (2.13)

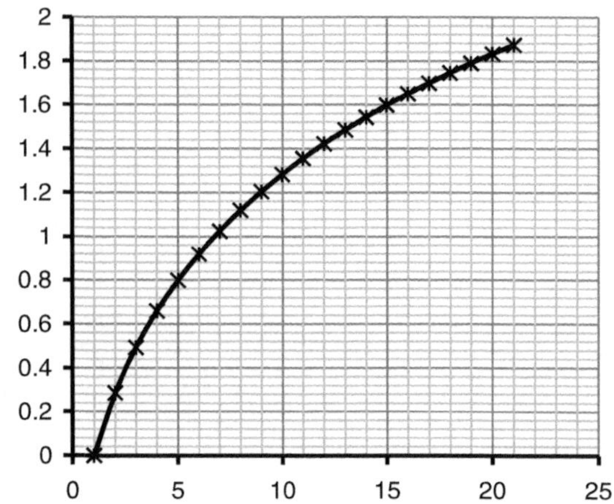

Fig. 2.4 Dynamics of imaginary part of complex trend (2.13)

The base of an exponential function may also be different, for example, a complex number, but we will not consider these variants.

In the exponential form function (2.14) may look like this:

$$y_r + iy_i = Ae^{b_0x}e^{i(\alpha+b_1x)}. \tag{2.15}$$

As we see, the module of this function varies according to the exponential law and variations in the polar angle are directly proportional to variations in the argument. Since the complex coefficient of an exponent can take various values, a modeled function can describe different variants of the dynamic whose details differ from function (2.7), but, similarly to that function, its mapping to the complex plane is spiral.

If we consider the real and imaginary parts of this complex function separately, we will have a system of equations:

$$\begin{cases} y_r = Ae^{b_0x}\cos(\alpha + b_1x), \\ y_i = Ae^{b_0x}\sin(\alpha + b_1x). \end{cases} \tag{2.16}$$

Now the differences between a complex exponential function of a real argument and a complex power function of a real argument are evident. The real and imaginary parts of an exponential function vary according to the cosine and sine laws with a constant period of oscillations, the oscillation range varying with the change in the argument. If $b_0 > 0$, then the oscillation range increases with the growth of the argument; if $b_0 < 0$, then the oscillation range decreases.

A complex trend model is a simple variant of this model.

For example, for low positive values of coefficients of a complex exponent like

$$y_{rt} + iy_{it} = e^{(0,15+i0,05)t} \tag{2.17}$$

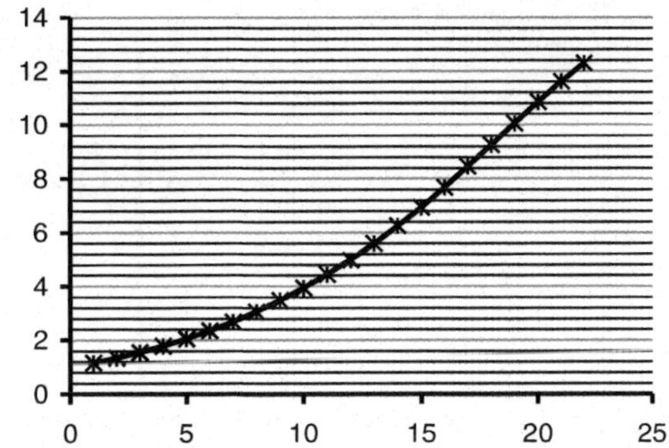

Fig. 2.5 Dynamics of real part of trend (2.17)

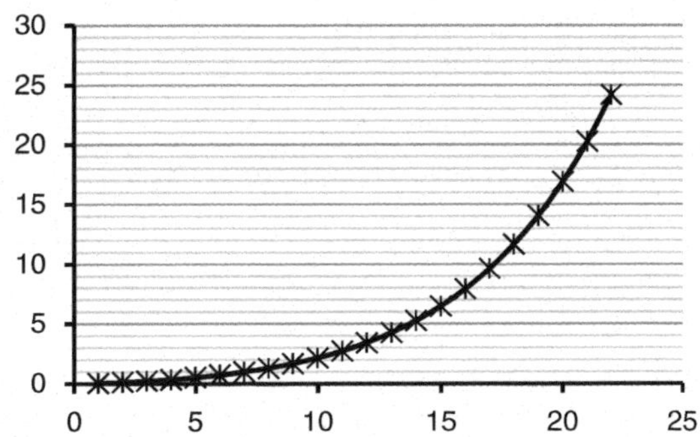

Fig. 2.6 Dynamics of imaginary part of trend (2.17)

each of the components is described for $t = 1,2,\ldots 22$ with an increasing area (Figs. 2.5 and 2.6).

For other coefficients a considerably more complex cyclical dynamics can be modeled, for example, if the model has the form

$$y_{rt} + iy_{it} = e^{(0{,}05+i60)t} \tag{2.18}$$

then the dynamics of the real and imaginary constituents of the complex-valued trend takes the form shown in Figs. 2.7 and 2.8.

It is seen from these figures that the function models the oscillation process with increasing amplitude at a constant oscillation frequency.

We could continue looking at similar elementary complex functions of real arguments, but this goes beyond the problems covered by our study. Thus, we will consider several special nonstandard functions that are subspecies of those mentioned previously.

First is an exponential-power function with an imaginary exponent:

$$y_r + iy_i = x^{ix}. \tag{2.19}$$

Fig. 2.7 Dynamics of real part of trend (2.18)

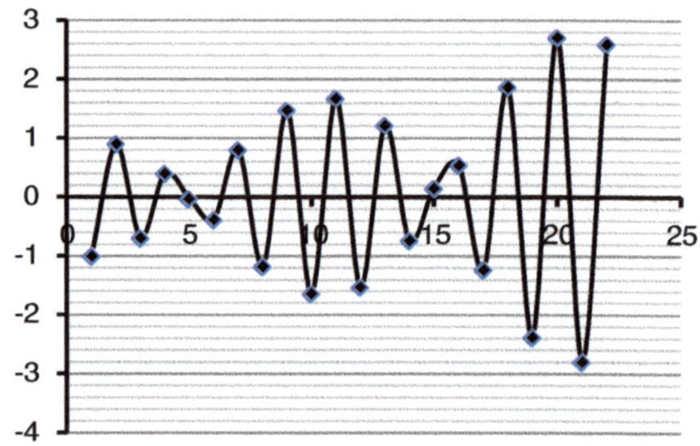

Fig. 2.8 Dynamics of imaginary part of trend (2.19)

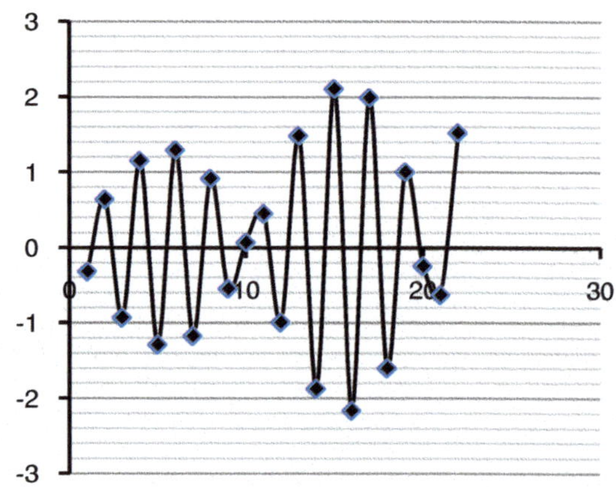

Let us present this function in exponential form:

$$y_r + iy_i = e^{ix \ln x}. \tag{2.20}$$

This helps us to determine a change in the real and imaginary parts with the growth of the argument:

$$\begin{cases} y_r = \cos(x \ln x), \\ y_i = \sin((x \ln x). \end{cases} \tag{2.21}$$

They change according to the cosine and sine laws with an increasing oscillation period. A logarithm limits the function domain; since a logarithm of zero does not exist, the zero point is not included in the function domain.

Since the module of this function is equal to one, in a complex plane the function represents a unit circumference.

An exponential-power function with a complex exponent is an expected development of this function:

$$y_r + iy_i = x^{(x+ix)} = x^{(1+i)x}. \tag{2.22}$$

In this case, the right-hand side of the equality is easily represented in exponential form:

$$y_r + iy_i = x^x e^{ix \ln x}. \tag{2.23}$$

This model has a domain in the positive part of real numbers since a logarithm of a negative number, as well as a logarithm of zero, does not exist.

Let us present the real and imaginary parts of this complex function separately:

$$\begin{cases} y_r = x^x \cos(x \ln x), \\ y_i = x^x \sin(x \ln x). \end{cases} \tag{2.24}$$

The module of this complex function increases sharply with an increase in the argument; this is why the real and imaginary parts of the function represent an oscillatory function with increasing oscillation period and sharply growing oscillation range. In a complex plane this function is shown as a sharply diverging spiral. This feature gives the function little applicability in economic modeling, though the initial part of the function could be of interest. The module of a function in the positive neighborhood of the zero point is close to one (any number to the zero power is equal to one); however, with an increase in the argument it will first decrease and then increase. The module of the complex function reaches its minimum value at the point where the first derivative is equal to zero:

$$\frac{dr}{dx} = (x^x)' = 0$$

After solving this equation and using the Leibniz-Bernoulli formula we have

$$r' = x^x(1 + \ln x) = 0.$$

Since $|x| > 0$, the module of the complex function reaches its minimum value at the point $x = e^{-1}$.

The dynamics of the polar angle u with changes in the argument within the interval [0;1) is complicated since it is determined by the following equality:

$$\theta = x \ln x.$$

The first derivative of this relationship with respect to the argument will have the following form:

$$\frac{d\theta}{dx} = (x)' \ln x + x(\ln x)' = \ln x + 1,$$

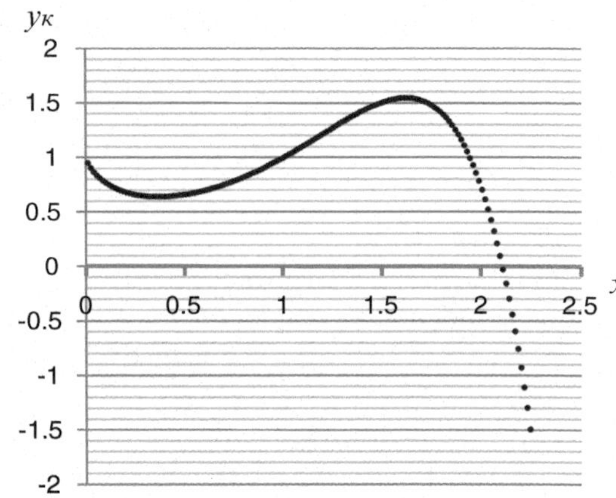

Fig. 2.9 Real parts of complex function (2.22)

which means that the polar angle reaches its minimum value at the same point as the module of the complex point $x = e^{-1}$.

Thus, for the argument $x{=}e^{-1}$ the complex function under consideration reaches its minimum values of both the module and polar angle. In the complex plane this will be shown with an increase in the argument as follows. The curve starts its movement in the clockwise direction from the neighborhood of a point with the coordinates $xr = 1, xi = 0$ until it reaches the point where both the module and the argument take their minimum values. The module then is equal to

$$r_{\min} = x^x = (e^{-1})^{e^{-1}} = \left(\frac{1}{e}\right)^{\frac{1}{e}}$$

and the polar angle to

$$\theta_{\min} = x \ln x = -\frac{1}{e}.$$

Then, with growth of the x module of the complex function, its polar angle starts growing, too. In the complex plane this growth is revealed in movement along the same line but in a counterclockwise direction.

Then the function module starts growing sharply, which leads to an increase in the values of the real and imaginary parts of the function under consideration. The separate dynamics of the real and imaginary parts of this complex function at low values of the real argument $x = [0;2)$ are of more interest. This dynamics is given in Fig. 2.9.

It is possible to narrow the spiral span and increase or reduce the rotation frequency by involution of the complex proportionality coefficient, which is different from the complex unit:

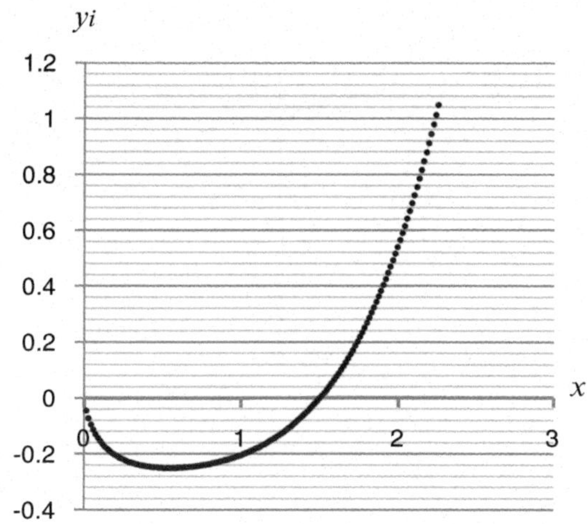

Fig. 2.10 Imaginary parts of complex function (2.22)

$$y_r + iy_i = x^{(b_0+ib_1)x}. \tag{2.25}$$

For different values of the real and imaginary parts of this coefficient, for the function under consideration there one obtains a great variety of spirals in the complex plane, as well as various types of dynamics of the real and imaginary parts of the complex function (Fig. 2.10).

We can continue the logic of the real argument transformation to the complex plane by suggesting a complex exponential-power function with a complex base:

$$y_r + iy_i = (x + ix)^x = (x(1 + i))^x. \tag{2.26}$$

Its exponential form will look like this:

$$y_r + iy_i = (\sqrt{2}x)^x e^{i\frac{\pi}{4}x}. \tag{2.27}$$

Then for the real and imaginary parts of this complex function we have

$$\begin{cases} y_r = (\sqrt{2}x)^x \cos\left(\dfrac{\pi}{4}x\right), \\ y_i = (\sqrt{2}x)^x \sin\left(\dfrac{\pi}{4}x\right). \end{cases}$$

The zero point of the real argument is included in the function domain. In general this function looks like a spiral; however, at low values its behavior is complicated as its polar angle increases with the growth of the real argument and its module first decreases, reaches its minimum, and starts increasing again.

The first derivative of the module is

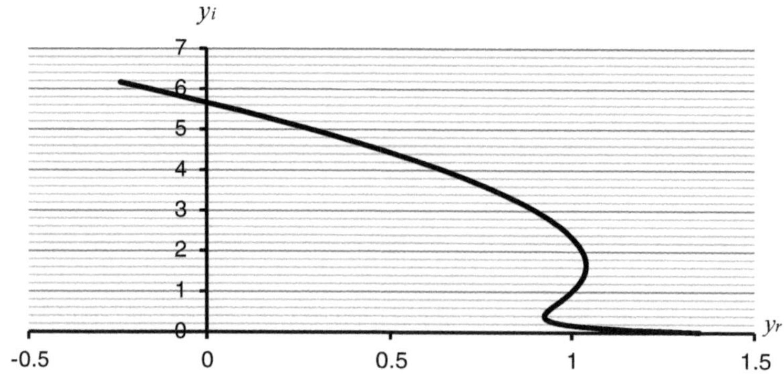

Fig. 2.11 Function (2.26) in complex plane at low values of real argument

$$r' = (\sqrt{2}x)^x\left(\frac{x}{\sqrt{2}x} + \ln(\sqrt{2}x)\right).$$

If we set it equal to zero we obtain a point where the module takes minimum values:

$$x = \frac{e^{-\frac{1}{\sqrt{2}}}}{\sqrt{2}} = 0.348652215$$

Taking into account these specifics, for initial values of the real argument the complex function will have nonlinear dynamics (Fig. 2.11).

The last elementary function of a real argument is a complex exponential-power function with a complex base and complex exponent:

$$y_r + iy_i = (x + ix)^{(x+ix)} = (x(1+i))^{(1+i)x}. \tag{2.28}$$

In exponential form it looks like this:

$$y_r + iy_i = (\sqrt{2}x)^{(1+i)x} e^{i\frac{\pi}{4}(1+i)x} = [(\sqrt{2}x)^x e^{-\frac{\pi}{4}x}][(\sqrt{2}x)^{ix} e^{i\frac{\pi}{4}x}]. \tag{2.29}$$

The module of this complex function of a real argument will be

$$R = (\sqrt{2}x)^x e^{-\frac{\pi}{4}x}. \tag{2.30}$$

And its polar angle will have the following form:

$$\theta = x\left[\ln(\sqrt{2}x) + \frac{\pi}{4}\right]. \tag{2.31}$$

It is easy to see that the zero value of the real argument is not included in the function domain.

The behavior of the module of this function is more complicated than that of the previous ones. For an argument close to zero the module will be close to one, then it gets lower up to a certain value, after which it starts increasing again, but not sharply as in the case of the previous function.

To determine the point where the module of complex function (2.30) takes its minimum value, we should find its first derivative:

$$R' = [(\sqrt{2}x)^x]'e^{-\frac{\pi}{4}x} + (\sqrt{2}x)^x(e^{-\frac{\pi}{4}x})' = (\sqrt{2}x)^x\left(\frac{1}{\sqrt{2}} + \ln(\sqrt{2}x)\right)e^{-\frac{\pi}{4}x} - \frac{\pi}{4}(\sqrt{2}x)^x e^{-\frac{\pi}{4}x},$$

which should be equal to zero. Then, solving the equation we find the point at which the module is at its minimum:

$$x = \frac{e^{\frac{\pi - 2^{\frac{3}{2}}}{4}}}{\sqrt{2}}.$$

The polar angle also varies nonlinearly – it decreases from values close but not equal to zero (the points are in the fourth quadrant of the complex value) and then grows. To determine the minimum value of the polar angle, let us find its first derivative of the real argument:

$$\frac{d\theta}{dx} = (x)'\left(\ln(\sqrt{2}x) + \frac{\pi}{4}\right) + x\left(\ln(\sqrt{2}x) + \frac{\pi}{4}\right)' = \ln(\sqrt{2}x) + \frac{\pi + 2^{\frac{3}{2}}}{4}.$$

After setting it equal to zero and solving the equation we find the value of the real argument for which the polar angle reaches its minimum value:

$$x = \frac{e^{-\frac{\pi + 2^{\frac{3}{2}}}{4}}}{\sqrt{2}}$$

With an increase in the argument, the conformal mapping of the function under consideration in a complex plane takes place in a spiral moving in a clockwise direction, as shown in Fig. 2.12.

With the further growth of the argument, the function module increases sharply, as does the polar angle, the function itself continuing its spiral movement in the clockwise direction.

The elementary complex exponential-power function with complex base and complex exponent (2.28) can be represented in a form more applicable to practical purposes, namely:

$$y_r + iy_i = (x(a_0 + ia_1))^{(b_0 + ib_1)x}. \tag{2.32}$$

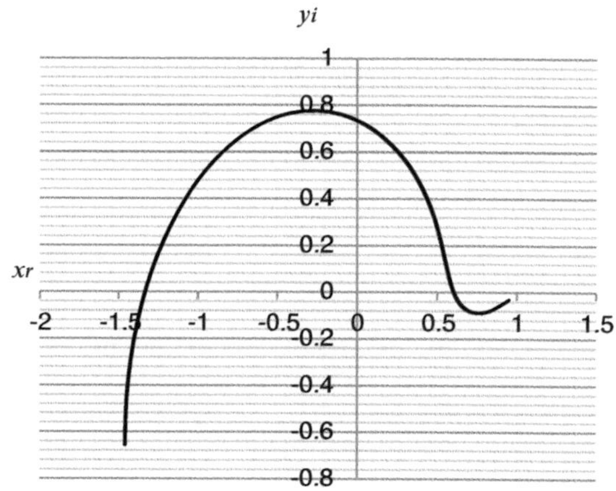

Fig. 2.12 Part of conformal mapping of function (2.28) for low value of argument

Its exponential form will have the form

$$y_r + iy_i = [(Ax)^{b_0 x} e^{-b_1 x\alpha}][(Ax)^{ib_1 x} e^{ib_0 x\alpha}]. \tag{2.33}$$

Hence, for the module of this complex function of a real argument

$$R = (Ax)^{b_0 x} e^{-b_1 x\alpha} \tag{2.34}$$

and for the polar angle

$$\theta = x[b_1 \ln(Ax) + b_0 \alpha]. \tag{2.35}$$

The function domain lies in the area of positive arguments, which clearly follows from (2.35).

Changing the values of function coefficients (2.32) we can obtain a great diversity of conformal mappings and variations of the real and imaginary parts of this function that have an oscillatory character.

As we see from (2.35), the polar angle of this complex function of a real argument depends largely on the constant b_1. The higher the values of this constant, the more rapid is the increase in the polar angle with the increase of the argument, and the faster is the turnover of the function values in the complex plane. This coefficient also influences the change in the module of the function under consideration, but for a low value of a_1 and high value of a_0 this influence decreases.

The coefficient b_0 is responsible for the growth in the function module. At its positive values the module increases sharply.

For various coefficient values the function behaves in a different way – it converges to zero and diverges, changes values around some circumference, changes chaotically, etc.

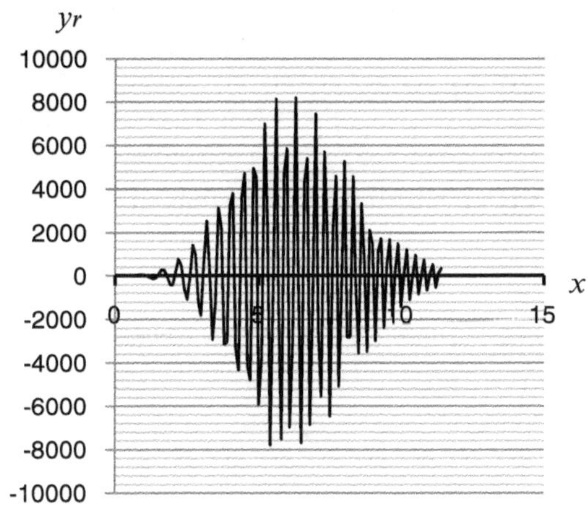

Fig. 2.13 Dynamics of real part of function (2.36)

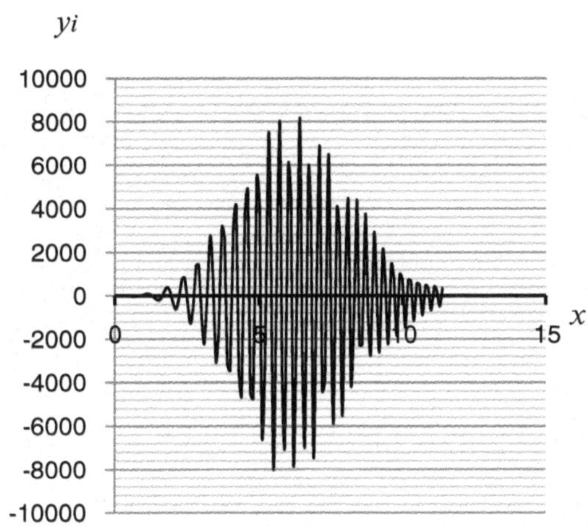

Fig. 2.14 Dynamics of imaginary part of function (2.36)

It is interesting that the function can also model the process of reaction of some system to an external influence with further stabilization at the previous level. This function behaves in this way, for example, for the following coefficients:

$$y_r + iy_i = (x(1 - i))^{(-1{,}5+i6)x}. \tag{2.36}$$

Subsequent change in the real and imaginary parts of this function with the growth of the argument within $0 < x \leq 10$ subject to the argument is shown in Figs. 2.13 and 2.14.

According to the results of this section we can draw the conclusion that complex functions of a real argument model a great diversity of cyclical dynamics.

Numerous functions of complex arguments are not limited at all to only the aforementioned types. However, it is not possible to consider all functions within the framework of this study; this would be at odds with the purpose of the present study, where we state only the basics of the application of the TFCV to solutions of economic problems.

The superposition of elementary complex functions provides vast possibilities for the generation of new functions. A simple example is the case where the complex power function of a real argument

$$z_r + iz_i = (c_0 + ic_1)x^{d_0+id_1}$$

is added to by a complex function of a real argument (2.32)

$$z_r + iz_i = (c_0 + ic_1)x^{d_0+id_1} + (x(a_0 + ia_1))^{(b_0+ib_1)x}. \tag{2.37}$$

If, for example, for the second term of this function we use coefficients like those proposed in (2.36), the resulting model will describe the dynamics of some nonlinear process, which may chaotically deviate from its previous trajectory on a certain segment under a certain external influence, but due to the stability of the object it returns to its former trajectory. It is evident that instead of the power function, the first term may be represented by other forms, for example, by a step function. With the proper selection of parameters, with the assistance of such a superposition, we model the transition from one stationary state to another.

The real argument itself can be presented in complex functions as a real function of a real argument, for example, *sinx* or *cosx*.

It is evident that the variety of complex functions of a real argument is enormous, and it is impossible to cover them in one section or chapter.

2.3 Functions of a Complex Argument: Linear Function

Since it is possible to transform a real argument to a complex plane using particular functions, a reverse transformation procedure is also possible – from the field of complex variables to the numerical axis of real variables. The relationship between a complex argument and a real result will represent a function of a complex argument:

$$y = f(x_r + ix_i) = f_r(x_r, x_i) + if_i(x_r, x_i). \tag{2.38}$$

Since there is a complex number in the right-hand side of this equality and a real one in the left-hand side, the function of the complex argument may be written as follows:

$$y + i0 = f_r(x_r, x_i) + if_i(x_r, x_i). \tag{2.39}$$

Hence we have a system of two real equations:

$$\begin{cases} y = f_r(x_r, x_i), \\ 0 = f_i(x_r, x_i). \end{cases} \quad (2.40)$$

The first equality of the system represents an equation of some surface in a three-dimensional space, the second one a line in the argument's plane. Since the problem is considered in a three-dimensional coordinate system, for the second equation of system (2.40) the equality is valid for any y value. This means a surface in three-dimensional space that is not crossed by the y-axis, i.e., the y-axis is parallel to this surface, the surface itself being perpendicular to the complex plane of the argument.

Since these two equations are combined in a system, they are simultaneously satisfied. Geometrically this means that system (2.40) and the initial function (2.38) represent an intersection of two planes in three-dimensional space – the first and second equations of system (2.40). The perpendicular nature of the second equation of system (2.40) means that the aggregate of the points lying on the surface of the first equation of system (2.40.) must be projected onto the plane of the complex argument as a line described by the second equation of system (2.41).

Let us sequentially consider the main functions of a complex argument and their graphical interpretation in order of increasing complexity, bringing each of them to the form (2.40).

The first such model to be used in economics is a linear function of a complex argument with a zero free term:

$$y = (b_0 + ib_1)(x_r + ix_i). \quad (2.41)$$

If we single out the real and imaginary parts of this function and group them, we have

$$\begin{cases} y = b_0 x_r - b_1 x_i, \\ 0 = b_1 x_r + b_0 x_i. \end{cases} \quad (2.42)$$

The first Eq. (2.42) is that of a plane in space passing through a zero point. The slope angle and position of the plane in space is fully determined by the signs and values of the coefficients of the complex proportionality coefficient.

The second equation of the system under consideration represents an equation of a line in the plane of the argument:

$$x_r = -\frac{b_0}{b_1} x_i. \quad (2.43)$$

This straight line originates from the zero point, and its location in the particular quadrant of the complex plane is determined by the values of the real and imaginary parts of the complex proportionality coefficient. Since the second equation of system (2.42) should be considered in space, it represents a plane parallel to the y-axis and perpendicular to the complex plane.

The two crossing planes form a line, meaning that the linear function of complex argument (2.41) represents a line in three-dimensional space $(0y;0x_r;0x_i)$ passing through the zero point.

If we now consider a linear function of a complex argument with a free complex coefficient:

$$y = (a_0 + ia_1) + (b_0 + ib_1)(x_r + ix_i). \tag{2.44}$$

then singling out the real and imaginary parts of this function and grouping them as in the previous case, we have

$$\begin{cases} y = a_0 + b_0x_r - b_1x_i, \\ 0 = a_1 + b_1x_r + b_0x_i. \end{cases} \tag{2.45}$$

It is clear that the nature of the function has not changed – both the first and the second equations are planar equations – only the location of the planes in space has changed, as has the location of the line resulting from the planes' intersecting. It follows from the first equation that the plane does not pass in space through the zero point and crosses the y-axis at the point $y = a_0$. The second equation shows that the line in space of the complex argument does not pass through the zero point either since

$$x_r = -\frac{a_1}{b_1} - \frac{b_0}{b_1}x_i. \tag{2.46}$$

And on the axis of real values of the complex argument this line passes through the point $x_r = -\frac{a_1}{b_1}$.

Thus, the linear function of a complex argument with free complex coefficient (2.44) represents the equation of a line in three-dimensional space "a complex plane of an argument – an axis of a real number." Or put another way, any line in three-dimensional space may be described by a linear function of complex argument (2.44).

It is appropriate to recall that in the Cartesian coordinate system the equation of a line is also determined by the intersection of two planes and may be written as follows:

$$\begin{cases} a_1y + b_1x_r + c_1x_i + d_1 = 0, \\ a_2y + b_2x_r + c_2x_i + d_2 = 0. \end{cases} \tag{2.47}$$

As follows from (2.47), a line in a Cartesian coordinate system is defined by eight coefficients; the same line in the form of a linear function of a complex argument, as follows from (2.44), is defined by only four coefficients and is represented in the form of a linear equation. We can again see that to actions with complex numbers correspond actions with real numbers, and functions of complex variables often represent a more convenient form of notation than those of real numbers.

It should be noted that in the Cartesian coordinate system the equation of a line passing through two different points $P_1(y_1,x_{r1},x_{i1})$ and $P_2(y_2,x_{r2},x_{i2})$ will be written as follows:

$$\frac{y - y_1}{y_2 - y_1} = \frac{x_r - x_{r1}}{x_{r2} - x_{r1}} = \frac{x_i - x_{i1}}{x_{i2} - x_{i1}}. \tag{2.48}$$

With reference to the line described by the function of the complex argument (2.44), the equation of the line for these two points will be written as follows:

$$\frac{y - y_1}{y_2 - y_1} = \frac{(x_r + ix_i) - (x_{r1} + ix_{i1})}{(x_{r2} + ix_{i2}) - (x_{r1} + ix_{i1})}. \tag{2.49}$$

The specifics of a linear function of a complex argument with reference to some economic problems will be considered in other chapters of this book.

2.4 Power Function of a Complex Argument with a Real Exponent

The linear function of a complex argument can be applied in many cases of economic modeling since in accordance with the general scientific principle "from the simple to the complex," to study some complex object, first simple models including linear ones are used, after which models become increasingly complex as the object's properties become clearer for a more adequate description of complex processes.

The power function of a complex argument is more complex than a linear one, its general form being

$$y = (a_0 + ia_1)(x_r + ix_i)^{(b_0+ib_1)}. \tag{2.50}$$

It is easy to see that if the exponent of the model is equal to one, it is turned into an elementary linear model of the complex argument (2.41).

Let us consider function (2.50) sequentially in order of increasing complexity depending on the exponent – real, imaginary, or complex.

The first of the possible models determined by the equality (2.50) is one with a real exponent:

$$y = (a_0 + ia_1)(x_r + ix_i)^{b_0}. \tag{2.51}$$

To understand the properties of this function, let us represent the complex proportionality coefficient and complex resource variable in exponential form. Then we have

$$y = ae^{i\alpha}\left(re^{i\varphi}\right)^{b_0} = ar^{b_0}e^{i(\alpha+b_0\varphi)}, \tag{2.52}$$

$$a = \sqrt{a_0^2 + a_1^2}, \alpha = arctg\frac{a_1}{a_0}, r = \sqrt{x_x^2 + x_i^2}, \varphi = arctg\frac{x_i}{x_r}.$$

Hence we have a system of equations for the real and imaginary parts of the function under consideration:

$$\begin{cases} y = ar^{b_0}\cos(\alpha + b_0\varphi), \\ 0 = ar^{b_0}\sin(\alpha + b_0\varphi). \end{cases} \tag{2.53}$$

It follows from the last equality that it holds for the following conditions:

$$\sin(\alpha + b_0\varphi) = 0 \rightarrow \alpha + b_0\varphi = \pm\pi k, \tag{2.54}$$

where k is a whole number.

It should be noted that for values of the polar angle of the function determined by these conditions, its cosine takes the following values:

$$\begin{cases} \cos(\alpha + b_0\varphi) = 1, \quad \forall\ k = 0, \pm 2, \pm 4, \ldots \\ \cos(\alpha + b_0\varphi) = -1, \forall\ k = \pm 1, \pm 3, \pm 5, \ldots \end{cases} \tag{2.55}$$

If, for example, we consider the polar angle in the complex plane of the argument from 0 to 2π, at б = 0 and $b_0 = 1$, we have that y is positive for $\varphi = 0$ and $\varphi = 2\pi$ and negative for $\varphi = \pi$. For any α and for $b_0 \neq 0$ (with exponent $b_0 = 0$, the function is turned into the point $y = a\cos\alpha = a_0$) function (2.51), subject to the values of coefficients α and b_0 and polar angle φ, takes both positive and negative values.

Since it follows from the second equation of system (2.53) that the relationship between the real and imaginary parts of the complex argument in this function is a constant value regardless of the values of y, in three-dimensional space this equation is represented by a plane parallel to the axis of the real variable 0y and perpendicular to the plane of the complex argument.

The first Eq. (2.53) determines the change in the y depending on the change in the two factors x_r and x_i, which may be represented in three-dimensional space in the form of some surface.

The complex proportionality coefficient changes the surface scale and slopes, which is why its influence on the result is negligible and we may consider this coefficient to be equal to one. For this reason let us consider a simplified analog of function (2.51):

$$y = (x_r + ix_i)^{b_0}. \tag{2.56}$$

Then the real and imaginary parts of this function will have the form

$$\begin{cases} y = r^{b_0} \cos b_0\varphi, \\ 0 = r^{b_0} \sin b_0\varphi. \end{cases} \tag{2.57}$$

Let us consider the influence on the real part of model (2.56) of exponent b_0, i.e., equation of the first function of system (2.57), since the second equation describes the linear relationship between the real and imaginary parts of the complex argument, as

$$0 = r^{b_0} \sin b_0\varphi \to \sin b_0\varphi = \pi k \to \varphi = arctg\frac{x_i}{x_r} = const.$$

Here three variants of the function's behavior are possible:

1. When the exponent is negative, $b_0 < 0$;
2. When the exponent lies within the range: $0 < b_0 < 1.$;
3. When the exponent is higher than 1, $b_0 > 1$.

Let us consider the first case where the exponent of the power function of complex argument b_0 is less than zero. For convenience, let us give the full form of the first equation of the system:

$$y = \left(\sqrt{x_r^2 + x_i^2}\right)^{b_0} \cos\left(b_0 arctg\frac{x_i}{x_r}\right). \tag{2.58}$$

Since the second equation of system (2.57) states that the function under consideration is projected in the complex plane of the argument to a line passing through the zero point and having a constant value of the polar angle, let us consider first what values the factor containing this constant polar angle can take. Cosine is a periodic function, but the argument of this function does not change; it is always constant due to the constant nature of the polar angle. The surface described in three-dimensional space by the first equation of system (2.57) is nonlinear and its character is determined by coefficient b_0, since it characterizes the frequency of oscillations – the higher it is by module, the more uneven ("corrugated") is the surface.

Since subsequently we will not need the type of this surface, we will consider only the nature of the lines in the space made by system (2.57).

Let us first consider the situation where the polar angle of the argument in the complex plane is equal to zero. This is possible when $x_r > 0$, $x_i = 0$. Since the cosine of zero is one, (2.58) will look as follows:

$$y = x_r^{b_0}. \tag{2.59}$$

In the (y, x_r) plane, this function will represent a hyperbola that decreases from plus infinity to zero with the growth of the module of the argument.

Now let us suppose that the polar angle of the complex argument is $\pi/4$, i.e., have the following form:

$$y = \left(x_r\sqrt{2}\right)^{b_0} \cos\left(b_0 \frac{\pi}{4}\right). \tag{2.60}$$

The function is positive for $-2 < b_0 < 0$, equal to zero for $b_0 = -2$, negative for $-6 < b_0 < -2$, etc. Absolute values of the function for this case also decrease with the growth of the module of the complex argument, as previously, but in the case of negative values of the function, it decreases from minus infinity to zero.

Let us consider another case where the real part of the complex argument is equal to zero: $x_r = 0$, and its imaginary part is positive: $x_i > 0$. The polar angle of the complex argument is equal to $\pi/2$ and the function looks like this:

$$y = x_i^{b_0} \cos\left(b_0 \frac{\pi}{2}\right). \tag{2.61}$$

If the coefficient b_0 lies within the range $-1 < b_0 < 0$, then the cosine of function (2.61) will be positive, which means the function has a positive character. If the exponent is equal to $b_0 = -1$, then cosine becomes zero and the function also becomes equal to zero. If the values of this coefficient are within the range $-3 < b_0 < -1$, then the cosine of the function becomes negative like the function itself. Since the cosine is a periodical function, then with the subsequent increase of the module of the values of the exponent b_0, the function becomes both positive and negative. With the growth in the values of the function argument (2.61), the absolute values of the function also decrease with the hyperbola.

Continuing on, it is clear that the power function of a complex argument with negative real exponent decreases in its absolute values according to the hyperbolic law with an increase in the argument's module. With this, the sign of the function is determined by the result of multiplying the exponent by the polar angle. In some cases the function is equal to zero.

In the second case, where the exponent of the function of the complex argument lies within the range $0 < b_0 < 1$, the function behaves slightly differently:

$$y = x_r^{b_0}. \tag{2.62}$$

However, since the exponent is positive and not greater than one, with the growth of the module, the function increases nonlinearly from zero to plus infinity according to the exponential law with a negative second derivative.

If the polar angle of the complex argument is $\pi/4$, then when the variables are in the first quadrant of the complex plane and $x_r = x_i$, then function (2.58) takes the following value:

$$y = (x_r\sqrt{2})^{b_0} \cos\left(b_0\frac{\pi}{4}\right). \tag{2.63}$$

The function will be positive for an exponent lying within $0 < b_0 < 2$ and equal to zero at $b_0 = 2$. It will be negative for $2 < b_0 < 6$, etc. With the growth of the module of the complex argument, absolute values will behave similarly to (2.62).

It makes no sense to examine this case further since it is clear that the function will behave just like this – its absolute values will increase nonlinearly from zero, and the function sign will be determined by the exponent of the function.

In the third case of the power function with a real exponent, the exponent $b_0 > 1$, function (2.58) will take negative or positive values, as well as values equal to zero, depending on the result of multiplying the exponent by the polar angle since the cosine of an angle may be both positive and negative and be equal to zero. However, by its absolute value, with the growth of the complex argument, the function will tend from zero to infinity according to the exponential law with positive second derivative.

We can now go back to the model with a complex proportionality coefficient under consideration (2.51):

$$y = (a_0 + ia_1)(x_r + ix_i)^{b_0}$$

and pay more attention to the influence of this coefficient on the function behavior.

Values of this complex proportionality coefficient influence both the module of the function and the polar angle.

For various values of the proportionality coefficient, the module of the function of the complex argument equal to $R = ar^{b_0}$ is presented on various scales.

When the values of this proportionality coefficient vary, the polar angle also turns in the plane of the complex argument:

$$\varphi = \frac{\pm\pi k}{b_0} - \alpha.$$

This is why variations in the proportionality coefficient move the power function curve in various parts of space symmetrically to the y-axis and change the row curve scale.

2.5 Exponential Function of Complex Argument with Imaginary Exponent

Having considered the case where the power function of complex argument (2.50) is represented by a real exponent and the function represents a line of the power function in three-dimensional space, let us move on to a more complicated case where the real part of this function is equal to zero and the exponent is imaginary:

$$y = (a_0 + ia_1)(x_r + ix_i)^{ib_1}. \tag{2.64}$$

In this case another relationship besides the power one will be modeled, though the complex argument is subject to involution.

Since the influence of the proportionality coefficient on the result in this case will remain the same, let us consider it to be equal to one:

$$a_0 + ia_1 = 1.$$

Then the function in exponential form will look like this:

$$y = \left(re^{i\varphi}\right)^{ib_1} = e^{-\varphi b_1} e^{ib_1 \ln r}, \tag{2.65}$$

where $r = \sqrt{x_x^2 + x_i^2}, \varphi = arctg\frac{x_i}{x_r}$.

From this follows a system of equations for the real and imaginary parts of the function under consideration:

$$\begin{cases} y = e^{-b_1 arctg\frac{x_i}{x_r}} \cos\left(b_1 \ln\sqrt{x_r^2 + x_i^2}\right), \\ 0 = e^{-b_1 arctg\frac{x_i}{x_r}} \sin\left(b_1 \ln\sqrt{x_r^2 + x_i^2}\right). \end{cases} \tag{2.66}$$

The first equation of this system represents a description of some nonlinear function in three-dimensional space, to be discussed later.

The second equation of this system represents a nonlinear surface perpendicular to the plane of a complex argument, where all the lines lying in that plane are parallel to the 0*y*-axis.

The intersection of these two surfaces is simply a graphical interpretation of function (2.64) in the space.

Let us consider what the second equation looks like in the plane of a complex argument:

$$0 = e^{-b_1 arctg\frac{x_i}{x_r}} \sin\left(b_1 \ln\sqrt{x_r^2 + x_i^2}\right). \tag{2.67}$$

The first factor can be equal to zero only if its exponent is equal to infinity. Variants when b_1 is equal to infinity are not considered here because they are meaningless. The arctangent is known to lie within a range of $-\pi/2$ to $+\ \pi/2$. Therefore, the first factor (2.67) will never be equal to zero and the equality holds when the second factor is equal to zero:

$$\sin\left(b_1 \ln \sqrt{x_r^2 + x_i^2}\right) = 0. \tag{2.68}$$

This equality holds when

$$b_1 \ln \sqrt{x_r^2 + x_i^2} = \pi k. \tag{2.69}$$

Hence,

$$x_r^2 + x_i^2 = e^{2\frac{\pi k}{b_1}}. \tag{2.70}$$

This means that the imaginary part (2.64) is equal to zero when the values of the complex argument in the complex plane lie on a circumference with radius $e^{2\frac{\pi k}{b_1}}$. In particular, if $k = 0$, then the equality holds when

$$x_r^2 + x_i^2 = 1, \tag{2.71}$$

i.e., when the points in the complex plane lie on a one-unit circumference.

k may take any whole values, which means a family of circumferences in a complex plane of arguments. In the three-dimensional space it is a cylindrical surface perpendicular to the plane of the complex argument.

Now, let us consider the first equation of system (2.66) referring to the real part of the function:

$$y = e^{-b_1 arctg\frac{x_i}{x_r}} \cos\left(b_1 \ln \sqrt{x_r^2 + x_i^2}\right). \tag{2.72}$$

This equation describes a nonlinear surface in space, but since the type of this surface will not be used subsequently, we should consider what line on this surface is cut off by the cylinder, since the function of complex argument (2.64) is an intersection of two nonlinear surfaces one of which is a cylinder (2.70). Thus, let us consider the behavior of (2.72) in the case where the variables x_r and x_i lie on some circumference.

In this case the logarithm argument is a constant, which is why the nature of this curve is completely determined by the first factor (2.72), which represents the exponent.

Since a complex argument changes its values in a circumference, in the initial point where the minimal component is equal to zero, the first factor is equal to one since any number to the zero power equals one. Then, for a polar angle equal to zero, the function will take the following values:

$$y(0) = \cos(b_1 \ln r). \tag{2.73}$$

With the growth of the values of the imaginary component x_i on the circumference and the respective decrease in the real component x_r (increasing polar angle), the polar angle of the complex argument tends from zero to π/2. In the extreme point, when the real component of the complex argument is equal to zero, the function will have the following form:

$$y\left(\frac{\pi}{2}\right) = e^{-b_1 \frac{\pi}{2}} \cos(b_1 \ln r). \tag{2.74}$$

In the interval between these two points the function will vary exponentially from points with coordinates determined by (2.73) to points determined by (2.74).

Further movement of the complex argument on the circumference corresponds to a variation in the polar angle from π/2 to π. In the extreme point, when the polar angle is equal to π, which means that the imaginary component is equal to zero and the real constituent $x_r = -r$, the function will take the following values:

$$y(\pi) = e^{-b_1 \pi} \cos(b_1 \ln r). \tag{2.75}$$

Continuing along the circumference and arriving at the point where the real part is equal to zero and the imaginary $x_i = -r$, we obtain a function value equal to

$$y\left(\frac{3}{2}\pi\right) = e^{-b_1 \frac{3\pi}{2}} \cos(b_1 \ln r). \tag{2.76}$$

Completing the movement along the circumference in the point where the imaginary part is equal to zero and the real one is equal to the radius, the function takes the following value:

$$y(2\pi) = e^{-b_1 2\pi} \cos(b_1 \ln r). \tag{2.77}$$

Now it is clear what the power function of a complex argument represents if the exponent is an imaginary number – this exponent is located on the cylinder surface. Completing the full circle equal to 2π we see that the function differs from its initial point by $e^{-b_1 2\pi}$ times.

If the coefficient b_1 is positive, then the function tends to zero; if the exponent is negative, then the function tends to infinity, making turn after turn on the cylinder surface, if the complex argument makes rotational movements in the complex plane. However, since in economics we do not observe such rotational movements, meaning variations in the polar angle of the complex argument within a range $0 \le \varphi \le 2\pi$, function (2.64) should be considered an exponent on the cylinder surface making one complete turn thereon.

2.6 Power Function of Complex Argument with Complex Exponent

When a complex argument of a power function is raised to a real power, in three-dimensional space this represents a curve described by an exponential function and lying in space in a plane perpendicular to the plane of the complex argument.

If the exponent of this function is an imaginary number, then it represents an exponent varying with the increase in the polar angle of the complex variable and lying in the space on the cylinder surface perpendicular to the complex argument plane.

Now let us consider the nature of a power function of a complex argument where the exponent is complex:

$$y = (a_0 + ia_1)(x_r + ix_i)^{(b_0+ib_1)}. \tag{2.78}$$

This function, taking into account previously introduced designations and the assumption that the complex proportionality coefficient is equal to one, may be written in exponential form as follows:

$$y = r^{b_0} e^{ib_0\varphi} e^{-\varphi b_1} e^{ib_1 \ln r} = r^{b_0} e^{-\varphi b_1} e^{i(b_0\varphi + b_1 \ln r)}. \tag{2.79}$$

The real and imaginary parts of this function may be written as a system of equations:

$$\begin{cases} y = r^{b_0} e^{-b_1\varphi} \cos(b_0\varphi + b_1 \ln r), \\ 0 = r^{b_0} e^{-b_1\varphi} \sin(b_0\varphi + b_1 \ln r). \end{cases} \tag{2.80}$$

Again, we have equations of complex nonlinear surfaces in the space, the second equation describing the surface perpendicular to the complex argument plane. As in the previous cases, let us examine the properties of function (2.78) with the condition that the imaginary part of the complex argument function is equal to zero.

The second equation will be equal to zero when the argument is equal to zero and when the sine is equal to zero:

$$\sin(b_0\varphi + b_1 \ln r) = 0. \tag{2.81}$$

In the zero point the function itself is equal to zero, which is why (2.81) is of interest and may be written as follows:

$$b_0\varphi + b_1 r = \pi k. \tag{2.82}$$

Here, as in the previous case, k is a whole number.

Let us take $k = 0$. Then (2.82) can be written as follows:

$$r = \frac{-b_0}{b_1}\varphi. \tag{2.83}$$

It is evident that we have obtained Archimedes' spiral, where with the change in the polar angle within the range $0 \le \varphi \le 2\pi$ the coefficient before the polar angle should always be positive since the module of the complex argument cannot be negative. This means that the signs of the real and imaginary parts of the complex exponent should be different.

We do not consider rotational processes that practically do not exist in economics, which is why in the space under consideration the second equation of system (2.80) indicates one turn of Archimedes' spiral, that is, a nonlinear surface in Archimedes' spiral perpendicular to the complex argument plane. This surface "cuts off" a nonlinear curve on the other surface represented by the first equation of system (2.80).

We are not interested in the type of the surface described in the space by the first equation of system (2.80), but in the line on this surface that is cut off by Archimedes' spiral.

To understand this, we substitute (2.83) into the first equation of system (2.80) and get

$$y = \left(\frac{-b_0}{b_1}\varphi\right)^{b_0} e^{-b_1\varphi} \cos\left(b_0\varphi + b_1 \ln\left(\frac{-b_0}{b_1}\varphi\right)\right). \tag{2.84}$$

Since it was shown previously that b_0 and b_1 have different signs, let us first take $b_0 > 0$ and $b_1 > 0$. For this case with a growing polar angle:

- The first factor $(\frac{-b_0}{b_1}\varphi)^{b_0}$ increases according to the power law;
- The second factor $e^{-b_1\varphi}$ increases according to the exponential law;
- The third factor $\cos(b_0\varphi + b_1 \ln(\frac{-b_0}{b_1}\varphi))$ varies nonlinearly depending on the modules of the values of coefficients b_0 and b_1. If the module of the complex cosine argument increases with the growth of the polar angle, this factor decreases up to zero, after which it becomes negative.

Thus, on the whole, (2.84) describes a function increasing up to a certain limit with its subsequent decrease to zero and further to the negative range. This line is located in the space on the nonlinear surface of Archimedes' spiral.

If we now change signs of the coefficients to the opposite ones and set $b_0 < 0$, $b_1 > 0$, then the picture will look as follows:

- The first factor $(\frac{-b_0}{b_1}\varphi)^{b_0}$ decreases according to the power law;
- The second factor $e^{-b_1\varphi}$ decreases according to the exponential law;

– The third factor $\cos(b_0\varphi + b_1 \ln(\frac{-b_0}{b_1}\varphi))$ behaves in the same way as in the first case as the cosine is a symmetrical function.

On the whole, with such signs of the coefficients, the function decreases with the growth of the argument and becomes negative as it travels along Archimedes' spiral.

Various combinations of coefficients give various forms of a curve in space. If the imaginary part of a complex exponent is equal to zero, then the function represents points lying on a line of the exponential function in space in a plane perpendicular to the plane of the complex argument. When the real part is close to zero, then the curve represents an exponent lying on the cylinder surface. If the exponent is equal to one, then we have a linear function of a complex argument.

To conclude our study of the properties of this function, it should be noted that the coefficients of a function may be easily estimated by two points.

Let there be two points (x_{r1},x_{i1},y_1) and (x_{r2},x_{i2},y_2) available to an economist disposaltwo , and she thinks that there is a relationship between these variables that may be described by a model in the form of a power function of complex argument (2.78). Substituting these values into the function and dividing the left- and right-hand sides by each other we obtain the following equation:

$$\frac{y_1}{y_2} = \frac{x_{r1} + ix_{i1}}{x_{r2} + ix_{i2}}^{(b_0+ib_1)}. \tag{2.85}$$

Here we can derive the exponent

$$b_0 + ib_1 = \frac{\ln\frac{y_1}{y_2}}{\ln\frac{x_{r1}+ix_{i1}}{x_{r2}+ix_{i2}}}. \tag{2.86}$$

Knowing this value we can easily find the value of the proportionality coefficient $(a_0 + ia_1)$.

Thus, for example, if an economist wants to build a model in the form of a power function of a complex argument and she has two points at her disposal in three-dimensional space – (2; 3; 5) and (2.5; 4.7; 15), then she can easily do this using (2.85) and (2.86), and model (2.78) passing through these two points in three-dimensional space will have the following form:

$$y = (-0,014 - i0,082)(x_r + ix_i)^{(2,648-i0,674)}.$$

2.7 Exponential Function of a Complex Argument

It is clear from the aforementioned properties of the power function of a complex argument that it can be used for modeling various complex nonlinear relationships in three-dimensional space. But this model hardly covers the entire possible variety of functions of a complex argument. One of the simple nonlinear functions of a complex argument that differ in their properties from the power function is the exponential function

$$y = (a_0 + ia_1)e^{(b_0+ib_1)(x_r+ix_i)}. \tag{2.87}$$

To study its properties let us first consider a situation where the proportionality coefficient is represented as a real coefficient, then when it is the imaginary part, we can consider the whole function (2.87).

The exponential function of a real argument with a real exponent coefficient will look as follows:

$$y = (a_0 + ia_1)e^{b_0(x_r+ix_i)}. \tag{2.88}$$

In exponential form it will be written as follows:

$$y = ae^{b_0x_r}e^{i(\alpha+b_0x_i)}, \tag{2.89}$$

where the following equalities hold for the real and imaginary parts:

$$\begin{cases} y = ae^{b_0x_r}\cos(\alpha + b_0x_i), \\ 0 = ae^{b_0x_r}\sin(\alpha + b_0x_i). \end{cases} \tag{2.90}$$

The imaginary part can be equal to zero in two cases –with a positive exponent coefficient $x_r \to$ -∞ and when

$$\alpha + b_0x_i = \pi k. \tag{2.91}$$

Situations where one or all of the factors tend to infinity do not exist in economics, so we will concentrate on equality (2.91).

The last condition represents a combination of equations of a line in a complex plane of the argument parallel to the axis of real values of the complex argument since it follows from (2.91):

$$x_i = \frac{\pi k - \alpha}{b_0}. \tag{2.92}$$

In a simple case where $k = 0$ there is a line

$$x_i = -\frac{\alpha}{b_0} \tag{2.93}$$

in the complex plane.

Since we are looking at the problem of presenting a function in three-dimensional space, (2.93) represents a plane perpendicular to the plane of a complex argument and parallel to the axes of the real part of the complex argument x_r and function y.

The first Eq. (2.90) describes a nonlinear surface in three-dimensional space. Let us consider it.

If the real part of a complex argument is a constant value $x_r = d = const$, then the function varies by the cosine law:

$$y = ae^{b_0 d} \cos(\alpha + b_0 x_i). \tag{2.94}$$

If the imaginary part of the complex argument is constant $x_i = g = const$, then the function varies according to the exponential law:

$$y = ae^{b_0 x_r} \cos(\alpha + b_0 g). \tag{2.95}$$

Since the last condition is a restriction (2.93) that follows from the fact that the imaginary part of the function under consideration is equal to zero, then in three-dimensional space, the exponential function of a complex argument with a real exponent coefficient represents an exponent:

$$y = ae^{b_0 x_r} \cos\left(\alpha + b_0 \frac{\pi k - \alpha}{b_0}\right) = ae^{b_0 x_r}. \tag{2.96}$$

Let the exponential function of a real argument have an imaginary coefficient of the exponent:

$$y = (a_0 + ia_1)e^{ib_1(x_r + ix_i)}. \tag{2.97}$$

In exponential form this will look like

$$y = ae^{-b_1 x_i} e^{i(\alpha + b_1 x_r)}, \tag{2.98}$$

where for the real and imaginary parts of the function

$$\begin{cases} y = ae^{-b_1 x_i} \cos(\alpha + b_1 x_r), \\ 0 = ae^{-b_1 x_i} \sin(\alpha + b_1 x_r). \end{cases} \tag{2.99}$$

If the imaginary part of the function under consideration equals zero, then

$$\alpha + b_1 x_r = \pi k \tag{2.100}$$

or

$$x_r = \frac{\pi k - \alpha}{b_1}. \tag{2.101}$$

This means we have again obtained an equation of lines that, in a plane of a complex argument, are parallel to the imaginary axis. We can limit ourselves to the case where $k = 0$. In the space under consideration this equation means a plane perpendicular to the plane of a complex argument and passing through line (2.101).

This means that on a complex nonlinear surface described by the first equality (2.99) there is a curve with a constant value of x_r. It is clear from the first equation of system (2.99) that this curve is described by the exponent

$$y = Ce^{-b_1 x_i}, \quad C = a\cos(\alpha + b_1 x_r), \quad x_r = const.$$

Now it is clear what will represent the full exponential function of a complex argument. Let us present again the complex values of the model – the proportionality coefficient and the complex argument – in exponential form. Grouping the constituents of the module and the polar angle we get

$$y = ae^{b_0 x_r - b_1 x_i} e^{i(\alpha + b_1 x_r + b_0 x_i)}. \tag{2.102}$$

Let us represent this model as an equality system of real and imaginary parts:

$$\begin{cases} y = ae^{b_0 x_r - b_1 x_i} \cos(\alpha + b_1 x_r + b_0 x_i), \\ 0 = ae^{b_0 x_r - b_1 x_i} \sin(\alpha + b_1 x_r + b_0 x_i). \end{cases} \tag{2.103}$$

From the last equality we can easily get

$$x_i = \frac{1}{b_0}(\pi k - \alpha - b_1 x_r). \tag{2.104}$$

This equation describes a family of parallel lines in a complex plane of the argument. In the simple case, where $k = 0$, it is a line in the plane and surface perpendicular to the complex plane of the argument in the space. Both the line and the plane are defined in the whole range of the problem.

This plane cuts off some line on the surface defined by the first equation of system (2.103):

$$y = ae^{b_0 x_r - b_1 x_i} \cos(\alpha + b_1 x_r + b_0 x_i). \tag{2.105}$$

Substituting (2.104) into this equation we get

$$y = ae^{b_0 x_r - \frac{b_1}{b_0}(\pi k - \alpha - b_1 x_r)} \cos(\alpha + b_1 x_r + (\pi k - \alpha - b_1 x_r)) = ae^{\frac{b_1}{b_0}(\alpha - \pi k) + \frac{b_0^2 + b_1^2}{b_0} x_r}. \tag{2.106}$$

This means that we have an exponent in the space located in a plane perpendicular to the complex plane of the argument.

2.8 Logarithmic Function of a Complex Argument

Let us now examine the properties of the logarithmic function of a complex argument. The logarithm of a complex variable is known as a periodical function, which is why when we study it we should specify what part of the function is being studied. It was determined in the first chapter of this book that from the entire combination of logarithmic values we will consider only the main values.

The logarithmic function of a complex argument may be presented in its general form as follows:

$$y = (a_0 + ia_1) + (b_0 + ib_1)\ln(x_r + ix_i). \tag{2.107}$$

If we apply the formula of a logarithm of a complex argument to the model under consideration, we get

$$y = (a_0 + ia_1) + (b_0 + ib_1)\left(\ln\sqrt{x_r^2 + x_i^2} + iarctg\frac{x_i}{x_r}\right). \tag{2.108}$$

Let us consider the variant where the imaginary part of the complex proportionality coefficient is equal to zero:

$$y = (a_0 + ia_1) + b_0\left(\ln\sqrt{x_r^2 + x_i^2} + iarctg\frac{x_i}{x_r}\right). \tag{2.109}$$

Opening the brackets and grouping the real and imaginary parts of this equation we get

$$y = a_0 + b_0\ln\sqrt{x_r^2 + x_i^2} + i\left(a_1 + b_0 arctg\frac{x_i}{x_r}\right). \tag{2.110}$$

Two equalities for the real and imaginary parts follow from the preceding equation:

$$\begin{cases} y = a_0 + b_0\ln\sqrt{x_r^2 + x_i^2}, \\ 0 = a_1 + b_0 arctg\dfrac{x_i}{x_r}. \end{cases} \tag{2.111}$$

The second equation requires a constant polar angle in the plane of the complex argument:

$$arctg\frac{x_i}{x_r} = -\frac{a_1}{b_0}. \tag{2.112}$$

This indicates an equation of the line passing through the neighborhood of the zero point but not including it. The zero point does not exist for the first equation as well since a logarithm of zero does not exist.

The first equation of system (2.111) describes a nonlinear surface in three-dimensional space. We are interested in the location of the line on this surface that satisfies condition (2.112). Thus, let us consider the equation

$$y = a_0 + b_0 \ln\sqrt{x_r^2 + x_i^2}$$

for $x_i = dx_r$.

If we substitute this into the equation, we get

$$y = a_0 + b_0 \ln x_r\sqrt{1 + d^2}. \tag{2.113}$$

Therefore, the logarithmic function of a complex argument with a real proportionality coefficient represents in three-dimensional space a logarithmic function passing through the zero point and lying in a plane perpendicular to the complex plane of the argument.

Now let us consider the second extreme version, when the real part of the complex proportionality coefficient is equal to zero:

$$y = (a_0 + ia_1) + ib_1\left(\ln\sqrt{x_r^2 + x_i^2} + iarctg\frac{x_i}{x_r}\right). \tag{2.114}$$

Grouping the real and imaginary parts of this function we get the following system:

$$\begin{cases} y = a_0 - b_1 arctg\frac{x_i}{x_r}, \\ 0 = a_1 + b_1 \ln\sqrt{x_r^2 + x_i^2}. \end{cases} \tag{2.115}$$

The fact that the imaginary part equals zero means that in the complex plane of the argument the function is determined on a circumference since this equality can easily be made as follows:

$$\sqrt{x_r^2 + x_i^2} = e^{-\frac{a_1}{b_1}} = const = d. \tag{2.116}$$

The zero point does not serve to define the function because a logarithm of zero does not exist. In the space under consideration the second equation of system (2.115) represents a cylinder surface. This cylinder surface cuts off a curve in the plane of the first equation, which we are interested in.

Since the polar angle in the plane of a complex argument varies on a circle, on the surface described by the first equation of system (2.115) a curve is defined that represents an arctangent function lying on the surface of the cylinder perpendicular to the complex plane of the argument.

The general logarithmic function of complex argument (2.107) represents a complex superposition of these two functions. After opening the brackets in the right-hand side of equality (2.107) and grouping the real and imaginary parts we get

$$y = a_0 + b_0 \ln \sqrt{x_r^2 + x_i^2} - b_1 arctg \frac{x_i}{x_r} + i\left(a_1 + b_1 \ln \sqrt{x_r^2 + x_i^2} + b_0 arctg \frac{x_i}{x_r}\right). \tag{2.117}$$

This equality holds only when the real and imaginary parts are equal to each other:

$$\begin{cases} y = a_0 + b_0 \ln \sqrt{x_r^2 + x_i^2} - b_1 arctg \dfrac{x_i}{x_r} \\ 0 = a_1 + b_1 \ln \sqrt{x_r^2 + x_i^2} + b_0 arctg \dfrac{x_i}{x_r} \end{cases}. \tag{2.118}$$

The second equation of the system describes a curve in the plane of the complex argument that does not include the zero point:

$$0 = a_1 + b_1 \ln \sqrt{x_r^2 + x_i^2} + b_0 arctg \frac{x_i}{x_r}. \tag{2.119}$$

The approximate form of the function can be imagined from the location of a function with these coefficients in the plane:

$$0 = -3 \ln \sqrt{x_r^2 + x_i^2} + 0,2 arctg \frac{x_i}{x_r}. \tag{2.120}$$

This function is given in Fig. 2.15.

The first equation of system (2.118) represents a complex surface. Its general form may look like this: in space there are a great number of lines like those shown in Fig. 2.15 that are parallel to each other and increase on the S-axis or decrease with the growth of the argument, depending on the function coefficient. This surface intersects with another one perpendicular to that of the complex argument and passing in the plane of the complex argument through the points determined by

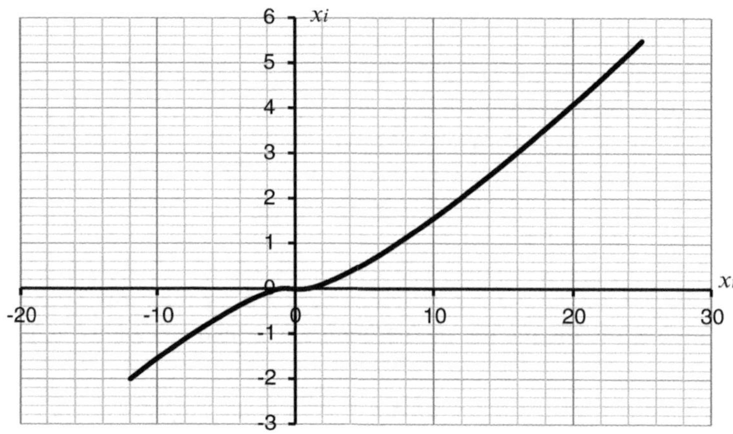

Fig. 2.15 Line (2.120) in plane of complex argument

line (2.119). Intersection of these two planes gives a line in space that has the form of the line in Fig. 2.15.

The functions of a complex argument studied here do not exhaust their full range, but of those that can be used in economic practice, the previously mentioned function are fundamental.

Reference

1. Kasana HS (2005) Complex variables: theory and applications. PHI Learning, New Delhi

Chapter 3
Conformal Mappings of Functions of a Complex Variable

To understand the meaning of a model applied in economics, one should know its basic properties including graphical characteristics. Conformal mapping shows graphically how one complex variable is mapped to another complex variable by means of a particular complex function. This chapter not only considers the corresponding section of the complex variable function theory from the point of view of the economy but presents conformal mappings of functions that are not considered or practically applied in the TFCV but that can, however, be widely applied in economic modeling, for example, the power complex-value function with complex exponent.

3.1 Power Functions of a Complex Variable

Since functions of a complex variable have analogs in the domain of real variables, models of complex variables are often considered a convenient notation of complex relationships y of the domain of real numbers. This property of describing in a simple way complex relationships in the domain of real numbers is clearly demonstrated by functions of a complex variable that are to be considered in this chapter [1]. We are interested in those properties of functions of a complex variable and those parts of the TFCV that may be applied in economic modeling. From this point of view we will consider the basic functions of a complex variable.

The power function of a complex variable is the most popular in the practice of economic modeling. As we did previously, let us denote the explanatory complex variable as follows:

$$z = x_r + ix_i = re^{i\varphi}, \tag{3.1}$$

where

x_r – real part of the complex variable;

S. Svetunkov, *Complex-Valued Modeling in Economics and Finance*,
DOI 10.1007/978-1-4614-5876-0_3,

x_i – imaginary part of the complex variable;
r – module of the complex variable, $r = \sqrt{x_r^2 + x_i^2}$
φ – polar angle of this variable (argument of a complex variable $Argz$):

$$\varphi = Argz = arctg\frac{x_i}{x_r},$$

i – imaginary unit.

Let us represent the resulting variable in the form of another complex variable:

$$w = y_r + iy_i = \rho e^{i\theta}, \tag{3.2}$$

where

y_r – real part of resulting complex variable;
y_i – imaginary part of resulting complex variable;
r – module of complex variable, $\rho = \sqrt{y_r^2 + y_i^2}$;
u – polar angle of this variable;

$$\theta = Argw = arctg\frac{y_i}{y_r}.$$

We will consider that, according to general axiomatic assumptions, all the variables considered in this study are nondimensional or have one dimension and one scale. In addition, they are defined in the entire domain of the complex plane. Taking into account the introduced symbols, the power function of complex variables will have the following form:

$$y_r + iy_i = (a_0 + ia_1)(x_r + ix_i)^{b_0 + ib_1}. \tag{3.3}$$

In exponential form, using (3.1) and (3.2), a power model of complex variables (3.3) may be represented as follows:

$$\rho e^{i\theta} = (a_0 + ia_1)(re^{i\varphi})^{b_0 + ib_1}. \tag{3.4}$$

Let us first consider a simple form of the power function of complex variables, when all the coefficients are real numbers, subsequently making the model increasingly complex. Let the coefficient of proportionality be equal to one, $a = 1$, and the exponent is a positive integer $b = n$. The model will have the form

$$y_r + iy_i = (x_r + ix_i)^n. \tag{3.5}$$

It is more convenient to consider the properties of this function in exponential form:

$$\rho e^{i\theta} = r^n e^{in\varphi},$$

where due to the properties of the equality of complex numbers

$$\begin{cases} \rho = r^n, \\ \theta = n\varphi. \end{cases}$$

Then it is evident that the conformal mapping provided by function (3.5) results in stretching of the module of the complex variable z to the nth power and an increase in the polar angle of the complex variable z n times. If we consider the complex variable z as a vector in polar coordinates, the vector will rotate by an angle

$$(n-1)\varphi.$$

Since the polar angle is defined up to a period, all of its functions are periodical. From this it obviously follows that complex numbers z_1 and z_2 with equal modules and arguments that differ from one another by a multiple of $2\pi/n$ move at mapping (3.5) to a single point.

In the TFCV, there is a concept of mutually monosemantic or univalent mapping. It has the following definition: if the function $w = f(z)$ is monosemantic or single-valued on the set M and two different points on this set M of the complex plane w are always correlated with different points N in the complex plane z, then this mapping is called univalent. Otherwise, the mapping is multivalent. For the exponential function under consideration, multivalency is valid only in particular sectors of the complex plane z, namely, in those sectors where a condition is true restricting the value of the initial variable's polar angle:

$$k2\pi/n < \varphi < (k+1)2\pi/n,$$

where k is a positive integer.

If the exponent of function (3.5) is not a whole number $b \neq n$, which is expected for most economic problems, then the nature of the arguments will not change. The module of the mapped variable z grows to a power b and the angle turns counterclockwise b times, except that under conditions of univalency one should insert nonwhole b instead of positive integer n:

$$k2\pi/b < \varphi < (k+1)2\pi/b.$$

With the values of the complex factor z, when the polar angle u goes beyond the restrictions (3.6), the function becomes multivalent, i.e., there is a set of points in a

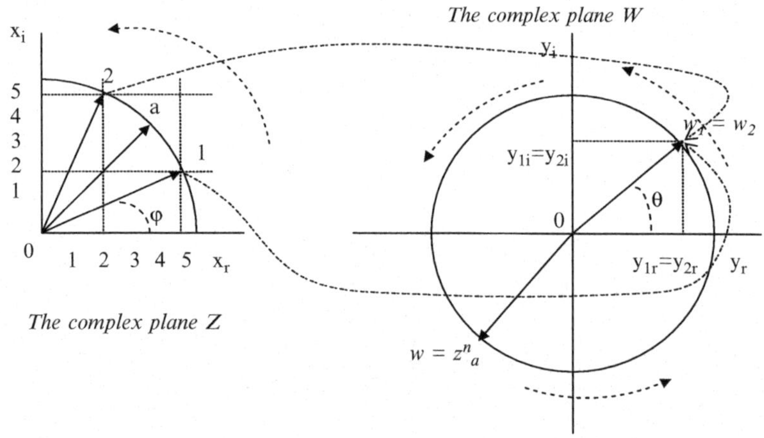

Fig. 3.1 Model (3.5) for $b = 3{,}128$ gives the same result $y_{1i} = y_{2i}$ and $y_{1r} = y_{2r}$ for various $x_{1r} = 5$; $x_{1i} = 2$ and $x_{2r} = 2$; $x_{2i} = 5$

complex plane of explanatory variables z that, by means of (3.5), will be mapped in the complex plane of results w to the same point. For example, with the exponent $b = 3{,}128$, the same result $w(y_r, y_i)$ may be obtained if we take five x_r and two x_i units or take a smaller number of x_r units, for example, two, but increase the number of x_i units to five (Fig. 3.1).

In Fig. 3.1 the first point of the complex plane $z(5,2)$ is mapped in the complex plane of results to point w_1. Another point in the complex plane $z(2,5)$ is mapped in the complex plane of results to point w_2. The coordinates of points w_1 and w_2 coincide with each other. This is why the complex variable z coming from the first point $z(5,2)$ around the circumference to point $z(2,5)$ is mapped, by means of an exponential function with the exponent $b = 3{,}128$, to a circumference of a complex plane of the results w with another radius, forming a complete cycle, beginning from point w_1 and ending at point w_2.

If we continue the counterclockwise movement in the complex plane z around the circumference shown in the figure, the conformal mapping of this movement to the plane of the results w will again correspond to a counterclockwise movement around the circumference from the point $w_1 = w_2$. And again, the points in the resulting plane will pass through the complete cycle while in plane z the points pass only the sector of the circumference.

This is what the multivalence of conformal mapping means – in a complex plane z there is a set of points that are mapped to another complex plane w at the same point. That is, the same result may be obtained for various combinations of the real and imaginary components of the complex factor.

Since model (3.5) represents the equality of the real and imaginary parts, they comprise a system of two real equations:

$$\begin{cases} y_r = \left(\sqrt{x_r^2 + x_i^2}\right)^b \cos\left(barctg\dfrac{x_i}{x_r}\right), \\ y_i = \left(\sqrt{x_r^2 + x_i^2}\right)^b \sin\left(barctg\dfrac{x_i}{x_r}\right). \end{cases} \tag{3.6}$$

It is virtually impossible to imagine a situation in economics where a complex argument would change cyclically around a circumference that has been set for good. More often are situations either involving the growth of a complex argument that is becoming linear or nonlinear dynamics.

In the first case, where the complex argument shows linear changes, we expect two variants:

- When a line passes through the zero point in the argument's complex plane, and the polar angle of the complex argument is a constant value;
- When the line does not pass through the zero point and the polar angle of the complex argument changes.

In the first case with the growth of the argument's module the real and imaginary parts of the complex function will change nonlinearly according to the exponential law, depending on exponent b, but in the complex plane the result will be seen on the line.

In the second case the real and imaginary components change nonlinearly according to a more complicated law, as they represent the result of multiplication of two components – the power function and the periodical component. In the complex plane they will represent smooth nonlinear functions monotonically increasing or decreasing depending on the argument's values and on how the polar angle changes: it will decrease (if the line passes above the zero point) or increase (if the line passes below the zero point).

If the argument changes nonlinearly, then the function under consideration will model various nonlinear relationships in the complex plane of the result. Figure 3.2 shows how the transition of the argument from point 1 to point 2 will be displayed in the complex plane of the result.

Let us consider particular features of the behavior of power function (3.5) with exponent b lying within the limits

$$0 < b < 1. \tag{3.7}$$

Let $b = 1/n$, where n is any positive whole number. This function is very well studied in the TFCV; its behavior has been studied for closed curves lying in a complex plane having (Fig. 3.3) or not having (Fig. 3.4) the point $z = 0$ inside [2].

In the first case, mapping of the complex variable by means of function (3.5) gives n continuous and single-valued functions called branches of a multivalued function (3.5) depending on condition (3.7), each of which takes one of the values $\sqrt[n]{z}$ since, in circumvention of a closed curve having point $z = 0$ inside, the argument of the complex variables of the factors receives an increment 2π, and

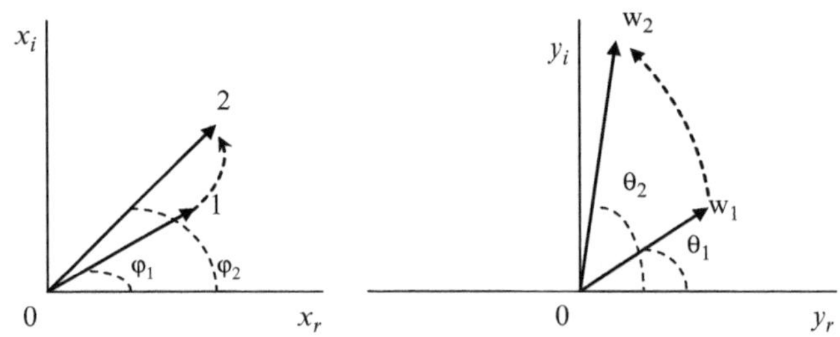

Fig. 3.2 Conformal mapping of $w = zb$ at $1 < b < 1.5$ when argument moves from point 1 to point 2

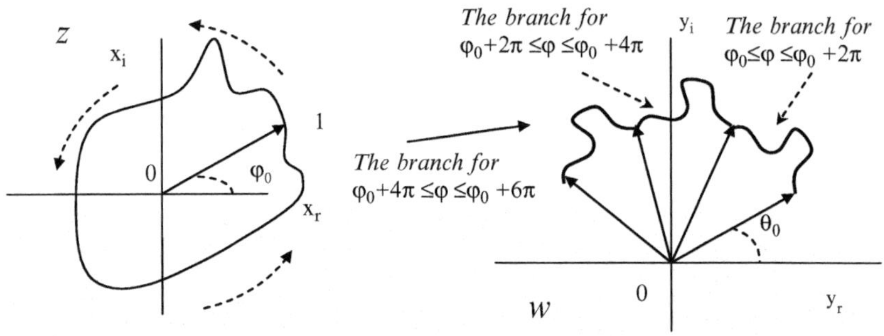

Fig. 3.3 Conformal mapping of closed curve in complex plane (x_r,x_i) having zero point inside by means of function $w = z^b$ to complex plane (y_r,y_i) at $0 < b < 1$

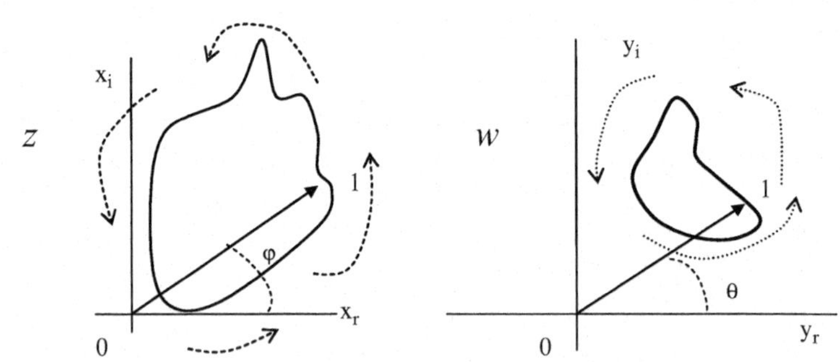

Fig. 3.4 Conformal mapping of closed curve in complex plane (x_r,x_i) without zero point inside by function $w = z^b$ to complex plane (y_r,y_i) at $0 < b < 1$

the point $\sqrt[n]{z}$ with this argument in the plane of a complex result does not return to its initial position until the vector on the complex plane of the factor makes an n-times circumvention of the closed curve.

In other words, a vector in the factor z plane, having made a complete circumvention, is mapped to the complex plane of the result w only in the form of one nonclosed curve. If a vector of factor z makes another circumvention, then in the complex plane of the result w this will be shown by points lying on the next curve similar in form to the first one but following it in the complex plane in a counter-clockwise direction. In this way branches in the complex plane of the result will grow until they finally close. But this can occur only when the vector of factor z makes a circumvention in its plane exactly n times.

In the second case, where the closed curve does not have the zero point inside (Fig. 3.4), the argument of the complex variable of resources does not receive an increment 2π because it does not make a full circumvention around the origin of coordinates, which is why function (3.5) under condition (3.7) correlates each point in the factor plane with only one point in the plane of complex results.

This means that in the second case the closed curve in the resulting complex plane w will correspond to the closed curve in the plane z.

Without loss of generality, these ideas may be extended to other cases of condition (3.7), i.e., when the exponent $b = 1/n$ takes any values on the specified interval including the cases when n is not a whole number.

Which of these two possible variants should be attributed to economic practice when complex power functions can be used? Certainly, a considerable, if not the largest, part of economic indicators is considered only in the first quadrant of a Cartesian coordinate system since they are essentially nonnegative (prime cost, volume of output, quantity of energy consumed, labor productivity, number of employees, etc.).

Gross profit as an indicator of production efficiency may also be negative if the company operates at a loss, but cases of negative economic variables are quite rare.

Thus, economists are most interested in the nature of conformal mapping using a model of a nonclosed curve lying in the first quadrant of a complex plane of factors, i.e., when the initial variables of the model are positive.

The real and imaginary parts of the function under consideration will represent a system of two equations:

$$\begin{cases} y_r = \left(\sqrt{x_r^2 + x_i^2}\right)^{\frac{1}{n}} \cos\left(\frac{1}{n} arctg \frac{x_i}{x_r}\right), \\ y_i = \left(\sqrt{x_r^2 + x_i^2}\right)^{\frac{1}{n}} \sin\left(\frac{1}{n} arctg \frac{x_i}{x_r}\right). \end{cases}$$

If n tends to infinity, then the real part of the function tends to one and the imaginary part to zero. In all other cases the function models nonlinear dynamics of the real and imaginary parts with an increase in the argument. For example, if in a complex plane of an argument its values are on a line passing below the zero point described, for example, by the equation

$$x_i = x_r - 3.55, \tag{3.8}$$

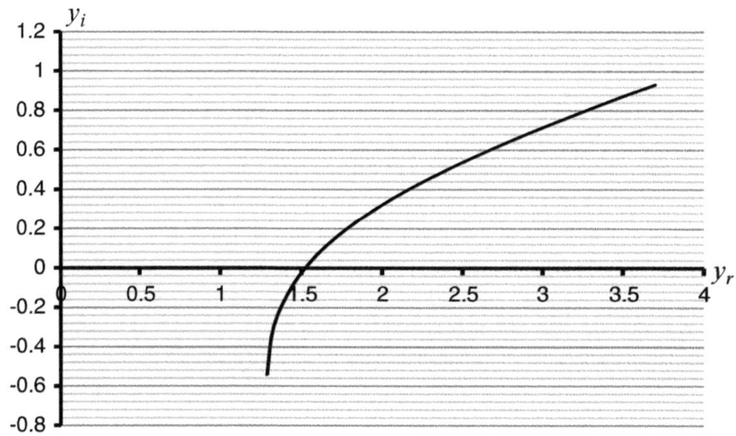

Fig. 3.5 Function (3.5) for argument (3.8) and exponent $b = 1/3$

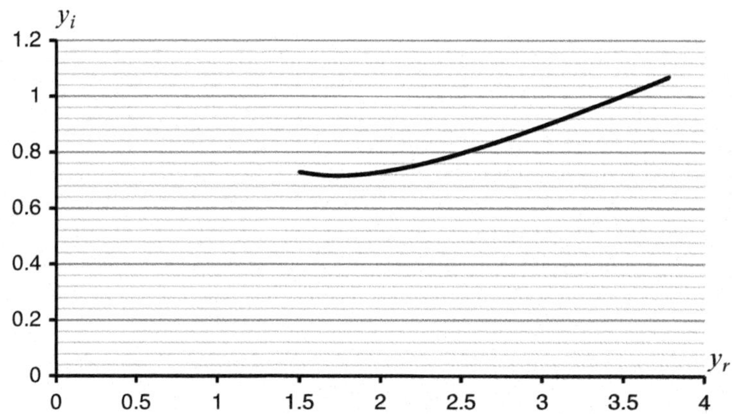

Fig. 3.6 Function (3.5) for argument (3.9) and exponent $b = 1/3$

then for $n = 3$ and $x_r = 1,2,3,\ldots 40$ on the complex plane the points of the function will lie on a curve shown in Fig. 3.5.

For the same exponent but in the case where the argument is described by a line passing above the zero point, for example,

$$x_i = x_r + 3.55, \tag{3.9}$$

a line will describe an increasing curve of another form, shown in Fig. 3.6.

If the exponent is negative, then any increase in argument z will definitely lead to a decrease in function w values – an increase in the argument r module leads to a decrease in the function c module, and an increase in the polar angle φ leads to a rotation of the conformal mapping in the opposite direction, as for the case

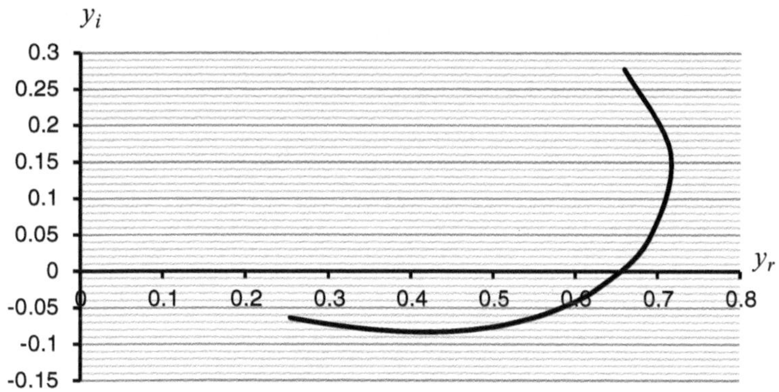

Fig. 3.7 Function (5.5) for argument (3.8) and exponent $b = -1/3$

$$\rho = r^{-b}, \quad \theta = -\mathrm{b}\varphi. \tag{3.10}$$

Are there cases in economics when a complex power function has a negative exponent? Such cases are quite possible in economic practice. For example, in production, when additional involvement of labor, resources, and capital only diminishes production results (the module of the complex variable of production results decreases) or, in other words, the redundancy of employees and the closure of some production sites (for example, noncore divisions) has a positive effect on production results, the volume of gross output increases and the gross margin grows. These cases are not rare in economics, especially in times of crisis.

This means that a complex power function with negative exponent has the right to exist in a complex economy.

The conformal mapping of a power function with negative exponent will again be nonlinear. If an argument increases according to a linear law (3.8) and within the same limits, and the exponent $b = -1/3$, then the function will change its values in the clockwise direction from $y_r = 0.658$, $y_i = 0.278$ to a point with the coordinates $y_r = 0.254$, $y_i = -0.064$ (Fig. 3.7).

When the argument line passes above the zero point, as described by relationship (3.9), a decrease in the module will be shown by another curve (Fig. 3.8) in a counterclockwise direction from a point with the coordinates $yr = 0.539$, $yi = -0.261$ up to a point with the coordinates $yr = 0.245$, $yi = -0.069$.

Let us consider a power function of complex variables with imaginary exponent and coefficient of proportionality equal to one:

$$y_r + iy_i = (x_r + ix_i)^{ib}. \tag{3.11}$$

Its exponential form is as follows:

$$\rho e^{i\theta} = \left(re^{i\varphi}\right)^{ib} = e^{-b\varphi}r^{ib} = e^{-b\varphi}e^{ib\ln r}. \tag{3.12}$$

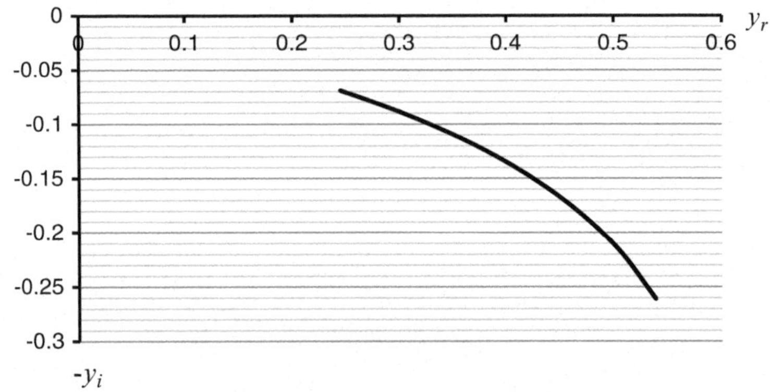

Fig. 3.8 Function (5.5) for argument (3.9) and exponent $b = -1/3$

Here exponent b is any real number.

Then the module of the function under consideration will represent the following relationship:

$$\rho = e^{-b\varphi}, \text{ the argument being } \theta = b \ln r.$$

It is easy to see that in this model, the module of function c depends on the correlation between factors x_r and x_i (of argument φ), and the argument of the function changes depending only on changes in the module of the complex variables of the factors.

Therefore, when the module of the complex variable of argument r is a constant value with only the polar angle changing, i.e., the argument changes its values along a circle, by means of function (3.11) it is mapped to the complex plane w in the form of a line at an angle equal to $b\ln r$. The points on the line increase with the increase in the polar angle if the exponent b is negative and decrease if it is positive.

If the complex argument changes linearly and this line passes through the zero point but does not include it within, then that indicates stability of the polar angle and the growth of the argument module. Conformal mapping of these points to the complex plane w takes place in the form of a circumference since the module is not changed and the polar angle changes with changes of the argument's module.

Of greater interest is the situation where an argument represents a line passing either above or below the zero point. In this case, both the polar angle of the argument and its module will change. Then function (3.11) will model the nonlinear dynamics of both the real and imaginary parts of the complex function, since they may be presented as a system:

$$\begin{cases} y_r = e^{-b\,\mathrm{artg}\frac{x_i}{x_r}} \cos\left(b \ln \sqrt{x_r^2 + x_i^2}\right), \\ y_i = e^{-b\,\mathrm{artg}\frac{x_i}{x_r}} \sin\left(b \ln \sqrt{x_r^2 + x_i^2}\right). \end{cases} \tag{3.13}$$

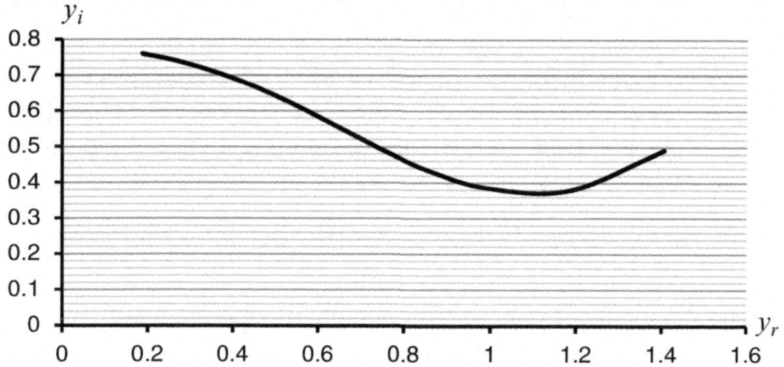

Fig. 3.9 Function (3.11) for argument (3.8) and exponent $b = 1/3$

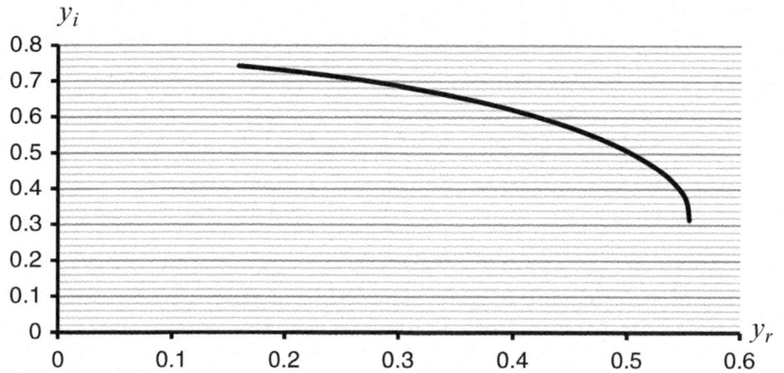

Fig. 3.10 Function (3.11) for argument (3.9) and exponent $b = 1/3$

If the argument is described by a line passing below the zero point as defined by (3.8), then with the exponent $b = 1/3$ the conformal mapping of function (3.11) will have the form of a nonlinear function, shown in Fig. 3.9. The function changes its values from right to left in a counterclockwise direction from a point with the coordinates $y_r = 1.407$, $y_i = 0.491$ to a point with the coordinates $y_r = 0.186$, $y_i = 0.759$.

If a line in the plane of the complex argument passes above the zero point as defined by (3.9), then, within the same range of values of the argument and with the exponent $b = 1/3$, the function will change again, decreasing the module with an increase of the polar angle. However, the form of this nonlinear change will differ from Fig. 3.9: it is smoother, which is clearly seen from Fig. 3.10.

For negative values of exponent (3.11) other forms of the conformal mapping will be modeled.

Let us now consider a general model of a complex power function with complex coefficients:

$$y_r + iy_i = (a_0 + ia_1)(x_r + ix_i)^{b_0 + ib_1}, \tag{3.14}$$

which combines all the aforementioned variants of a conformal mapping and includes a complex coefficient of proportionality.

The effect of the complex coefficient of proportionality on the behavior of this function is easy to understand if it is represented in exponential form:

$$a_0 + ia_1 = ae^{i\alpha}.$$

This coefficient changes the function module a times and shifts the polar angle by α.

Let us present model (3.14) in exponential form:

$$w = ae^{b_0 \ln r - b_1 \varphi} e^{i(\alpha + b_0 \varphi + b_1 \ln r)}. \tag{3.15}$$

Then, for the real and imaginary parts of the complex function under consideration

$$\begin{cases} y_r = ae^{b_0 \ln r - b_1 \varphi} \cos(\alpha + b_0 \varphi + b_1 \ln r), \\ y_i = ae^{b_0 \ln r - b_1 \varphi} \sin(\alpha + b_0 \varphi + b_1 \ln r). \end{cases} \tag{3.16}$$

It is evident from the equalities obtained that for various combinations of the complex coefficients ($a_0 + ia_1$) and ($b_0 + ib_1$) function (3.14) will describe the most varied forms of relationship including cyclical ones. From this point of view complex model (3.14) is universal and can be used for numerous economic applications.

To get an idea of the possible forms of a conformal mapping of this function, let us consider the following model:

$$y_r + iy_i = (x_r + ix_i)^{-0.5 + i}. \tag{3.17}$$

In the first case line (3.8) passing in the plane of the complex argument below the zero point (Fig. 3.11) is mapped to the complex plane w, and in the second case line (3.9) passes in the plane of the complex argument above the zero point. This second case is shown in Fig. 3.12.

Comparing these two figures we see that the position of the line in the plane of the complex argument considerably affects the form of the conformal mapping.

Thus, it should be stated that a complex power function with complex exponent may be used to model the most varied nonlinear processes in an economy.

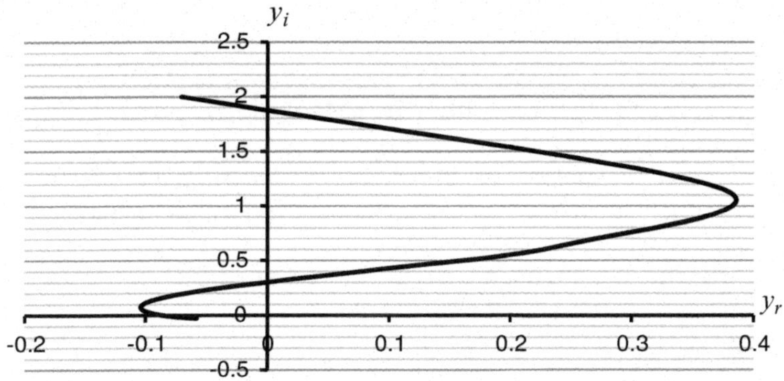

Fig. 3.11 General form of conformal mapping of exponential complex function for argument (3.8) and complex exponent (−0.5 + *i*)

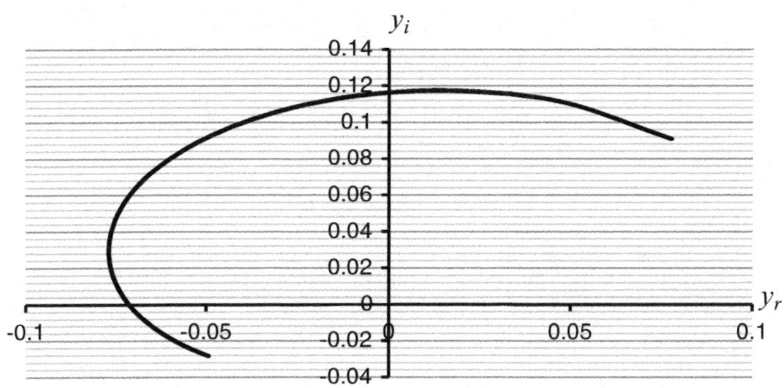

Fig. 3.12 General form of conformal mapping of exponential complex function for argument (3.9) and complex exponent (−0.5 + *i*)

3.2 Exponential Functions of Complex Variables

Exponential functions of real variables are not the only class of functions used in modern economics, though they are dominant among economic models of real numbers. In particular, it is exponential functions that are widely applied in the theory of production functions since exponents have a clear economic interpretation and the differentiability of functions makes it possible to judge the modeled production processes by the values of the first and second derivatives. These models are fundamental for modeling economic dynamics, for parametric models in price formation, etc. However, since other functions are also used in the economics of real variables, in a complex economy it is necessary to study the properties of complex functions that are similar to those of real variables and the possibility of using them in the practice of mathematical economic modeling.

Let us consider exponential functions taking the following exponential complex function as an example:

$$y_r + iy_i = (a_0 + ia_1)e^{(b_0+ib_1)(x_r+ix_i)}. \tag{3.18}$$

The properties of other exponential functions will be similar to this one, which is why we can extrapolate them to the entire class of exponential functions of complex variables.

Let us agree that initially the exponent is a real number, i.e., $b1 = 0$. Then the function in exponential form will look like this:

$$y_r + iy_i = ae^{i\alpha}e^{b_0(x_r+ix_i)} = ae^{b_0x_r}e^{i(\alpha+b_0x_i)}. \tag{3.19}$$

That is, the function module depends only on the changes of the real part of the complex argument:

$$\rho = ae^{b_0x_r}. \tag{3.20}$$

The polar angle is wholly defined by changes in the imaginary part of the complex argument:

$$\theta = \alpha + b_0x_i. \tag{3.21}$$

Let us first consider the simplest situation, where one of the factors x_r or x_i is a constant value, with only the other factor changing.

If the real part of the complex argument remains constant, i.e., $x_r = X = const$, but the imaginary part x_i increases, then in the plane of the complex argument this indicates a line parallel to the axis of imaginary values and crossing the line of real numbers in point X. A conformal mapping of this line will represent a circumference with constant radius:

$$\rho = ae^{b_0X} \tag{3.22}$$

and a polar angle (3.21) that increases with an increase in the imaginary part of the positive coefficient b_0 and decreases if it is negative.

In the second case, where the line in the complex plane of the argument is perpendicular to the axis of the imaginary numbers at $x_i = X = const$ and changing x_r, function (3.19) is mapped in the plane w as a line since it is characterized by a constant value of the polar angle

$$\theta = \alpha + b_0x_i = \alpha + b_0X,$$

with the radius changing according to exponent (3.20).

When the complex argument changes its values linearly and is not parallel to any axis of the plane of the complex argument, function (3.19) models a more complex line. The real and imaginary parts of this function will look as follows:

$$\begin{cases} y_r = ae^{b_0 x_r}\cos(\alpha + b_0 x_i), \\ y_i = ae^{b_0 x_r}\sin(\alpha + b_0 x_i). \end{cases} \tag{3.23}$$

The nature of the conformal mapping depends on the sign of exponent b_0. If $b > 0$, a linear increase in the values of the complex argument will be followed by a considerable increase in the module of the function. Since its polar angle also increases, in the complex plane w, this will mean movement along a divergent spiral, the real and imaginary parts of the complex function changing by sine and cosine respectively with a nonlinear increase of the oscillations.

When the exponent $b_0 < 0$, the module of the function decreases exponentially, and in the complex plane it will look like a spiral that tends to zero, and the real and imaginary parts considered separately will represent a damped oscillatory process.

Now let us consider the case where the exponent is multiplied by the imaginary coefficient b_I and the real one is equal to zero, $b_0 = 0$:

$$y_r + iy_i = (a_0 + ia_1)e^{ib_1(x_r + ix_i)} = (a_0 + ia_1)e^{-b_1 x_i}e^{ib_1 x_r}. \tag{3.24}$$

The module of this function represents an exponential dependence on the imaginary part of the complex argument

$$\rho = ae^{-b_1 x_i}, \tag{3.25}$$

and the polar angle represents a linear dependence on the real part of the argument:

$$\theta = \alpha + b_1 x_r. \tag{3.26}$$

If we compare (3.25) with (3.20), and (3.26) with (3.21), we will see that the function has varied "symmetrically" – now the imaginary part of the argument influences the module of the function and the real part influences the polar angle.

This means that exponential function (3.24) will map a linear increase of the complex argument to the complex plane w in the form of a spiral, which is divergent if coefficient $b_1 = 0$ and convergent to zero if $b_1 > 0$.

Now we can consider function (3.18) with a complex exponent:

$$y_r + iy_i = (a_0 + ia_1)e^{(b_0 + ib_1)(x_r + ix_i)} = (a_0 + ia_1)e^{(b_0 x_r - b_1 x_i)}e^{i(b_1 x_r + b_0 x_i)}. \tag{3.27}$$

Presenting all the complex variables and coefficients of function (3.25) in exponential form and grouping the components of the polar angle and those of the module we have

$$\rho e^{i\theta} = ae^{b_0 x_r - b_1 x_i} e^{i\alpha} e^{i(b_1 x_r + b_0 x_i)}. \tag{3.28}$$

Whence for the module of the complex function,

$$\rho = ae^{b_0 x_r - b_1 x_i} \tag{3.29}$$

and for the polar angle of this function,

$$\theta = \alpha + b_1 x_r + b_0 x_i. \tag{3.30}$$

The case is of interest when an argument represents a line lying in a complex plane, i.e.,

$$c_r x_r + c_i x_i = d \to x_i = \frac{d - c_r x_r}{c_i}. \tag{3.31}$$

Let us substitute (3.31) into the module of function (3.29):

$$\rho = ae^{b_0 x_r - b_1 \frac{d - c_r x_r}{c_i}} = ae^{\frac{(c_i b_0 + b_1 c_r) x_r - b_1 d}{c_i}}. \tag{3.32}$$

The module of an exponential function with a complex proportionality coefficient varies by exponent again and may increase depending on the coefficient or decrease depending on its values.

If now we substitute (3.31) into (3.30), then it is easy to see how the polar angle of this function varies with linear variations of the complex argument:

$$\theta = \alpha + b_1 x_r + b_0 \frac{d - c_r x_r}{c_i} = \alpha + \frac{(b_1 c_i - c_r b_0) x_r + b_0 d}{c_i}. \tag{3.33}$$

It varies linearly.

Therefore, a general exponential function with complex exponent models a divergent spiral process or one that converges to zero (depending on the coefficient values) if the complex argument varies linearly.

For nonlinear variations of the argument the form of the line of the conformal mapping will change.

3.3 Logarithmic Functions of Complex Variables

Let is consider the possibility of applying logarithmic functions of complex variables in a complex economy. In general, the function of a complex variable will have the following form:

$$y_r + iy_i = (a_0 + ia_1) + (b_0 + ib_1) \ln(x_r + ix_i). \tag{3.34}$$

Obviously, other logarithm bases (not just natural but decimal, binary, etc.) are possible, but this does not influence our subsequent arguments or the basic properties of the function itself, which is why we will consider the natural basis of logarithms as the most convenient basis for our analysis.

It should be noted immediately that the logarithmic function of complex variables, as well as the function of real variables, does not exist in the point $x_r = 0, x_i = 0$, but, unlike the logarithmic function of real variables, it exists for negative values of the argument. This is why in our subsequent reasoning in this section we exclude this zero point from our consideration by default. If we need to consider the situation where some line displayed in a complex plane issues from the zero point, then we will think that it comes from the neighborhood of this point.

Since the free term on the right-hand side of the equality characterizes the initial conditions and its effect on the complex result does not go further, let us consider the function without it:

$$y_r + iy_i = (b_0 + ib_1)\ln(x_r + ix_i). \tag{3.35}$$

From the TFCV it is known that a logarithm of the complex number z may be represented as the sum

$$\ln z = \ln|z| + i\arg z = \ln r + i(\varphi + 2k\pi), \tag{3.36}$$

where k is a whole number.

This means that the complex number $z \neq 0$ has an infinite number of logarithms (functions with an infinite number of values), as the real part of a logarithmic function is defined by a single value and its imaginary part is defined up to a multiple of 2π.

It is clear from (3.36) that if the real or imaginary part of a complex argument of a logarithmic function is negative, then a logarithm of this function exists. This is an additional argument in favor of using functions of complex variables in economics since a logarithmic relationship may be extrapolated to a range of negative values of an argument, unlike functions of real variables.

In the theory of complex variables there is a concept known as the "main logarithm value," when $k = 0$. We will just use this main value in models of a complex economy.

Applying the property of a logarithm of a complex number to the function under consideration (3.35), we get

$$y_r + iy_i = (\mathrm{b}_0 + ib_1)\left(\ln\sqrt{x_r^2 + x_i^2} + iarctg\frac{x_i}{x_r}\right) = (\mathrm{b}_0 + ib_1)(\ln r + i\varphi). \tag{3.37}$$

Let us consider the case where the imaginary part of a complex coefficient is equal to zero. Then the function will look as follows:

$$y_r + iy_i = b_0 \ln r + b_0 i\varphi \tag{3.38}$$

or

$$y_r = b_0 \ln r \tag{3.39}$$

$$y_i = b_0 \varphi. \tag{3.40}$$

If a complex argument represents points lying on a circumference, then the module of the argument will also be a constant value. This indicates the constant nature of the real part of the function. The imaginary part will change linearly since it represents a linear function of the polar angle of the argument. That is, the circumference in a complex plane of an argument is mapped to the complex plane w in the form of a line perpendicular to the axis of real variables.

If a complex argument represents a line passing in the neighborhood of the zero point, then the polar angle of the argument will be a constant, $\varphi = const$, which will lead to the constant nature of the function $y_i = const$. Therefore, in this case the function is also mapped in the complex plane w as a line, though it is perpendicular to the axis of the imaginary part of the function.

But if a complex argument is represented as a line passing beyond the zero point, then the function in question will provide another mapping of this line. Let it be

$$c_r x_r + c_i x_i = d. \tag{3.41}$$

Then, according to (3.39), the real part of the function will represent

$$y_r = b_0 \ln \sqrt{x_r^2 + \left(\frac{d - c_r x_r}{c_i}\right)^2}. \tag{3.42}$$

This means that with a linear increase in the argument, the real part of the function increases logarithmically.

The imaginary part of the function, taking into account the linear nature of the change in the argument, will change as follows:

$$y_i = b_0 arctg \frac{x_i}{x_r} = b_0 arctg \frac{d - c_r x_r}{c_i x_r} = b_0 arctg \left(\frac{d}{c_i x_r} - \frac{c_r}{c_i}\right). \tag{3.43}$$

This change follows the arctangent of the inverse value.

A simultaneous change in the real and imaginary parts of a function in a complex plane will provide a picture of smooth curves.

Figure 3.13 shows this mapping for the case where the complex argument changes according to the linear law

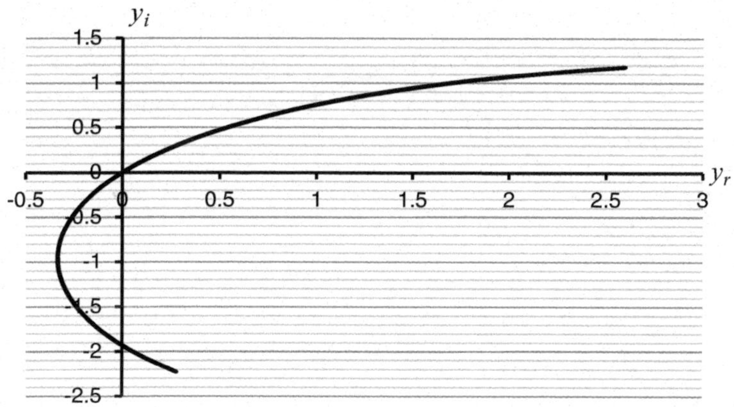

Fig. 3.13 Function (3.38) for linear change in argument

$$2x_r - 1,5x + 2 = 0,$$

where proportionality coefficient $b_0 = 1.5$.

Here the real part of the argument changes from $x_r = 0.1$ to $x_r = 4.0$, with an interval of 0.1.

Let us consider a situation where the real part of the complex proportionality coefficient of model (3.35) is equal to zero, i.e., the following function is used:

$$y_r + iy_i = ib_1 \ln(x_r + ix_i) = -b_1\varphi + ib_1 \ln r, \tag{3.44}$$

where

$$y_r = -b_1\varphi,$$

$$y_i = b_1 \ln r.$$

Here we get properties symmetrical to those of model (3.38) – the same circumference in the complex plane will be mapped to the complex plane w in the form of a line parallel to y_r, and the line issuing from the neighborhood of the zero point on the argument axis will be mapped as a line parallel to the y_i-axis.

Now we can consider the properties of model (3.35) where both the real and imaginary parts of the proportionality coefficient are not equal to zero. After opening the parentheses (3.37) we have

$$x_r + ix_i = b_0 \ln r - b_1\varphi + i(b_0\varphi + b_1 \ln r), \tag{3.45}$$

where for the real part of the equality

$$y_r = b_0 \ln r - b_1\varphi$$

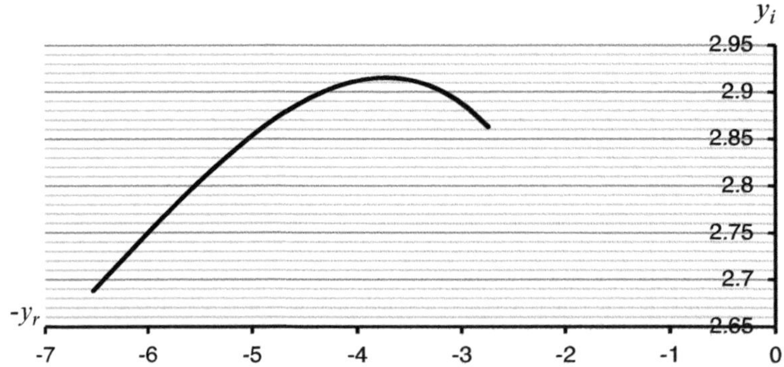

Fig. 3.14 Function (3.48) for linear changes in argument

and for the imaginary part

$$y_i = b_1 \ln r + b_0\varphi.$$

To understand the behavior of this model, let us represent it in exponential form. The tangent of the polar angle for the function under consideration will be

$$tg\theta = \frac{b_1 \ln r + b_0\varphi}{b_0 \ln r - b_1\varphi}. \tag{3.46}$$

The module of the complex result is

$$\rho = \sqrt{(b_0^2 + b_1^2)(\ln^2 r + \varphi^2)}. \tag{3.47}$$

The function in exponential form will look like this:

$$w = \sqrt{(b_0^2 + b_1^2)(\ln^2 r + \varphi^2)e^{-iarctg\frac{b_1 \ln r + b_0\varphi}{b_0 \ln r - b_1\varphi}}}.$$

As is clear, complex-form smooth curve is being modeled that depends on the proportionality coefficient and on the position of the argument points in the complex plane. An example of how the function

$$w = \sqrt{(b_0^2 + b_1^2)(\ln^2 r + \varphi^2)}e^{iarctg\frac{b_1 \ln r + b_0\varphi}{b_0 \ln r - b_1\varphi}} \tag{3.48}$$

maps linear changes in the complex argument $x_r - 2x_i - 2 = 0$ to the complex plane is given in Fig. 3.14.

3.4 Zhukovsky's Function and Trigonometric Complex Functions

The aforementioned simple complex functions in the TFCV are also supplemented by Zhukovsky's function and trigonometric functions. Since an economy is very diverse and the presence in that economy of relationships that can be described by these functions cannot be excluded, let us study their properties.

The fractional-rational function

$$w = \frac{1}{2}\left(z + \frac{1}{z}\right) \tag{3.49}$$

was thoroughly studied by N.E. Zhukovsky, which is why in the TFCV it is called Zhukovsky's function.

To determine the range of univalency of a conformal mapping of this function, we should first determine if there is a range of multivalency, i.e., if there are such z_1 and z_2 that, by means of Zhukovsky's function, transfer to one point w. To answer this question, we equate Zhukovsky's functions for these two complex variables:

$$z_1 + \frac{1}{z_1} = z_2 + \frac{1}{z_2} \rightarrow (z_1 - z_2)\left(1 - \frac{1}{z_1 z_2}\right) = 0. \tag{3.50}$$

If this equality is valid, then the function is multivalent; if not, then each of the values of the complex variable of factors is associated with only one complex value of the result.

Since by the problem condition $z_1 \neq z_2$, the function is univalent except for the points for which $z_1 z_2 = 1$. Let us see which points do not satisfy the univalency condition, i.e., where $z_1 z_2 = 1$. For that let us represent complex variables in exponential form:

$$z_1 z_2 = r_1 e^{i\theta_1} r_2 e^{i\theta_2} = 1 \leftrightarrow r_1 r_2 e^{i(\theta_1 + \theta_2)} = 1.$$

This equality will be equal to 1 if

- The sum of polar angles is equal to zero – $\theta_1 = -\theta_2$, which means symmetry of points with respect to the real number axis,
- Modules of complex variables z_1 and z_2 are equal to 1.

Thus, Zhukovsky's function is univalent in the entire complex plane, except for points of the unit circumference.

Since any complex number may be represented both in arithmetic and trigonometric forms, let us represent z as follows:

$$z = r\cos\varphi + ir\sin\varphi. \tag{3.51}$$

If we substitute this expression into Zhukovsky's function we have

$$w = \frac{1}{2}\left(z + \frac{1}{z}\right) = \frac{1}{2}\left(r\cos\varphi + ir\sin\varphi + \frac{1}{r\cos\varphi + ir\sin\varphi}\right)$$
$$= \frac{1}{2}\left(r\cos\varphi + ir\sin\varphi + \frac{r\cos\varphi - ir\sin\varphi}{r^2(\cos^2\varphi + \sin^2\varphi)}\right)$$

or

$$w = \frac{1}{2}\left(r + \frac{1}{r}\right)\cos\varphi + i\frac{1}{2}\left(r - \frac{1}{r}\right)\sin\varphi. \tag{3.52}$$

If we consider a circumference, $r = r_0 \neq 1$, in a plane, Zhukovsky's function is mapped in the resulting plane in the form of ellipses with semiaxes:

$$a = \frac{1}{2}\left(r_0 + \frac{1}{r_0}\right), \ \mathrm{b} = \frac{1}{2}\left(r_0 - \frac{1}{r_0}\right). \tag{3.53}$$

For $r_0 \to 1$ the ellipse shrinks to the interval $[-1,1]$ of the real axis of the complex plane, for $r_0 \to 0$, as well as for $r_0 \to \infty$, it tends to infinity. Only when $|z| < 1$, i.e., the points lie inside a unit circumference, with an increase in the polar angle of the argument, does the ellipse go g around in the negative direction, and when $|z| > 1$ – positive.

It is easy to see from (3.52) that for $r >> 1$ Zhukovsky's function almost coincides with the complex function, and on the interval $r < 1$ it becomes nonlinear. That is, if a complex argument varies linearly, then Zhukovsky's function will vary practically linearly, with the exception of an interval of values of the argument inside a unit circumference. Here the function becomes nonlinear.

With reference to economic problems, this function may be used in econometrics. Then it is convenient to represent it in the following way:

$$w = (a_0 + ia_1)\left(z + \frac{1}{z}\right) \tag{3.54}$$

since the complex coefficient of proportionality promotes the generation of conformal mapping in the form of ellipses of various scales and forms.

As an example, Fig. 3.15 shows a conformal mapping of Zhukovsky's function to the complex plane w for the linearly varying argument z when the imaginary component represents a dependence on the real component of the argument calculated by the formula

$$x_i = 0{,}4 + x_r, \ -7 < x_r < 7.$$

For other methods of linear variations in the argument inside a unit circumference, the function takes other nonlinear forms.

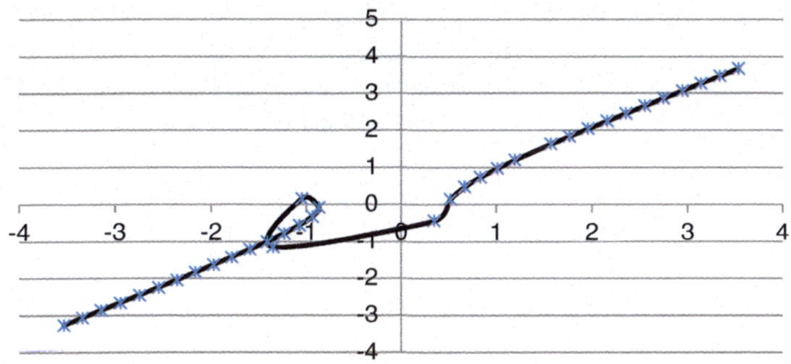

Fig. 3.15 Zhukovsky's function for linearly varying argument values ($-7 < x_r < 7$)

This means that Zhukovsky's function may be used in modeling particular economic processes when in the area of change in the real and imaginary parts of the argument from -1 to +1 the instability of linear dynamics is modeled.

Zhukovsky's function is convenient for studying trigonometric functions of complex variables.

Since the following two equalities follow from Euler's formula:

$$e^{ix} = \cos x + i \sin x; e^{-ix} = \cos x - i \sin x,$$

it is easy to derive formulae for the calculation of sines and cosines:

$$\sin x = \frac{e^{ix} - e^{-ix}}{2i}; \cos x = \frac{e^{ix} + e^{-ix}}{2}.$$

These formulae are used in the TFCV to determine the trigonometric functions of the complex variable z:

$$\sin z = \frac{e^{iz} - e^{-iz}}{2i}; \cos z = \frac{e^{iz} + e^{-iz}}{2}. \quad (3.55)$$

For these functions all trigonometric correlations are valid, they are periodical with period 2π, etc. To provide the essence of a conformal mapping of the first function, it can be represented as follows:

$$\sin z = \frac{e^{iz} - e^{-iz}}{2i} = \frac{1}{2}\left(\frac{e^{iz}}{i} + ie^{-iz}\right) = \frac{1}{2}\left(\frac{e^{iz}}{i} + \frac{i}{e^{iz}}\right) = \frac{1}{2}\left(z_1 + \frac{1}{z_1}\right). \quad (3.56)$$

Here $z_1 = \frac{e^{iz}}{i}$.

Thus, a mapping of the function sin*z* can be considered a superposition of other mappings – an exponential function of a complex variable and Zhukovsky's function. Since currently it is not clear what economic practices trigonometric functions of complex variables are applicable in, we can only state this fact.

In conclusion, it should be noted that the cosine of a complex variable differs from the sine only in the shift due to the obvious equality

$$\cos z = \sin\left(z + \frac{\pi}{2}\right).$$

The sine and cosine formulae serve as the basis for defining other trigonometric functions.

References

1. Kasana H.S. (2005), Complex Variables: Theory And Applications. New Delhi: PHI Learning Pvt. Ltd
2. Cohen H. (2007), Complex Analysis with Applications in Science and Engineering. Springer

Chapter 4
Principles of Complex-Valued Econometrics

Mathematical statistics has paid little attention to the processing of random complex values; however, such statistical processing is crucial for models and methods of complex variables to be used in practical economics. That is why in this chapter we propose principles of a new mathematical apparatus for statistical processing of economic data, namely, principles of complex-valued econometrics – regression and correlation analysis – and adapt the least-squares method to complex random variables. We also derive the formula for a pair correlation coefficient for two random complex variables and provide an interpretation of its values. The method of estimation of confidence limits of complex-valued econometric models is also provided.

4.1 Statistics of Random Complex Value: Standard Approach

Econometrics is one of the developed and most in-demand branches of economic and mathematical modeling. Economists at all levels of activity – from the macro to the micro level – continuously face problems associated with the description, explanation, and forecasting of various trends in socioeconomic dynamics or relationships between the factors. Econometrics is a branch of economics that involves methods of processing f statistical data to build economic models giving a quantitative description of the laws of economic relationships. Modern econometrics is based on methods of mathematical statistics, above all methods of regression and correlation analysis.

The main goal of these branches of mathematical statistics is to discern the interrelations between random factors, the estimation of these interrelations, the selection of a proper form of regression model, the estimation of model coefficient values, and the assessment of the reliability of the obtained results.

The set of regression models is determined by well-known mathematical functions of a variable whose number barely exceeds two tens. In terms of the types of socioeconomic dynamics this number is infinitely small since economic dynamics is

S. Svetunkov, *Complex-Valued Modeling in Economics and Finance*,
DOI 10.1007/978-1-4614-5876-0_4,

varied and one can count several tens of thousands of market factors. This is why the problem extending the range of models to be used in economics remains very relevant. It is this very problem that the complex-valued functions considered in previous solve providing an tool that describes the dependencies between variables in a different way compared to real-variable models. Various elementary functions of complex variables, with rare exceptions, allow economists to model such nonlinear interrelations that have no analogs in the econometrics of real variables, or these analogs are so complex that their practical application is pointless. Thus, using elementary models of complex variables we considerably extend the instrumental base of econometric research. Properties of elementary models of functions of complex variables and complex arguments are such that they can be used to describe the most varied economic processes – intensive and extensive, efficient and stagnating.

To make models of a complex-valued economy work in reality we appeal to econometrics. However, since models of complex variables are not practically used in the economy, modern econometrics will not help us solve this problem, so we should turn to mathematical statistics.

In practice, problems of processing statistical data of a random complex value are quite rare – they are more often in recognition of certain signals. This is why it is not easy to solve the problem of building econometric models of complex variables on the basis of the results of mathematical statistics.

Nevertheless, in mathematical statistics there are methods concerning just the statistics of a random complex value that can be applied to the solution of a given problem. According to the available literature, interest in the statistical processing of observations about variations of complex variables appeared in the 1950s and 1960s [1–4]. Later other properties of random value statistics were discovered, but these studies are insufficient and do not provide the necessary knowledge for purposes of complex-variable econometrics.

First of all let us mention the fundamental concept of the statistics of a complex variable – its mathematical expectation. The expectation of a complex random value $z = x + iy$ is a complex number:

$$M(z) = M(x) + iM(y). \tag{4.1}$$

Since any complex number may be represented as a pair of real numbers, mathematical statistics assumes that the complex representation of random functions is simply convenient for analyzing the mathematical form of a mapping, which may always be transferred to the real form. If is for that reason that functions of variance, correlation, and covariation are presented in mathematical statistics as simple and nonrandom real characteristics of random processes and functions regardless of the form of their mathematical representation.

Then, following the approach adopted in mathematical statistics, let us determine the variance of a complex random value as a real number. "The variance of a complex random value is the expectation of the square of the Absolute value of the corresponding centered value" [5]:

$$D(z) = M[|\dot{z}|] = M[\dot{x}^2 + \dot{y}^2] = M[\dot{x}^2] + M[\dot{y}^2] \tag{4.2}$$

where

$$D(x) = M[\dot{x}^2] = M[(x - \bar{x})^2], \tag{4.3}$$

$$D(y) = M[\dot{y}^2] = M[(y - \bar{y})^2]. \tag{4.4}$$

This means that the variance of a complex random value equals the sum of fvariances of its real and imaginary parts.

A correlation moment is an important characteristic used in mathematical statistics. Since according to the above-mentioned definition, a correlation moment is a real number, the expectation of a product of two random complex values

$$M[(x_1 + iy_1)(x_2 + iy_2)]$$

may not be considered a correlation moment since the product of two complex variables will be a complex number.

Incidentally, in mathematical statistics of real variables the correlation moment of two equal random values is equal to the variance; however, this does not follow from the preceding formula.

That is why it was decided to consider the correlation moment of a complex random value as the mathematical expectation of the multiplication of one variable by the conjugate of another variable:

$$\mu_{zz} = M[(x_1 + iy_1)(x_2 - iy_2)]. \tag{4.5}$$

If we perform the multiplication and group the terms, we get

$$\begin{aligned} \mu_{zz} = M[x_1x_2] + M[y_1y_2] + i(M[y_1x_2] - M[x_1y_2]) = \\ \mu_{x_1x_2} + \mu_{y_1y_2} + i(\mu_{y_1x_2} - \mu_{x_1y_2}). \end{aligned} \tag{4.6}$$

If $z_1 = z_2 = z$, then the last term of (4.6) with the imaginary unit becomes equal to zero and the correlation moment is equal to variance (4.2), which was necessary to obtain for a complete analogy with the statistics of real random variables.

Further, in mathematical statistics there is a concept of a complex random function:

$$z = x(t) + iy(t). \tag{4.7}$$

For this we calculate the variance in a similar way:

$$D_z(t) = D_x(t) + D_y(t) \tag{4.8}$$

and the correlation function

$$\mu_z(t_1, t_2) = \mu_x(t_1, t_2) + \mu_y(t_1, t_2) + i(\mu_{xy}(t_1, t_2) - \mu_{xy}(t_2, t_1)). \tag{4.9}$$

This is everything that mathematical statistics gives economists if they want to use its tools for complex-valued econometrics. As follows from the given materials, there is a rule according to which all the basic parameters characterizing random processes, such as variance, correlation moment, covariation, etc., are real values. This rule does not result from a strict mathematical proof but from the logical conclusion that all the characteristics of variability of a random value are real ones regardless of the form of presentation of random variables – real or complex.

4.2 Method of Least Squares of Complex Variables: Standard Approach

As we can see from the previous section, mathematical statistics does not answer those questions that can be answered using econometric tools, these questions are as follows:

- How can the relationship between two random complex variables be determined?
- If we reveal this relationship, how do we select the required model of complex variables and estimate its parameters on statistical data?

The first problem in the range of real variables is solved by correlation analysis, the second one by regression. Mathematical tools of regression-correlation analysis are unique and based on a uniform methodology.

We have not found any available works in mathematical statistics containing the results of complex-value-regression-correlation analysis. That is why we had to solve this problem independently [6]. The only scientific publication referring to the problem in question is a work published in 2007 [7], but, first of all, when there was a vital necessity to develop the tools of regression analysis for the evaluation of coefficients of complex-valued models this article was unavailable, and, second, it is based on the conditions stated in the previous sections, which, as we will see later, may not be axiomatic for the econometrics of complex variables.

Since the main computational coefficients of correlation analysis are based on regression analysis methods, let us start our examination of the conditions of econometrics of complex variables with the main task of regression analysis – the evaluation of regression model coefficients.

The least-squares method (LSM) is the most popular method in modern econometrics, as it possesses a number of remarkable properties and is highly standard. We are not going to speak here about the specifics of the distribution of a random complex variable, taking only its normal distributions. Let us assume a priori that in this chapter we will consider only stationary processes having a normal distribution. These and other assumptions usually made to substantiate regression-correlation analysis methods will be considered as given by default so as not to repeat them henceforth.

Let us designate the values of some complex variable varying in time:

$$y_{rt} + iy_{it}. \tag{4.10}$$

The reason for this dynamic set of complex variables is another random variable that is a complex argument:

$$x_{rt} + ix_{it}. \tag{4.11}$$

For example, the amount and price of an acquired product combined into a complex variable (4.10) are explained by a consumer's cash income and accumulated income (property), which may be presented in the form of complex variable (4.11). We can provide quite a few pairs of related socioeconomic indicators; some of them will be discussed in subsequent chapters of this study.

Values of the dynamic range (4.10) should be approximated by some regression model to calculate the computational values of this complex variable:

$$F(x_{rt} + ix_{it}) = \hat{y}_{rt} + i\hat{y}_{it}. \tag{4.12}$$

The method of approximation of the calculated values to actual ones is represented by the difference between them:

$$(y_{rt} + iy_{it}) - (\hat{y}_{rt} + i\hat{y}_{it}) = \varepsilon_{rt} + i\varepsilon_{it}. \tag{4.13}$$

This may be defined as a complex approximation error. Depending on the values taken by the model coefficients, error (4.13) can also be different. This is why the econometric model coefficient should provide the average minimum error (4.13). This requirement should be met on the entire range of values of t. However, this general wish should be made a strict form of mathematical criterion. The problem is that if the range of real numbers provides an easy comparison between two numbers, then the range of complex numbers makes it impossible.

In fact, it is no use looking for an answer to the question of which number is more, z_1 or z_2, if they take the following values:

$$z_1 = 2 + i3, \quad z_2 = 3 + i2.$$

It is possible to compare only real numbers, i.e., we can say which number has a higher real or imaginary part, or we can compare the absolute value of complex variables or their polar angles (which are real numbers), but it is not possible to compare two complex numbers.

Here we can use assumptions that all the measures of the variability of complex random variables are real numbers. In particular, the variance of the complex approximation error (4.13) will also be a real number calculated by variance formula (4.2). The least-squares method for real variables minimizes the variance of the approximation error of the actual values by the calculated ones. With reference to the complex approximation error, the minimum of its variance will be determined by the following criterion:

$$\Phi = \sum_t \left(\varepsilon_{rt}^2 + \varepsilon_{it}^2\right) \to \min. \tag{4.14}$$

If we write down the approximation error in exponential form, we get

$$\varepsilon_{rt} + i\varepsilon_{it} = \sqrt{\varepsilon_{rt}^2 + \varepsilon_{it}^2}\, e^{i\,arctg\frac{\varepsilon_{it}}{\varepsilon_{rt}}}.$$

Hence it is easy to see that the LSM criterion for the evaluation of coefficients of econometric complex-valued model (4.14) means a minimization of the absolute value of the complex approximation error. It is seen from (4.14) that the polar angle of the complex approximation error has no meaning, i.e., when we use (4.14) corresponding to the standard formulation of the problem, some information on the nature of the complex variable, its complex regression model, and complex approximation error gets lost.

It is evident that the sum (4.14) is determined by the values that the complex coefficients of the econometric model take. Let an econometric model have a pair of complex coefficients, as is characteristic for most analytical functions considered in Chap. 2, i.e.

$$a_0 + ia_1 \text{ и } b_0 + ib_1$$

the sum (4.14) represents a function of these parameters.

In the general case the LSM criterion for the evaluation of these coefficients of a complex-valued model will be written as follows:

$$\Phi(a_0, a_1, b_0, b_1) = \sum (y_{rt} - \mathrm{Re}[F(x_{rt} + ix_{it})])^2 + \sum (y_{it} - \mathrm{Im}F(x_{rt} + ix_{it})])^2 \to \min. \tag{4.15}$$

With reference to a linear complex-valued model:

$$y_{rt} + iy_{it} = (a_0 + ia_1) + (b_0 + ib_1)(x_{rt} + ix_{it}) \tag{4.16}$$

this criterion will look like this:

$$\begin{aligned}\Phi(a_0, a_1, b_0, b_1) = &\sum (y_{rt} - [a_0 + b_0 x_{rt} - b_1 x_{it}])^2 \\ &+ \sum (y_{it} - [a_1 + b_1 x_{rt} + b_0 x_{it}])^2 \to \min.\end{aligned} \tag{4.17}$$

To find the minimum of such a function of real variables it is necessary to calculate the first partial derivatives of the function for the variables, set them equal to zero, and solve the resulting system of equations. If there are doubts that the extremum is really the function minimum, it is necessary to build a Hesse matrix and make sure that it is positively determined. The system of normal equations for the evaluation of coefficients of a linear complex-valued model for a number of observations T looks like this:

$$\begin{cases} \sum y_{rt} = a_0 T + b_0 \sum x_{rt} - b_1 \sum x_{it}, \\ \sum y_{it} = a_1 T + b_1 \sum x_{rt} + b_0 \sum x_{it}, \\ \sum y_{rt} x_{rt} + \sum y_{it} x_{it} = a_0 \sum x_{rt} + b_0 (\sum x_{rt}^2 + \sum x_{it}^2) + a_1 \sum x_{it}, \\ \sum y_{it} x_{rt} - \sum y_{rt} x_{it} = -a_0 \sum x_{it} + b_1 (\sum x_{rt}^2 + \sum x_{it}^2) + a_1 \sum x_{rt}. \end{cases} \tag{4.18}$$

This is quite a bulky system of four equations with four unknowns. Since modern computation equipment makes it possible to work with complex variables – for example, Microsoft Excel has an "engineering calculations" section that allows you to perform basic operations with complex variables – we would like to convert the system of normal (4.18) to a system of complex equations. Then the calculations for practical purposes would be much easier. And this can be done.

It is easy to see that system (4.18) is equivalent to a system of two complex equations with two complex coefficients:

$$\begin{cases} \sum y_{rt} + i y_{it} = (a_0 + i a_1) T + (b_0 + i b_1) \sum (x_{rt} + i x_{it}), \\ \sum (y_{rt} + i y_{it})(x_{rt} - i x_{it}) \quad = (a_0 + i a_1) \sum (x_{rt} - i x_{it}) \\ + (b_0 + i b_1) \sum (x_{rt} + i x_{it})(x_{rt} - i x_{it}). \end{cases} \tag{4.19}$$

It is a system of two linear complex-valued equations with two complex coefficients that has a simple solution. But it can also be simplified, as in the case where all the initial variables are centered with respect to their averages, the free complex coefficient is equal to zero. In this case from (4.19) we can derive a formula for the complex proportionality coefficient:

$$b_0 + i b_1 = \frac{\sum (y_{rt} + i y_{it})(x_{rt} - i x_{it})}{\sum (x_{rt} + i x_{it})(x_{rt} - i x_{it})}. \tag{4.20}$$

Let us see how the LSM is applied to the data given in Table 4.1.

After centering these variables with respect to their averages we obtain the numerator (4.20)

171,373-i41,862 and its denominator: *130,819*.

Table 4.1 Data of conditional example

t	x_{rt}	x_{it}	y_{rt}	y_{it}
1	1.0	2	4.250	−5.100
2	1.2	3	4.832	−3.854
3	1.3	7	6.243	1.354
4	1.3	1	4.323	−6.506
5	1.2	2	4.512	−5.164
6	1.5	−2	3.625	−10.500
7	1.4	3	5.094	−3.918
8	1.6	1	4.716	−6.602
9	1.4	8	6.694	2.632
10	1.7	−3	3.567	−11.874
11	1.5	−3	3.305	−11.810
12	1.8	2	5.298	−5.356
Average	1.40	1.75	4.705	−5.558

Substituting these values into the formula for finding complex coefficients, we get

$$b_0 + ib_1 = \frac{171.373 - i41.862}{130.819} = 1.31 - 0.32i$$

Now it is easy to calculate the values of the free complex term of the linear regression relationship because for evaluations of the LSM of linear relationships an equality with respect to the averages always takes place:

$$\bar{y}_r + i\bar{y}_i = (a_0 + ia_1) + (b_0 + ib_1)(\bar{x}_r + i\bar{x}_i).$$

The complex free coefficient for this model built on the data of Table 4.1 looks like this:

$$a_0 + ia_1 = \bar{y}_r + i\bar{y}_i - (b_0 + ib_1)(\bar{x}_r + i\bar{x}_i) =$$
$$4,705 - i5,558 - (1,31 - i0,32)(1,4 + i1,75) = 2,3 - i7,4.$$

The linear complex-valued model looks like this:

$$y_{rt} + iy_{it} = (2,3 - i7,4) + (1,31 - i0,32)(x_{rt} + ix_{it}).$$

The obtained evaluations of the standard LSM will provide, on the considered set of initial data, values of approximation errors with minimum variance, the latter being a real number in the case under consideration.

4.3 Correlation Analysis of Complex Variables: Contradictions of the Standard Approach

With respect to the expansion of econometrics by including functions of complex variables, we should add to that all the other econometric attributes: not only regression but also correlation analysis.

It should be noted that in mathematical statistics correlation analysis means a combination of approaches, methods, and techniques aimed at determining the degree and nature of the relation between two random factors. If this combination is intended for studying multiple relationships, then this involves multiple correlations.

Let us consider the possibility of transforming and adapting the basic conditions of correlation analysis to the econometrics of complex variables. As follows from the literature on the application of the TFCV in various branches of science, this problem has not been posed or solved in its full scope.

The point is that to solve problems of hydrodynamics or gas dynamics, the theory of cumulative charge, the theory of elasticity, to calculate electric contours and other practical calculations, where the TFCV has been successfully applied, the real and imaginary parts of complex variables have clear semantic interpretations, and the models used are deterministic. For example, if electrical engineering studies alternating currents, then active resistance R refers to the real part and reactive resistance of, for example, coil L refers to the imaginary part of full resistance. Then the model of full resistance Z for these two elements joined successively will look like this:

$$Z = R + iL$$

The values of active and reactive resistance are easily measured using the appropriate tools, taking into account the measurement error. And there is no need to reveal the degree of correlation between the complex resistance and consumed complex power, as this relationship is uniquely determined by the objectively existing Kirchhoff's law. If complex resistances are connected to a complex network, the processing of the network characteristics will depend on the connection of the elements – successive or parallel. A model of the most complicated electrical contour is deterministic since all the elements and characteristics – resistances, currents, voltages, power, etc. – are defined therein. It is only when we move to huge systems, influenced by many random factors, are the calculated characteristics resulting from the properties of the mathematical model of an electrical contour observed with some error.

However, the task of correlation analysis in these cases is to determine the relationship between the errors of the calculated characteristics of electrical circuits and random external factors. In this case, there can be no talk of a correlation between complex variables, for example, active and reactive current.

Economics is another matter; here, deterministic functional relationships practically do not exist, and even revealed relationships change their force, characteristics, and even direction with time; in economics, laws do not have a quantitative interpretation, and models describing them are not universal and do not take into account the actual variety of factors existing in the economy. The discovery of these relationships at any particular period of time does not mean the discovery of any law since at some point this law may change its force and instead of a linear dependence we might be faced with a nonlinear one. This is why there is an objective necessity to regularly revise previous research on the relationship among the same economic indicators. In this situation regression and correlation analysis is practically the only tool for studying real economic phenomena and the relationships among them, since in a changing economy there are no and there may be no frozen relationships. The results of regression and correlation analysis are regularly reviewed because, once revealed, they can and do change.

Therefore, expanding the instrumental base of econometrics by including in it models of complex variables and considering the proper mathematical tools to estimate the indicators of econometric models of complex variables (LSM), we must give thought to the problem of discovering the interrelations among random complex variables. Based on the principles and approaches of correlation analysis, it is necessary to derive the coefficients applicable to complex variables that could characterize the interrelations among complex factors.

It should be noted that modern correlation analysis does not provide an exhaustive solution to this problem, even with respect to random real variables. The most popular tool in this area of mathematical statistics is the pair correlation coefficient, which shows how closely the relationship between two random values approaches a linear one, if, of course, this relationship exists. The researcher should put forward and substantiate a hypothesis on the existence of a linear relationship between the factors and then, using the pair correlation coefficient, confirm or refute it. This is why the proposed correlation coefficients between complex variables may only serve as an additional argument confirming (or refuting) the linear relationship between complex variables.

We were unable to find methods and techniques of application of correlation analysis to complex variables, though there are general approaches, some of which were stated in the first section of this chapter. It was shown that in modern mathematical statistics, the variance of a complex random value equals the sum of variances of its real and imaginary parts, i.e., is a real value like the variability measure. The expectation of the product of deviation of one of the values from its expectation by the conjugate deviation of the other is called the correlation momentum of two complex random values (covariation).

The correlation momentum on whose calculation we must rely was presented in the first section in the following form (4.5):

$$\mu_{zz} = M[(x_1 + iy_1)(x_2 - iy_2)]$$

or

$$\mu_{zz} = \mu_{x_1x_2} + \mu_{y_1y_2} + i(\mu_{y_1x_2} - \mu_{x_1y_2}). \tag{4.21}$$

For the discrete case, which we must deal with in econometrics,

$$\mu_{zz} = \frac{1}{n}\left[\sum_t (y_{rt} - \overline{y_r})(y_{it} - \overline{y_i}) + \sum_t (x_{rt} - \overline{x_r})(x_{it} - \overline{x_i}) \right.$$
$$\left. + i(\sum_t (x_{rt} - \overline{x_r})(y_{it} - \overline{y_i}) - \sum_t (y_{rt} - \overline{y_r})(x_{it} - \overline{x_i})\right]. \tag{4.22}$$

It is known that the pair correlation coefficient for real variables can be found using the correlation momentum in the following way:

$$r_{XY} = \frac{\mu_{XY}}{\sigma_X \sigma_Y}. \tag{4.23}$$

One can use this formula for complex random values. To do this, we should substitute the correlation momentum of two random values into the numerator of formula (4.22).

To calculate the pair coefficient correlation (4.23), it is necessary to determine its denominator. Since the standard approach in mathematical statistics assumes that the variance of a complex random value equals the sum of the variances of its real and imaginary parts, we get

$$\sigma_X^2 = \frac{1}{n}(\sum_t (x_{rt} - \overline{x_r})^2 + \sum_t (x_{it} - \overline{x_i})^2), \tag{4.24}$$

$$\sigma_Y^2 = \frac{1}{n}(\sum_t (y_{rt} - \overline{y_r})^2 + \sum_t (y_{it} - \overline{y_i})^2). \tag{4.25}$$

Substituting these values into the denominator (4.23) and taking into account the fact that the standard deviation represents the square root of the variance, we obtain a formula for calculating a sample value of the coefficient of correlation of two complex random values:

$$r_{XY} = \frac{\sum_t (y_{rt} - \overline{y_r})(y_{it} - \overline{y_i}) + \sum_t (x_{rt} - \overline{x_r})(x_{it} - \overline{x_i}) + i(\sum_t (x_{rt} - \overline{x_r})(y_{it} - \overline{y_i}) - \sum_t (y_{rt} - \overline{y_r})(x_{it} - \overline{x_i})}{\sqrt{[\sum_t (x_{rt} - \overline{x_r})^2 + \sum_t (x_{it} - \overline{x_i})^2][\sum_t (y_{rt} - \overline{y_r})^2 + \sum_t (y_{it} - \overline{y_i})^2]}}. \tag{4.26}$$

It is clear that the obtained coefficient is a complex number since its imaginary component will be equal to zero only in particular cases for a small number of phenomena.

Since all the variables used in the calculation of the pair correlation coefficient (4.26) represent values centered with respect to their averages, we will use a simplified form, taking into consideration that all the variables are preliminarily centered. Then formula (4.26) may be represented like this:

$$r_{XY} = \frac{\sum_t (y_{rt}y_{it} + x_{rt}x_{it}) + i(\sum_t (x_{rt}y_{it} - y_{rt}x_{it}))}{\sqrt{\sum_t ({x_{rt}}^2 + {x_{it}}^2)\sum_t ({y_{rt}}^2 + {y_{it}}^2)}}. \tag{4.27}$$

Here we will not examine the properties of the derived pair correlation coefficient; we would like to draw attention to its substance. It is complex! The degree of approximation of the relationship between two random complex variables to a linear one turns out to be a complex number, though the standard approach of mathematical statistics to random complex variables assumes axiomatically that all the measures for complex random variables are real ones. This is the first paradox of the standard approach to the econometrics of complex random variables.

It is known that the pair correlation coefficient between two random real variables was derived by another method [8], and later scientists discovered that it may be represented via the correlation momentum and variances. Originally, the pair correlation coefficient was determined via regression coefficients. Let us see how this coefficient can be derived via regression coefficients with respect to random real variables. Regression coefficients of real variable Y_t to X_t and X_t to Y_t can be found using LSM, which for the model

$$Y_t = a_0 + a_1X_t \tag{4.28}$$

suggests solving a system of normal equations:

$$\begin{cases} \sum_t Y_t = na_0 + a_1\sum_t X_t \\ \sum_t Y_tX_t = a_0\sum_t X_t + a_1\sum_t X_t^2 \end{cases} \tag{4.29}$$

and for the model

$$X_t = b_0 + b_1Y_t \tag{4.30}$$

it suggests solving another system of normal equations:

$$\begin{cases} \sum_t X_t = nb_0 + b_1\sum_t Y_t, \\ \sum_t Y_tX_t = a_0\sum_t Y_t + a_1\sum_t Y_t^2. \end{cases} \tag{4.31}$$

To find regression coefficients a_1 and b_1 initial variables are centered with reference to their averages:

$$X_t - \bar{X};\ \ Y_t - \bar{Y}.$$

Taking account the fact that the sum of deviations of any variable from its average is equal to zero, the equalities for the first equations of systems of normal equations will hold only if their free terms (a_0 and b_0, respectively) are equal to zero. Then the first system of normal equations for centered variables will become one equation:

$$\sum_t (Y_t - \bar{Y})(X_t - \bar{X}) = a_1 \sum_t (X_t - \bar{X})^2$$

as well as the second one:

$$\sum_t (Y_t - \bar{Y})(X_t - \bar{X}) = b_1 \sum_t (Y_t - \bar{Y})^2$$

Hence it is easy to calculate the values of regression coefficients a_1 and b_1 via centered variables. The geometric mean of regression coefficients a_1 and b_1

$$r = \pm\sqrt{a_1 b_1} \tag{4.32}$$

will represent the pair correlation coefficient:

$$r = \frac{\sum_t (Y_t - \bar{Y})(X_t - \bar{X})}{\sqrt{\sum_t (X_t - \bar{X})^2 \sum_t (Y_t - \bar{Y})^2}}. \tag{4.33}$$

It is clear that a coefficient value whose absolute value equals zero is obtained only when $a_1 = 1/b_1$, i.e., when there is functional linear relationship between the variables.

Let us now derive a formula for the pair correlation coefficient between two random complex variables as the geometric mean of two complex regression coefficients. For that let us substitute into the appropriate formulae the values of the complex coefficients of regression obtained previously with the variance minimization if we consider to be a real characteristic of the process (4.2). This approach allowed us in Sect. 4.2. to obtain a system of equations that can help us calculate complex coefficients that minimize the values of this variance. Since we use centered variables and therefore get rid of the free term, we can directly use formula (4.20) obtained previously.

The complex regression coefficient of linear dependence of complex random variable Y on another complex random variable X will be calculated as follows, by means of LSM:

$$a = \frac{\sum (y_{rt} + iy_{it})(x_{rt} - ix_{it})}{\sum (x_{rt} + ix_{it})(x_{rt} - ix_{it})} = \frac{\sum (y_{rt} + iy_{it})(x_{rt} - ix_{it})}{\sum (x_{rt}^2 + x_{it}^2)}. \quad (4.34)$$

If we consider the inverse dependence, i.e., that of complex random variable X on another complex random variable Y, then the complex coefficient of this regression will be found similarly by the standard LSM:

$$b = \frac{\sum (x_{rt} + ix_{it})(y_{rt} - iy_{it})}{\sum (y_{rt} + iy_{it})(y_{rt} - iy_{it})} = \frac{\sum (x_{rt} + ix_{it})(y_{rt} - iy_{it})}{\sum (y_{rt}^2 + y_{it}^2)}. \quad (4.35)$$

Let us now substitute formulae (4.34) and (4.35) into formula (4.32) to calculate the complex pair correlation coefficient:

$$r = \pm \sqrt{\frac{\sum (x_{rt} + ix_{it})(y_{rt} - iy_{it}) \sum (y_{rt} + iy_{it})(x_{rt} - ix_{it})}{\sum_t ({x_{rt}}^2 + {x_{it}}^2) \sum_t ({y_{rt}}^2 + {y_{it}}^2)}}. \quad (4.36)$$

The first factor of the numerator may be put in the following form:

$$\sum (x_{rt} + ix_{it})(y_{rt} - iy_{it}) = i \sum (x_{it} - ix_{rt})(y_{rt} - iy_{it}). \quad (4.37)$$

The second factor of the coefficient numerator (4.36) may also be transformed in the following way:

$$\sum (x_{rt} - ix_{it})(y_{rt} + iy_{it}) = i \sum (x_{rt} - ix_{it})(y_{it} - iy_{rt}). \quad (4.38)$$

Now we can be sure that these two factors of the numerator radicand (4.36) are equal to each other, which is why this numerator may be written as follows:

$$i \sum (x_{rt} - ix_{it})(y_{it} - iy_{rt}). \quad (4.39)$$

Taking this into account, the formula for calculating the pair correlation coefficient of two random complex variables (4.36) may be written as follows:

$$r = \frac{i \sum (x_{rt} - ix_{it})(y_{it} - iy_{rt})}{\sqrt{\sum (y_{rt}^2 + y_{it}^2) \sum (x_{rt}^2 + x_{it}^2)}}. \quad (4.40)$$

The imaginary unit in the numerator is not very convenient for understanding the properties of this coefficient. However, if we expand the expression in the numerator under the summation sign and multiply the resulting expression by an imaginary unit, grouping the real and the imaginary parts, we obtain the following formula for the numerator:

$$i\sum(x_{rt} - ix_{it})(y_{it} - iy_{rt}) = \sum(x_{rt}y_{rt} + x_{it}y_{it}) + i\sum(x_{rt}y_{it} - x_{it}y_{rt}). \quad (4.41)$$

Substituting it into the formula, we obtain for the pair correlation coefficient

$$r = \frac{\sum(x_{rt}y_{rt} + x_{it}y_{it}) + i\sum(x_{it}y_{rt} - x_{rt}y_{it})}{\sqrt{\sum(y_{rt}^2 + y_{it}^2)\sum(x_{rt}^2 + x_{it}^2)}}. \quad (4.42)$$

Since this coefficient was derived from the premises of the standard approach, we should expect its complete identity with that derived previously via the correlation momentum of two complex random variables (4.28). This formula is given here to simplify comparison of the two variants of the same formula:

$$r = \frac{\sum_t(y_{it}y_{rt} + x_{it}x_{rt}) + i(\sum_t(x_{rt}y_{it} - y_{rt}x_{it}))}{\sqrt{\sum_t(y_{rt}{}^2 + y_{it}{}^2)\sum_t(x_{rt}{}^2 + x_{it}{}^2)}}. \quad (4.43)$$

The denominators of the two coefficients (4.42) and (4.43) coincide. The imaginary parts of the numerators differ from each other by minus one. But their real parts differ from each other in principle – in formula (4.42) we sum up the product of the real parts of the two variables and that of the imaginary parts of the variables. In formula (4.43) we sum up the product of the real and imaginary parts of one complex variable with that of another complex variable.

This means that the standard approach, which assumes that all the statistical measures of variability of complex variables should be real, lead us to the *second paradox* – the formulae of the pair correlation coefficient of two random variables do not coincide, though this is the case for the domain of real variables.

There is a well-known scientific principle that says that if there are paradoxes in a theory, one should pay special attention to the consistency of the theory's axioms, which should be done in our case.

4.4 Consistent Axioms of the Theory of Mathematical Statistics of Random Complex Variables

It is evident from the discussion of the previous section that the rule introduced in mathematical statistics with respect to complex variables does not hold in full. As shown in the first section of this chapter, it is accepted in mathematical statistics that a complex representation of random functions is simply a mathematical form convenient for analysis and may always be transformed into the form of real functions. This is why it is recommended that when we examine random complex variables, we should consider the parameters of variation of random complex variables to be real. And it for this very reason that functions of variance, correlation, and covariation represent simple and nonrandom real characteristics of complex random processes and functions.

In the domain of real numbers, there is no operation for extracting the square root of a negative number – it was this operation that gave birth to the TFCV as a branch of mathematics. Then why should we look back to the rules adopted in the area of real variables to deal with operations with complex ones? Why can we not consider complex variance, complex correlation momentum, etc.? Simply because this very variance means a particular measure of variability, and the meaning of the measure itself is demonstrated fully when it is a real number? But for a complex series represented by a complex variable series it is not a simple but a complex measure that is an adequate measure of variability, which may always be considered a convenient form of representing two real measures of variability. The real part of a complex variance characterizes one side of the variance of a complex random variable and the imaginary part characterizes the other one.

This view changes the axiomatic core of the theory of mathematical statistics of complex random variables – all the measures of variability of these variables can be complex, showing the complex nature of the modeled processes.

With this axiomatic basis, the correlation momentum is considered in a different way than is accepted in standard mathematical statistics, which leads to conflicting results, i.e., as follows from the substance of the correlation momentum:

$$\mu_{zz} = M[(x_1 + iy_1)(x_2 + iy_2)]. \tag{4.44}$$

It is evident that new axioms of the theory will entail results that are different from the standard formulation of the problem. However, the consistency of the results and recommendations of the new theory should be verified.

First of all, let us redefine a basic concept that will be used subsequently, namely, the concept of variance. Let us define the variance of a complex random value as a complex number. To differentiate the newly introduced concept from the generally accepted one we will add to it the word "complex" in this and all other cases. Thus, the complex variance of a complex random value will be the expectation of the square of the corresponding centered value:

$$D(z) = M[\dot{z}^2] = M[\dot{x}^2 - \dot{y}^2 + i2\dot{x}\dot{y}] = M[\dot{x}^2] - M[\dot{y}^2] + 2iM[\dot{x}\dot{y}], \tag{4.45}$$

where

$$D(x) = M[\dot{x}^2] = M[(x - \bar{x})^2], \tag{4.46}$$

$$D(y) = M[\dot{y}^2] = M[(y - \bar{y})^2]. \tag{4.47}$$

This means that complex variance may also be a real negative value or it may be an imaginary one. Since in Minkowsky's geometry distances may be negative or even imaginary, there is nothing extraordinary in the result obtained; for complex variables that is how it should be if we consider them fully.

Obviously, complex correlation momentum (4.44) will take a different form:

$$\begin{aligned}\mu_{zz} &= M[x_1x_2] - M[y_1y_2] + i(M[y_1x_2] + M[x_1y_2]) \\ &= \mu_{x_1x_2} - \mu_{y_1y_2} + i(\mu_{y_1x_2} + \mu_{x_1y_2})\end{aligned}$$

It is clear again that the complex correlation momentum may also be both negative and imaginary, and complex.

We will not give an interpretation of the values of variance and correlation momentum here.

We are interested in applied matters: how does one use these basic concepts to solve a problem of evaluating coefficients of complex-valued economic models and what form should a complex pair correlation coefficient have? Could there be any contradictions that might refute the new axioms of the theory?

Let us find answers to these questions.

4.5 Least-Squares Method from the Point of View of the New Axiomatic Theory

In regression analysis the problem of evaluating the coefficients of regression models is considered with respect to simple linear monofactor models, after which the problem gets complicated as one moves toward nonlinear functions. That is why, in the case of the econometrics of complex variables, we start with the solution of a problem for a simple linear model of complex variables. Its parameters are to be assessed using LSM. This model has the following form:

$$\hat{y}_{rt} + i\hat{y}_{it} = (a_0 + ia_1) + (b_0 + ib_1)(x_{rt} + ix_{it}). \tag{4.49}$$

The complex variance of the errors of approximation of the actual values of a random complex variable under this model will look like this:

$$\begin{aligned}f(z) &= \sum_t [y_{rt} + iy_{it} - (a_0 + ia_1) - (b_0 + ib_1)(x_{rt} + ix_{it})]^2. \\ &= \sum_t [Y_t - A - BX_t]^2 = \sum_t (Y_t^2 + A^2 + B^2X_t^2 - 2AY_t - 2BY_tX_t + 2ABX_t).\end{aligned} \tag{4.50}$$

Let us consider each of the terms on the right-hand side of the last equality separately taking into account the properties of complex numbers:

$$\sum_t Y_t^2 = \sum_t (y_{rt} + iy_{it})^2 = \sum_t y_{rt}^2 - \sum_t y_{it}^2 + i2\sum_t y_{rt}y_{it}, \tag{4.51}$$

$$\sum_t A^2 = \sum_t (a_0 + ia_1)^2 = \sum_t a_0^2 - \sum_t a_1^2 + i2\sum_t a_0 a_1, \tag{4.52}$$

$$\begin{aligned}\sum_t B^2 X_t^2 &= \sum_t (b_0 + ib_1)^2 (x_{rt} + ix_{it})^2 = \sum_t (b_o^2 - b_1^2 + i2b_0b_1)(x_{rt}^2 - x_{it}^2 + i2x_{rt}x_{it}) \\ &= b_0^2 \sum_t x_{rt}^2 - b_1^2 \sum_t x_{rt}^2 - b_0^2 \sum_t x_{it}^2 + b_1^2 \sum_t x_{it}^2 - 4b_0b_1 \sum_t x_{rt}x_{it} \\ &+ i2\left[b_0b_1 \sum_t x_{rt}^2 - b_0b_1 \sum_t x_{it}^2 + b_0^2 \sum_t x_{rt}x_{it} - b_1^2 \sum_t x_{rt}x_{it}\right],\end{aligned} \tag{4.53}$$

$$\begin{aligned}&-2\sum_t AY_t = -2\sum_t (a_0 + ia_1)(y_{rt} + iy_{it}) = \\ &-2a_0 \sum_t y_{rt} + 2a_1 \sum_t y_{it} + i\left[-2a_1 \sum_t y_{rt} - 2a_0 \sum_t y_{it}\right],\end{aligned} \tag{4.54}$$

$$\begin{aligned}&-2\sum_t BY_tX_t = -2\sum_t (b_0 + ib_1)(y_{rt} + iy_{it})(x_{rt} + ix_{it}) = \\ &-2b_0 \sum_t x_{rt}y_{rt} + 2b_1 \sum_t x_{rt}y_{it} + 2b_0 \sum_t x_{it}y_{it} + 2b_1 \sum_t x_{it}y_{rt} \\ &+ i\left[-2b_0 \sum_t x_{rt}y_{it} - 2b_1 \sum_t x_{rt}y_{rt} - 2b_0 \sum_t x_{it}y_{rt} + 2b_1 \sum_t x_{it}y_{it}\right],\end{aligned} \tag{4.55}$$

$$\begin{aligned}&2\sum_t ABX_t = 2\sum_t (a_0 + ia_1)(b_0 + ib_1)(x_{rt} + ix_{it}) \\ &= 2a_0b_0 \sum_t x_{rt} - 2a_1b_1 \sum_t x_{rt} - 2a_1b_0 \sum_t x_{it} - 2a_0b_1 \sum_t x_{it} \\ &+ i\left[2a_1b_0 \sum_t x_{rt} + 2a_0b_1 \sum_t x_{rt} + 2a_0b_0 \sum_t x_{it} - 2a_1b_1 \sum_t x_{it}\right].\end{aligned} \tag{4.56}$$

Substituting various values of coefficients of the complex-valued model into the complex variance (4.50), we obtain different values of complex variances. Since it is impossible to compare complex numbers with each other, it is impossible to suggest a combination of coefficients for which the complex variance is at a minimum. There is no concept of a minimum of a complex-valued function; therefore, there is no concept of a complex variance minimum.

One can find the minimum of the real part of a complex variance. One can find the minimum of its imaginary part. However, one cannot find the minimum of a complex variance. As a matter of fact, the situation is simpler since we are dealing with a complex-valued function of the dependence of the complex variance on the values of complex coefficients.

Let us first see what the application of the first criterion – the minimum of the real part of a complex variance – will lead to. For that we should calculate the first derivative of the real part of complex variance (4.50) and set it equal to zero.

Let us calculate the first derivatives of this part for each of the coefficients a_0, a_1, b_0, b_1. Using Riemann–Cauchy formulae it is easy to calculate partial derivatives for a_0, then for a_1, b_0, and b_1:

$$\frac{\partial \mathrm{Re}(f(z))}{\partial a_0} = 2\left(a_0 n - \sum_t y_{rt} + b_0 \sum_t x_{rt} - b_1 \sum_t x_{it}\right),$$

$$\frac{\partial \mathrm{Re}(f(z))}{\partial a_1} = 2\left(-a_1 n + \sum_t y_{it} - b_1 \sum_t x_{rt} - b_0 \sum_t x_{it}\right),$$

$$\frac{\partial \mathrm{Re}(f(z))}{\partial b_0} = 2\left(b_0 \sum_t x_{rt}^2 - b_0 \sum_t x_{it}^2 - 2b_1 \sum_t x_{rt} x_{it} - \sum_t x_{rt} y_{rt} + \sum_t x_{it} y_{it}\right.$$
$$\left. + a_0 \sum_t x_{rt} - a_1 \sum_t x_{it}\right),$$

$$\frac{\partial \mathrm{Re}(f(z))}{\partial b_1} = 2\left(-b_1 \sum_t x_{rt}^2 + b_1 \sum_t x_{it}^2 - 2b_0 \sum_t x_{rt} x_{it} + \sum_t x_{rt} y_{it} + \sum_t x_{it} y_{rt}\right.$$
$$\left. = -a_0 \sum_t x_{it} - a_1 \sum_t x_{rt}\right),$$

where n is the number of observations, $t = 1,2,3, \ldots, n$. If we set each of the partial derivatives equal to zero and group them, we get the following system of equations:

$$\begin{cases} \sum_t y_{rt} = n a_0 + b_0 \sum_t x_{rt} - b_1 \sum_t x_{it}, \\ \sum_t y_{it} = n a_1 + b_1 \sum_t x_{rt} + b_0 \sum_t x_{it}, \\ \sum_t x_{rt} y_{rt} - \sum_t x_{it} y_{it} = a_0 \sum_t x_{rt} - a_1 \sum_t x_{it} + b_0 \sum_t (x_{rt}^2 - x_{it}^2) - 2b_1 \sum_t x_{rt} x_{it}, \\ \sum_t x_{rt} y_{it} + \sum_t x_{it} y_{rt} = a_0 \sum_t x_{it} + a_1 \sum_t x_{rt} + b_1 \sum_t (x_{rt}^2 - x_{it}^2) + 2b_0 \sum_t x_{rt} x_{it}. \end{cases} \tag{4.57}$$

This system can easily be solved since we have four equations with four unknowns.

Now let us study another criterion and minimize the imaginary part of the complex variance as a complex-valued function of complex coefficients. We will get four levels corresponding to four partial derivatives:

$$\frac{\partial \operatorname{Im}(f(z))}{\partial a_0} = 2\left(a_1 n - \sum_t y_{it} + b_1 \sum_t x_{rt} + b_0 \sum_t x_{it}\right),$$

$$\frac{\partial \operatorname{Im}(f(z))}{\partial a_1} = 2\left(a_0 n - \sum_t y_{rt} + b_0 \sum_t x_{rt} - b_1 \sum_t x_{it}\right),$$

$$\frac{\partial \operatorname{Im}(f(z))}{\partial b_0} = 2\left(b_1 \sum_t x_{rt}^2 - b_1 \sum_t x_{it}^2 + 2b_0 \sum_t x_{rt}x_{it} - \sum_t x_{rt}y_{it}\right.$$
$$\left. - \sum_t x_{it}y_{rt} + a_1 \sum_t x_{rt} + a_0 \sum_t x_{it}\right),$$

$$\frac{\partial \operatorname{Im}(f(z))}{\partial b_1} = 2\left(b_0 \sum_t x_{rt}^2 - b_0 \sum_t x_{it}^2 - 2b_1 \sum_t x_{rt}x_{it} - \sum_t x_{rt}y_{rt} + \sum_t x_{it}y_{it}\right.$$
$$\left. + a_0 \sum_t x_{rt} - a_1 \sum_t x_{it}\right).$$

When we set these partial derivatives equal to zero and group them, we obtain

$$\begin{cases} \sum_t y_{it} = na_1 + b_1 \sum_t x_{rt} + b_0 \sum_t x_{it}, \\ \sum_t y_{rt} = na_0 + b_0 \sum_t x_{rt} - b_1 \sum_t x_{it}, \\ \sum_t x_{rt}y_{it} + \sum_t x_{it}y_{rt} = a_0 \sum_t x_{it} + a_1 \sum_t x_{rt} + b_1 \sum_t (x_{rt}^2 - x_{it}^2) + 2b_0 \sum_t x_{rt}x_{it}, \\ \sum_t x_{rt}y_{rt} - \sum_t x_{it}y_{it} = a_0 \sum_t x_{rt} - a_1 \sum_t x_{it} + b_0 \sum_t (x_{rt}^2 - x_{it}^2) - 2b_1 \sum_t x_{rt}x_{it}. \end{cases} \tag{4.58}$$

It is clear that we have the same system as (4.57), only the sequence of the equations has changed in accordance with the order of calculation of the first derivatives with respect to the imaginary part of the complex-valued function of the variance (4.50).

It turns out that any of the criteria (of the minimum of the complex-valued function with respect to its real and imaginary parts) gives us the same result. It is this conclusion that follows from the Riemann–Cauchy rule (in some works on the TFCV it is called the d'Alembert-Euler rule).

Since in the first and second cases we used the criterion of minimization of the sum of the squares of the deviations, the result obtained can be called a complex least-squares method. The minimum of complex variance (4.50) for the complex coefficients of complex-valued model (4.49) agrees with the LSM criterion

$$f(z) \to \min \leftrightarrow \begin{cases} \dfrac{\partial f(z)}{\partial(a_0 + ia_1)} = 0, \\ \dfrac{\partial f(z)}{\partial(b_0 + ib_1)} = 0. \end{cases}$$

Similarly to (4.19), let us make the obtained system of equations convenient for practical application in software products that allow for direct operations with complex variables. It is easy to see that system (4.18) is equal to the system of two complex equations with two complex coefficients:

$$\begin{cases} \sum (y_{rt} + iy_{it}) = (a_0 + ia_1)n + (b_0 + ib_1)\sum (x_{rt} + ix_{it}), \\ \sum (y_{rt} + iy_{it})(x_{rt} + ix_{it}) = (a_0 + ia_1)\sum (x_{rt} + ix_{it}) \\ \quad + (b_0 + ib_1)\sum (x_{rt} + ix_{it})(x_{rt} + ix_{it}). \end{cases}$$

If we now compare the obtained system with the similar system obtained in Sect. 4.2, we will easily see the difference between them – in the second equation of the obtained system there is no multiplication by a conjugate variable, as there is in the standard approach:

$$\begin{cases} \sum (y_{rt} + iy_{it}) = (a_0 + ia_1)n + (b_0 + ib_1)\sum (x_{rt} + ix_{it}), \\ \sum (y_{rt} + iy_{it})(x_{rt} - ix_{it}) = (a_0 + ia_1)\sum (x_{rt} - ix_{it}) \\ \quad + (b_0 + ib_1)\sum (x_{rt} + ix_{it})(x_{rt} - ix_{it}). \end{cases}$$

Since the use of complex variables gives the researcher more diverse modeling options than real-variable models, the family of linear complex-valued models is not limited only to model (4.49). Options are possible when only a real coefficient or only an imaginary one is used instead of a complex coefficient, or perhaps a real argument instead of a complex argument, or vice versa – the model of a complex argument describes the behavior of a real variable. This diversity was used in the second and third chapters of the present study.

Let us show how this approach can be used to implement a complex LSM for a model of a complex argument:

$$y_t = (a_0 + ia_1) + (b_0 + ib_1)(x_{rt} + ix_{it}). \tag{4.59}$$

A complex-valued function whose minimum agrees with LSM evaluations of coefficients of a linear model of a complex argument will be written as follows:

$$f(z) = \sum_t \left[y_t - (a_0 + ia_1) - (b_0 + ib_1)(x_{rt} + ix_{it})\right]^2$$
$$= \sum_t \left[y_t - A - BX_t\right]^2 = \sum_t (y_t^2 + A^2 + B^2X_t^2 - 2Ay_t - 2By_tX_t + 2ABX_t)).$$

Let us consider each of the terms of the right-hand side of the last equality separately, except for the square of the real variable y_t^2 since this term has already been presented in a form convenient for the calculation of derivatives:

$$\sum_t A^2 = \sum_t (a_0 + ia_1)^2 = \sum_t a_0^2 - \sum_t a_1^2 + i2\sum_t a_0a_1, \tag{4.60}$$

$$\sum_t B^2X_t^2 = \sum_t (b_0 + ib_1)^2(x_{rt} + ix_{it})^2 = \sum_t (b_o^2 - b_1^2 + i2b_0b_1)(x_{rt}^2 - x_{it}^2 + i2x_{rt}x_{it})$$
$$= b_0^2\sum_t x_{rt}^2 - b_1^2\sum_t x_{rt}^2 - b_0^2\sum_t x_{it}^2 + b_1^2\sum_t x_{it}^2 - 4b_0b_1\sum_t x_{rt}x_{it} +$$
$$i2\left[b_0b_1\sum_t x_{rt}^2 - b_0b_1\sum_t x_{it}^2 + b_0^2\sum_t x_{rt}x_{it} - b_1^2\sum_t x_{rt}x_{it}\right],$$

$$-2\sum_t Ay_t = -2\sum_t a_0y_t - 2ia_1\sum_t y_t), \tag{4.62}$$

$$-2\sum_t By_tX_t = -2\sum_t (b_0 + ib_1)y_t(x_{rt} + ix_{it}) =$$
$$-2b_0\sum_t x_{rt}y_t + 2b_1\sum_t x_{it}y_t + i\left[-2b_1\sum_t x_{rt}y_t - 2b_0\sum_t x_{it}y_t\right], \tag{4.63}$$

$$2\sum_t ABX_t = 2\sum_t (a_0 + ia_1)(b_0 + ib_1)(x_{rt} + ix_{it}) =$$
$$2a_0b_0\sum_t x_{rt} - 2a_1b_1\sum_t x_{rt} - 2a_1b_0\sum_t x_{it} - 2a_0b_1\sum_t x_{it} + \tag{4.64}$$
$$i\left[2a_1b_0\sum_t x_{rt} + 2a_0b_1\sum_t x_{rt} + 2a_0b_0\sum_t x_{it} - 2a_1b_1\sum_t x_{it}\right].$$

Let us use the obtained components to calculate the first partial derivatives of the real part of complex-valued function (4.59) for each of the coefficients a_0, a_1, b_0, b_1. Then we obtain four levels corresponding to four partial derivatives and four variables of the problem under consideration:

$$\frac{\partial \mathrm{Re}(f(z))}{\partial a_0} = 2\left(a_0n - \sum_t y_t + b_0\sum_t x_{rt} - b_1\sum_t x_{it}\right),$$

$$\frac{\partial \mathrm{Re}(f(z))}{\partial a_1} = 2\left(- a_1 n - b_1 \sum_t x_{rt} - b_0 \sum_t x_{it} \right),$$

$$\frac{\partial \mathrm{Re}(f(z))}{\partial b_0} = 2\left(b_0 \sum_t x_{rt}^2 - b_0 \sum_t x_{it}^2 - 2b_1 \sum_t x_{rt}x_{it} - \sum_t x_{rt}y_t + a_0 \sum_t x_{rt} - a_1 \sum_t x_{it} \right),$$

$$\frac{\partial \mathrm{Re}(f(z))}{\partial b_1} = 2(-b_1 \sum_t x_{rt}^2 + b_1 \sum_t x_{it}^2 - 2b_0 \sum_t x_{rt}x_{it} + \sum_t x_{it}y_t - a_0 \sum_t x_{it} - a_1 \sum_t x_{rt}).$$

If we set each of the derivatives equal to zero and group them, we get

$$\begin{cases} \sum_t y_t = na_0 + b_0 \sum_t x_{rt} - b_1 \sum_t x_{it}, \\ 0 = na_1 + b_1 \sum_t x_{rt} + b_0 \sum_t x_{it}, \\ \sum_t x_{it}y_t = a_0 \sum_t x_{it} + a_1 \sum_t x_{rt} + b_1 \sum_t (x_{rt}^2 - x_{it}^2) + 2b_0 \sum_t x_{rt}x_{it}, \\ \sum_t x_{rt}y_t = a_0 \sum_t x_{rt} - a_1 \sum_t x_{it} + b_0 \sum_t (x_{rt}^2 - x_{it}^2) - 2b_1 \sum_t x_{rt}x_{it}. \end{cases} \quad (4.65)$$

Similar equalities can be obtained by finding partial derivatives of the imaginary part of complex-valued function (4.64), setting them equal to zero, and using other variants resulting from the d'Alembert–Euler condition.

If now we compare system (4.65) with system (5.4.19) we may be certain that system (4.65) is easily obtained from (4.57) if we substitute $y_{it} = 0$ into the latter.

In a whole series of cases, instead of linear model (4.58) we can use its simpler analog – a linear model of a complex argument without a free term:

$$y_t = (b_0 + ib_1)(x_{rt} + ix_{it}). \quad (4.66)$$

With reference to this case, a complex-valued function LSM will look as follows:

$$f(z) = \sum_t \left[y_t - \left(b_0 + ib_1)(x_{rt} + ix_{it}) \right]^2 = \sum_t (y_t^2 - 2By_tX_t + B^2X_t^2 \right). \quad (4.67)$$

Let us use (4.61) and (4.63) to find partial derivatives of the real part of this function for each of the coefficients:

$$\frac{\partial \mathrm{Re}(f(z))}{\partial b_0} = 2\left(b_0\sum_t x_{rt}^2 - b_0\sum_t x_{it}^2 - 2b_1\sum_t x_{rt}x_{it} - \sum_t x_{rt}y_t\right),$$

$$\frac{\partial \mathrm{Re}(f(z))}{\partial b_1} = 2\left(-b_1\sum_t x_{rt}^2 + b_1\sum_t x_{it}^2 - 2b_0\sum_t x_{rt}x_{it} + \sum_t x_{it}y_t\right),$$

where the LSM system is to be written in the following form:

$$\begin{cases} \sum_t x_{rt}y_t = b_0\sum_t (x_{rt}^2 - x_{it}^2) - 2b_1\sum_t x_{rt}x_{it}, \\ \sum_t x_{it}y_t = b_1\sum_t (x_{rt}^2 - x_{it}^2) + 2b_0\sum_t x_{rt}x_{it}. \end{cases} \tag{4.68}$$

If we compare this system of normal equations with system (4.57) we see that (4.68) can be obtained without calculating derivatives but setting the components missing in (4.66) equal to zero.

Let us now show the interrelations between the LSM problems for real and complex variables.

As is known, for a simple linear monofactor model of real variables

$$y = a + bx,$$

the system of normal LSM equations has the following form:

$$\begin{cases} \sum_t y_t = an + b\sum_t x_t, \\ \sum_t y_t x_t = a\sum_t x_t + b\sum_t x_t^2. \end{cases} \tag{4.69}$$

If now we substitute complex variables and complex coefficients into this system of normal equations instead of real variables and coefficients, then for a linear complex-valued monofactor function we get the same system of normal equations (4.57) as previously. In fact, the first equation of system (4.69), if we substitute complex variables and complex coefficients into it, will have the form

$$\sum_t y_{rt} + i\sum_t y_{it} = na_0 + b_0\sum_t x_{rt} - b_1\sum_t x_{it} + i\left(na_1 + b_1\sum_t x_{rt} + b_0\sum_t x_{it}\right).$$

Dividing the real and imaginary parts of the equality we get the first part of LSM system (4.57):

$$\begin{cases} \sum_t y_{rt} = na_0 + b_0\sum_t x_{rt} - b_1\sum_t x_{it}, \\ \sum_t y_{it} = na_1 + b_1\sum_t x_{rt} + b_0\sum_t x_{it}. \end{cases} \tag{4.70}$$

If we substitute complex variables and complex coefficients into the second equation of system (4.69), it will be a more complicated equation, which is why we divide it into separate components. Following substitution the left-hand side of the equality will have the form

$$\sum_t (y_{rt}+iy_{it})(x_{rt}+ix_{it}) = \sum_t y_{rt}x_{rt} - \sum_t y_{it}x_{it} + i\left(\sum_t y_{it}x_{rt} + \sum_t y_{rt}x_{ir}\right).$$

The first term of the right-hand side of the second (4.69) will have the form $a_0\sum_t x_{rt} - a_1\sum_t x_{ir} + i\left(a_1\sum_t x_{rt} + a_0\sum_t x_{it}\right)$, and the second will look like this:

$$b_0\left(\sum_t x_{rt}^2 - \sum_t x_{it}^2\right) - 2b_1\sum_t x_{rt}x_{it} + i\left(b_1\left(\sum_t x_{rt}^2 - \sum_t x_{it}^2\right) + 2b_0\sum_t x_{rt}x_{it}\right).$$

If we group the real components into one equality and the imaginary ones into another, we get two more equations:

$$\begin{cases} \sum_t x_{rt}y_{rt} - \sum_t x_{it}y_{it} = a_0\sum_t x_{rt} - a_1\sum_t x_{it} + b_0\sum_t (x_{rt}^2 - x_{it}^2) - 2b_1\sum_t x_{rt}x_{it}, \\ \sum_t x_{rt}y_{it} + \sum_t x_{it}y_{rt} = a_0\sum_t x_{it} + a_1\sum_t x_{rt} + b_1\sum_t (x_{rt}^2 - x_{it}^2) + 2b_0\sum_t x_{rt}x_{it}. \end{cases} \tag{4.71}$$

Combining the systems of equations (4.70) and (4.71) into one system it is easy to see that we have a system of normal LSM equations for linear functions of complex variables (4.57).

The direct parallels between the method suggested in this section and LSM applied to models of real variables constitute one of the arguments in favor of the method under consideration, but not in favor of the one that follows from the standard formulation of the problem adopted in mathematical statistics.

Since the solution of a system of four equations with four variables is not the most enjoyable problem, the procedure for assessing the coefficients of linear complex-valued function (4.49) by means of LSM can and should be simplified.

To do this, after preliminary centering of the initial variables of the problem with respect to their averages:

$$y_r' = y_r - \bar{y}_r;\, y_i' = y_i - \bar{y}_i;\, x_r' = x_r - \bar{x}_i;\, x_i' = x_i - \bar{x}_i,$$

let us simplify model (4.49):

$$y_{rt}' + iy_{it}' = (b_0 + ib_1)(x_{rt}' + ix_{it}'). \tag{4.72}$$

To find the coefficients, it is sufficient to solve the following system of two equations with two variables:

$$\begin{cases} \sum y'_{rt}x'_{rt} - \sum y'_{it}x'_{it} = b_0 \sum (x'^2_{rt} - x'^2_{it}) - 2b_1 \sum x'_{rt}x'_{it}, \\ \sum y'_{it}x'_{rt} + \sum y'_{rt}x'_{it} = b_1 \sum (x'^2_{rt} - x'^2_{it}) + 2b_0 \sum x'_{rt}x'_{it}, \end{cases} \tag{4.73}$$

or, for work with complex variables,

$$b_0 + ib_1 = \frac{\sum (y_{rt} + iy_{it})(x_{rt} + ix_{it})}{\sum (x_{rt} + ix_{it})(x_{rt} + ix_{it})}. \tag{4.74}$$

Let us show the applicability of LSM to an example whose initial data are given in Table 4.1. The numerator of the preceding fraction for finding the complex proportionality coefficient is equal to
$-163,905 + i31,048$ and the denominator is equal to

$$-129,681 - i6,350.$$

The complex proportionality coefficient will be equal to

$$b_0 + ib_1 = 1{,}249 - \mathrm{i}0{,}301.$$

It should be noted that the standard approach used in Sect. 4.2 provided other values of the complex proportionality coefficient:

$$b_0 + ib_1 = 1{,}257 - \mathrm{i}0{,}295$$

Clearly, the differences are not as great as might be expected, but they do exist, since the minimization criteria are different.

4.6 Complex Pair Correlation Coefficient

An attempt to calculate the coefficient of a pair correlation between two random complex variables based on the standard axiomatic core of the theory of mathematical statistics of complex random variables showed contradictory results and became the basis for the creation of a new axiomatic core of the theory of complex-valued econometrics.

It was shown in Sect. 4.3 how to derive a formula for calculating a pair correlation coefficient using two methods. First we used a well-known formula in mathematical statistics:

$$r_{XY} = \frac{\mu_{XY}}{\sigma_X \sigma_Y}. \tag{4.75}$$

Let us substitute into this formula values of the complex correlation momentum and complex variances from Sect. 4.4 that result from the new axiomatic theory. To simplify the form, let us assume that all the initial variables are centered with respect to their variables:

$$r_{XY} = \frac{\sum (y_{rt} + iy_{it})(x_{rt} + ix_{it})}{\sqrt{\sum (x_{rt} + ix_{it})^2 \sum (y_{rt} + iy_{it})^2}}. \tag{4.76}$$

Expanding the numerator and grouping the real and imaginary parts we obtain

$$\sum (y_{rt}x_{rt} - y_{it}x_{it}) + i \sum (x_{rt}y_{it} + y_{rt}x_{it}). \tag{4.77}$$

Now in the denominator, let us write the expression under the summation symbol to the second power and group the real and imaginary parts:

$$\sqrt{\sum (x_{rt} + ix_{it})^2 \sum (y_{rt} + iy_{it})^2} = \sqrt{\sum (x_{rt}^2 - x_{it}^2 + i2x_{rt}x_{it}) \sum (y_{rt}^2 - y_{it}^2 + i2y_{rt}y_{it})}. \tag{4.78}$$

Substituting the numerator and denominator into the initial formula (4.76) we obtain a formula for a complex pair correlation coefficient that may be used in the absence of a way to work with complex numbers:

$$r_{XY} = \frac{\sum (y_{rt}x_{rt} - y_{it}x_{it}) + i \sum (x_{rt}y_{it} + y_{rt}x_{it})}{\sqrt{\sum (x_{rt}^2 - x_{it}^2 + i2x_{rt}x_{it}) \sum (y_{rt}^2 - y_{it}^2 + i2y_{rt}y_{it})}}. \tag{4.79}$$

It is easy to determine the differences of the obtained formula from the one following from the standard formulation of problem (4.28) if we write this formula here:

$$r_{XY} = \frac{\sum_t (y_{rt}y_{it} + x_{rt}x_{it}) + i(\sum_t (x_{rt}y_{it} - y_{rt}x_{it}))}{\sqrt{\sum_t ({x_{rt}}^2 + {x_{it}}^2) \sum_t ({y_{rt}}^2 + {y_{it}}^2)}}. \tag{4.80}$$

Both numerators and denominators differ from each other.

Let us now derive a formula for calculating the pair correlation coefficient of complex variables via the geometric mean of the product of complex coefficients of regressions (4.32).

With respect to the complex regression coefficient of X to Y designated as a, we obtain

$$a = \frac{\sum (y_{rt} + iy_{it})(x_{rt} + ix_{it})}{\sum (x_{rt} + ix_{it})(x_{rt} + ix_{it})}. \tag{4.81}$$

Now let us consider a regression inverse to the given one, i.e., a complex regression of Y to X:

$$X = X_0 + bY, \tag{4.82}$$

where X_0 and b are complex coefficients of the equation of regression.

The complex proportionality coefficient b may also be found by means of complex LSM:

$$b = \frac{\sum (y_{rt} + iy_{it})(x_{rt} + ix_{it})}{\sum (y_{rt} + iy_{it})(y_{rt} + iy_{it})}. \tag{4.83}$$

Since the pair correlation coefficient represents the geometric mean of regression coefficients (4.32), let us find the geometric mean of regression coefficients (4.81) and (4.83):

$$r_{XY} = \pm\sqrt{a_1 b_1} = \frac{\sum (y_{rt} + iy_{it})(x_{rt} + ix_{it})}{\sqrt{\sum (x_{rt} + ix_{it})^2 \sum (y_{rt} + iy_{it})^2}}. \tag{4.84}$$

It is evident that we get the same formula as (4.76) derived via the complex correlation momentum and complex variance. It may be transformed into (4.79) if the economist has no way of working with complex numbers and must work with real numbers only.

It is crucial that the same result was obtained using two different approaches to the derivation of a formula for calculating a complex pair correlation coefficient between two random complex variables, in contrast to attempts made for the standard formulation of the problem.

Neither of the paradoxes found in Sect. 4.3 has been revealed for the new axiomatic core of the theory of complex-valued econometrics. This is why subsequently we will just on this axiomatic core – all the measures of variability of random complex variables are complex, showing the complex nature of the process.

4.7 Interpretation of Values of Complex Pair Correlation Coefficient

Having determined a method for correctly calculating a complex pair correlation coefficient, we should provide an interpretation of the values it can take. A linear relationship between two complex variables in the domain of real variables means that both the real and imaginary parts of one complex variable function as two-factor linear dependencies on the real and imaginary parts of another complex variable. This is why if one variable varies nonlinearly, then the other will do the same, and it will be difficult to determine this dependence visually. If the dependence under consideration is not functional but regression, the scattering of the points in the complex planes is to an even lesser extent associated with a linear dependence. This is why a visual analysis of the dependence between variables is hard, and one can judge the linear relationship of two complex variables only by the calculated characteristics, primarily the complex pair correlation coefficient.

As follows from the discussion of the previous section, a complex pair correlation coefficient represents the geometric mean of two complex coefficients of regression:

$$r_{XY} = \pm\sqrt{ab}. \tag{4.85}$$

This is why, for a strictly functional linear complex-valued relationship

$$Y = aX, \ \ X = bY,$$

the following equality will be evident:

$$a = \frac{1}{b}, \tag{4.86}$$

where

$$ab = \frac{1}{b}b = 1.$$

That is, a complex pair correlation coefficient for a linear functional relationship is

$$r_{XY} = \pm(1 + i0). \tag{4.87}$$

This means that for a linear functional relationship between two complex variables the absolute value of the real part of the complex pair correlation coefficient is equal to one, and its imaginary component is equal to zero.

Therefore, the square of a complex pair correlation coefficient (complex coefficient of determination) for a linear relationship will always be equal to a real unit. But in what cases of a linear functional relationship between two complex variables will the square root of the determination coefficient take values of "plus one" and in what cases "minus one"?

To answer this question, let us present the complex proportionality coefficients in arithmetic and exponential forms:

$$a = a_0 + ia_1 = ae^{i\alpha}, \tag{4.88}$$

$$b = b_0 + ib_1 = be^{i\beta}. \tag{4.89}$$

Their product will be

$$ab = r_a r_b e^{i(\alpha+\beta)} = r_a r_b \cos(\alpha+\beta) + ir_a r_b \sin(\alpha+\beta). \tag{4.90}$$

Since for a linear functional relationship we have (4.87), i.e., the imaginary part of the complex pair correlation coefficient is equal to zero,

$$\alpha + \beta = 2\pi k, ab = 1. \tag{4.91}$$

Let us consider a simple case where $k = 0$. Then the complex pair correlation coefficient is determined as the square root of

$$r_{XY} = \sqrt{ab} = \sqrt{r_a r_b cos(\alpha+\beta)}. \tag{4.92}$$

Since the absolute value of each proportionality coefficient is positive by default, whether the pair correlation coefficient is equal to "plus one" or "minus one" will be determined by the cosine of α.

We are interested in the case where the radicand may be

$$\sqrt{(-1)(-1)}. \tag{4.93}$$

Then it is possible to determine the characteristics of the complex proportionality coefficient. Based on equality (4.91), the cosine of radicand (4.92) may be written as follows:

$$\cos(\alpha + (-\alpha)) = \cos\alpha\cos(-\alpha) - \sin\alpha\sin(-\alpha).$$

Then it is easy to determine that the case we are interested in, (4.93), is determined by a polar angle of the complex proportionality coefficient $a_0 + ia_1$ lying within the range

$$\frac{3}{4}\pi \le \alpha \le \frac{5}{4}\pi. \tag{4.94}$$

For this case the real component of the complex proportionality coefficient is always nonpositive:

$$a_0 \le 0, \tag{4.95}$$

and its imaginary part is always not less than the real one:

$$a_0 \leq a_1. \tag{4.96}$$

These conditions are valid for the following coefficients:

$$-1 + i; -1 + 0i; -10 - i9,999\ldots$$

What does a linear complex-valued relationship with such a complex pair correlation coefficient mean? To answer this question, let us present a linear functional complex-valued relationship as a system of equalities of real and imaginary parts:

$$y_r = a_0 x_r - a_1 x_i, \tag{4.97}$$

$$y_i = a_1 x_r + a_0 x_i. \tag{4.98}$$

According to conditions (4.95) and (4.96), the coefficient *a0* is always nonpositive, and the imaginary part *a1* can take both positive (1) and negative (2) values. Considering variations of the complex argument in the first quadrant of the complex plane, we obtain that with the simultaneous growth of the real and imaginary parts of the argument, the real part of the complex result *Yr* decreases, and the imaginary part *Yi* can both increase and decrease due to (4.96).

Thus, the real part of the complex pair correlation coefficient *rr* testifies to the degree of approximation of the dependence of two random complex variables to a linear dependence, and the interpretation of its values is similar to that of the values of the pair correlation coefficient in the domain of real numbers.

Now it is necessary to determine the meaning of the imaginary component of the complex pair correlation coefficient *ri*. It definitely follows from (4.87) that the imaginary component will be equal to zero only if there is a linear functional relationship between two complex variables. In all other cases it will not be equal to zero.

The case where

$$r_{XY} = \pm(0 + i) \tag{4.99}$$

is the extreme manifestation of this component of the complex pair correlation coefficient.

It follows then that

$$ab = r_a r_b e^{i(\alpha+\beta)} = r_a r_b \cos(\alpha + \beta) + i r_a r_b \sin(\alpha + \beta) = -1 + i0. \tag{4.100}$$

This means that

$$\alpha + \beta = (2k - 1)\pi, r_a r_b = 1. \tag{4.101}$$

What does this equality mean? If we consider the situation where $k = 1$, then the complex proportionality coefficient $a_0 + ia_1 = 1 + i0$ should be associated with coefficient assessments such as $b_0 + ib_1 = -1 + i0$, so that the complex pair correlation coefficient could take values (4.99) by the LSM as applied to the inverse dependence of the argument on the result. Or the coefficient $b_0 + ib_1 = -1 - i$ should correspond to the coefficient $a_0 + ia_1 = 1 + i$, a vector whose direction is opposite to that of the first one in the complex plane.

It is clear now in what case the real part of the complex pair correlation coefficient will be equal to zero and the absolute value of its imaginary part will be equal to one. If we find the regression of the complex argument to the complex result

$$y_r + iy_i = (a_0 + ia_1)(x_r + ix_i),$$

then the proportionality coefficient $a_0 + ia_1$, found by LSM, will model some linear sequence $\hat{y}_r + i\hat{y}_i$. If we find the inverse regression of the complex result to the complex argument

$$x_r + ix_i = (b_0 + ib_1)(y_r + iy_i),$$

then LSM will give a complex coefficient $b_0 + ib_1$ such that its application to the regression

$$y_r + iy_i = \frac{x_r + ix_i}{b_0 + ib_1}$$

will model a series of points $\hat{y}'_r + i\hat{y}'_i$, turned with respect to the initial series $\hat{y}_r + i\hat{y}_i$ at an angle π, i.e., backward.

This is possible if there is no linear relationship whatsoever.

To understand the essence of the intermediate values (from zero to one by a absolute value) of the imaginary component of the complex pair correlation coefficient, one can use the results of empirical studies. The main idea was as follows.

A linear functional relationship between two complex variables was plotted. The real part of the complex pair correlation coefficient was one, and the imaginary one – zero.

The error variance of this linear relationship, determined in fractions of the modeled value Y, was subsequently increased. In other words, the functional relationship was replaced by a regression complex-valued one with an increasing variance. The higher the variance of the approximation error between the calculated and actual values of complex Y, the higher the imaginary component of the complex pair correlation coefficient and the lower its real part. This means that the growth of the imaginary component indicates a growth in the variance of the approximation of a random variable by a linear complex-valued model.

Since the coefficient itself characterizes the degree of approximation of the relationship under study to the linear complex-valued function, in practice we should stick to the following rule. If the absolute value of the real part of a complex pair correlation coefficient is close to one and the imaginary part is close to zero, then one can safely use a linear complex-valued regression relationship. If the absolute value of the imaginary part is far from zero, a linear relationship will give a high variance.

If the absolute value of a complex pair correlation coefficient is lower than 0.8 and the imaginary part is too low, then one should choose a nonlinear complex-valued regression model since the linear model will describe the relationship badly. A low value of the absolute value of the imaginary part shows that there is some nonlinear regression relationship between the variables. However, if the imaginary part of the complex pair correlation coefficient is rather high, for example, greater than 0.5, then the existence of any relationship between random complex variables is quite doubtful.

4.8 Assessments of Parameters of Nonlinear Econometric Models of Complex Variables

The general principles of LSM, determined in Sect. 4.5 of this chapter on econometric complex-valued models, when applied to problems related to the assessment of parameters of econometric models, require individual consideration with respect to each complex-variable model. LSM adapted for a simple linear model of complex variables gives rise to general principles applicable to linear complex-valued econometric models. This could underlie an approach that would make it possible to use LSM for the assessment of selected coefficients of nonlinear econometric models of complex variables if they are represented in a linear form.

In the general case, in the domain of real variables, the procedure for assessing the coefficients of nonlinear models is not easy. For additive nonlinear models the creation of a system of normal equations is not difficult, for example, for a nonlinear additive model like

$$y = a_1 x^2 + a_2 \ln x + a_3 \cos x.$$

The system of normal equations by LSM will have the form

$$\begin{cases} \sum y x^2 = a_1 \sum x^4 + a_2 \sum x^2 \ln x + a_3 \sum x^2 \cos x, \\ \sum y \ln x = a_1 \sum x^2 \ln x + a_2 \sum \ln^2 x + a_3 \sum \ln x \cos x, \\ \sum y \cos x = a_1 \sum x^2 \cos x + a_2 \sum \ln x \cos x + a_3 \sum \cos^2 x. \end{cases}$$

Assessments of coefficients of this model will possess all the remarkable properties of LSM assessments. However, it is very hard to use LSM for assessing the parameters of a simple power model.

Thus, to assess the parameters of a simple power model

$$y_t = a_0 x_t^{a_1}$$

using LSM, we should solve the system of nonlinear equations

$$\begin{cases} \dfrac{\partial (Y_t - a_0 x_t^{a_1})^2}{\partial a_0} = 0, \\ \dfrac{\partial (Y_t - a_0 x_t^{a_1})^2}{\partial a_1} = 0. \end{cases}$$

Calculating the derivatives we get

$$\begin{cases} \sum_t y_t e^{a_1 x_t} - a_0 \sum_t e^{2a_1 x_t} = 0, \\ a_0 \sum_t y_t x_t e^{a_1 x_t} - a_0^2 \sum_t x_t e^{2a_1 x_t} = 0. \end{cases}$$

Although nowadays it is not hard at all to solve such a system of two nonlinear equations – it is rather easy using numerical methods – when econometrics was first created, it was practically unsolvable. Nowadays many practicing economists who do not have proper expertise in the use of mathematical methods find it difficult to solve similar problems, unless, of course, they have the software applications to do it.

It is for this precise reason that another method was proposed to solve the problem set – the linearization of the initial nonlinear model. It was less precise but much simpler. For the power function under consideration, we take a logarithm of the left- and right-hand sides of the initial equality to any base and obtain a linear model (here we do it to the natural base):

$$\ln y_t = \ln a_0 + a_1 \ln x_t \Leftrightarrow y_t' = a_0' + a_1 x_t'.$$

The coefficients of this linear model can be found easily using LSM. After assessing the parameters of this model it is easy to return to the power model – find the coefficient a_0 using the known value of a_0':

$$a_0 = e^{a_0'}.$$

It has long been known that assessments of the parameters of an original power model found in this way will be shifted, as the minimized squared deviations are not of the power function but of its linear analog. However, in the vast majority of cases this is not a big problem since models in which the parameters were found by a similar method describe the original series quite well.

It is evident that if we use LSM to assess parameters of econometric models of real variables, direct application of this method will create certain computational burdens connected with the need to solve systems of nonlinear equations, whereas for the case of nonlinear econometric models of complex variables the situation is exacerbated due to the properties of nonlinear complex-valued functions considered in Chaps. 2 and 3.

This is why, if we want to apply LSM in the econometrics of complex variables, we should develop practical techniques for assessing the parameters of each of the previously considered econometric models of complex variables using an approach similar to that used in the econometrics of real variables.

Let us consider this possibility and sequentially examine the functions of complex variables according to the same procedure that was used to study the conformal mapping of these functions in Chap. 3, bearing in mind that the complex models of a real argument and models of a complex argument are a special case.

A complex-valued power function with real coefficients will be the first to consider. It has the following form:

$$y_{rt} + iy_{it} = a_0(x_{rt} + ix_{it})^{b_0}. \tag{4.102}$$

Before we show one way of applying LSM to this function, one interesting feature of this and similar models should be noted. If we present each of the complex variables in exponential form and substitute them into model (4.102), we obtain

$$R_{yt}e^{i\theta_{yt}} = a_0 R_{xt}^{b_0} e^{ib_0\theta_{xt}}, \tag{4.103}$$

where

$$R_{yt} = \sqrt{y_{rt}^2 + y_{it}^2},\ R_{xt} = \sqrt{x_{rt}^2 + x_{it}^2},\ \theta_{yt} = arctg\frac{y_{it}}{y_{rt}},\ \theta_{xt} = arctg\frac{x_{it}}{x_{rt}}.$$

Two complex variables presented in exponential form are equal to each other only if their moduli and arguments are. This is why

$$R_{yt} = a_0 R_{xt}^{b_0},\ \textbf{and}\ \theta_{yt} = b_0\theta_{xt}.$$

Therefore, for each observation t one can find the value of the exponent b_0:

$$b_{0t} = \frac{\theta_{yt}}{\theta_{xt}} = \frac{arctg\frac{y_{it}}{y_{rt}}}{arctg\frac{x_{it}}{x_{rt}}} \tag{4.104}$$

and of the proportionality coefficient a_0:

$$a_{0t} = \frac{R_{yt}}{R_{xt}^{b_{0t}}} = \frac{R_{yt}}{R_{xt}^{\frac{\theta_{yt}}{\theta_{xt}}}}. \tag{4.105}$$

Thus, to assess the values of the coefficients of a complex-valued power function, it is not necessary to use some set of observations – the coefficients are assessed at each observation!

This possibility again demonstrates the difference in the properties of models of complex variables from those of real ones – there is a unique possibility to assess the coefficients of nonlinear models at each observation and give an economic interpretation of the modeled process, not generally for a period but for each observation, if only the coefficients carry economic meaning and the modeled relationship is described by this very function.

This property makes it possible to analyze two series of values $\{a_{0t}\}$ and $\{b_{0t}\}$, which in the case of reversible random process represent values of coefficients selected from some universe of coefficients of model (4.102) describing the expectation of the relationship under study. Since econometrics a priori assumes that we are dealing with a normal distribution of probabilities and the selected values of coefficients a_0 and b_0 found according to (4.104) and (4.105) vary around some expectations a_0 and b_0, the best assessment of the searched parameters of model (4.102) will be their averages:

$$b_0 = \frac{1}{T}\sum_{t=1}^{T}\frac{arctg\frac{y_{it}}{y_{rt}}}{arctg\frac{x_{it}}{x_{rt}}} \tag{4.106}$$

and

$$a_0 = \frac{1}{T}\sum_{t=1}^{T}\frac{(y_{rt}^2+y_{it}^2)^{\frac{1}{2}}}{(x_{rt}^2+x_{it}^2)^{\frac{1}{2}\frac{arctg\frac{y_{it}}{y_{rt}}}{arctg\frac{x_{it}}{x_{rt}}}}}. \tag{4.107}$$

For these assessments it is easy to calculate their variance and confidence limits, which is why nonlinear model (4.102) seems to be quite convenient and easy for modeling various economic processes.

Since this section aims at looking for approaches to using LSM in application to this model, let us show its solution. This problem may occur if researchers must not only calculate the parameters of an econometric model but build this model to describe some average trend. Then LSM is essential. First of all, in this case the original model should be linearized. Then we get

$$\ln(y_{rt} + iy_{it}) = \ln a_0 + b_0 \ln(x_{rt} + ix_{it}).$$

Since we are working with the principal values of logarithms, the expression obtained may be represented as follows:

$$\ln R_{yt} + i\varphi_{yt} = \ln a_0 + b_0(\ln R_{xt} + i\varphi_{xt}).$$

In the linear form the model has the form

$$\ln R_{yt} + i\varphi_{yt} = (\ln a_0 + i0) + (b_0 + i0)(\ln R_{xt} + i\varphi_{xt}).$$

Taking into account this form and the previously obtained LSM system for linear model (4.58) we can substitute these values into the LSM system (including zero ones) and obtain

$$\begin{cases} \sum_t \ln R_{yt} = n \ln a_0 + b_0 \sum_t \ln R_{xt} \\ \sum_t \varphi_{yt} = b_0 \sum_t \varphi_{it} \end{cases}. \tag{4.108}$$

Hence

$$b_0 = \sum_t \varphi_{yt} \Big/ \sum_t \varphi_{xt}, \tag{4.109}$$

$$\ln a_0 = \frac{1}{n}\left(\sum_t \ln R_{yt} - \frac{\sum_t \varphi_{yt}}{\sum_t \varphi_{xt}} \sum_t \ln R_{xt} \right). \tag{4.110}$$

In the general case assessments (4.108) will differ from (4.106) and (4.107). In the present work we are not going to compare them to each other or give recommendations. We need only specify that use of LSM for a linearized model will result in a shift of assessments (4.109) and (4.110) compared to (4.106) and (4.107).

Now let us show how to find the parameters of a power model of complex variables with complex coefficients that represents the most general form of power functions of complex variables:

$$y_{rt} + iy_{it} = (a_0 + ia_1)(x_{rt} + ix_{it})^{(b_0 + ib_1)}. \tag{4.111}$$

It will not be possible to find coefficients at each observation since the model has four coefficients and not two as in model (4.102). It is appropriate to state that any complex-valued function with two coefficients (either real or imaginary) may be built only for one observation. The explanation for this is quite simple. Complex-valued functions are simply systems of two real functions, and if at some observation t economists have data for x_{rt}, x_{it}, y_{rt} и y_{it}, then they can substitute them into the function, set the left- and right-hand sides of the equality equal to each other, and obtain two equalities with two unknown coefficients, which can be calculated immediately.

As for the power function, its coefficients are found at one observation for such of its variants as a model with real coefficients (4.102) and such models as

$$y_{rt} + iy_{it} = a_0 (x_{rt} + ix_{it})^{ib_1},$$

$$y_{rt} + iy_{it} = ia_1(x_{rt} + ix_{it})^{b_0},$$

$$y_{rt} + iy_{it} = ia_1(x_{rt} + ix_{it})^{ib_1}.$$

After obtaining the coefficients of these models for each observation t it is easy to calculate their averages, variances, etc.

Let us return to the complex-valued power model with complex coefficients (4.108) and present it in linear form taking natural logarithms of the left- and right-hand sides of the equality:

$$\ln(y_{rt} + iy_{it}) = \ln(a_0 + ia_1) + (b_0 + ib_1)\ln(x_{rt} + ix_{it}). \tag{4.112}$$

As previously, we use the principal values of the logarithms. For the complex argument the logarithm will have the following form:

$$\ln(x_{rt} + ix_{it}) = \ln R_x + i\varphi_x, \tag{4.113}$$

where R_x is the absolute value of the complex variable of the determinant and φ_x is its polar angle. For a complex variable of the modeled result the logarithm will be written as follow:

$$\ln(y_{rt} + iy_{it}) = \ln R_y + i\varphi_y, \tag{4.114}$$

where R_y is the absolute value of the complex variable of the result and φ_y is its polar angle. For the complex proportionality coefficient of model (4.108) the logarithm will be written as follows:

$$\ln(a_0 + ia_1) = \ln R_a + i\varphi_a, \tag{4.115}$$

where R_a is the absolute value of the complex proportionality coefficient and φ_a is its polar angle.

It is not necessary to write the complex coefficient (b_0 + ib_1) in exponential form. With these designations we obtain the following linear model of the power function of complex variables with complex coefficients:

$$\ln R_y + i\varphi_y = (\ln R_a + i\varphi_a) + (b_0 + ib_1)(\ln R_x + i\varphi_x). \tag{4.116}$$

To make subsequent operations simple, let us introduce the following designations:

$$\ln(a_0 + ia_1) = \ln R_a + i\varphi_a = A_0 + iA_1. \tag{4.117}$$

It is clear that the model obtained is a linear model of complex variables with complex coefficients:

$$\ln R_y + i\varphi_y = (A_0 + iA_1) + (b_0 + ib_1)\,(\ln R_x + i\varphi_x).$$

The coefficients of this model can be found using LSM adapted to linear complex-valued models. Substituting the values of model (4.116) into the system of normal equations (4.58), taking into account the designations (4.117), we have

$$\begin{cases} \sum_t \ln R_{yt} = TA_0 + b_0 \sum_t \ln R_{xt} - b_1 \sum_t \varphi_{xt}, \\ \sum_t \varphi_{yt} = TA_1 + b_1 \sum_t \ln R_{xt} + b_0 \sum_t \varphi_{xt}, \\ \sum_t \ln R_{yt} \ln R_{xt} - \sum_t \varphi_{yt}\varphi_{xt} = A_0 \sum_t \ln R_{xt} - A_1 \sum_t \varphi_{xt} \\ \quad + b_0 \sum_t (\ln^2 R_{xt} - \varphi_{xt}^2) - 2b_1 \sum_t \varphi_t \ln R_t \\ \sum_t \varphi_{yt} \ln R_{xt} + \sum_t \varphi_{xt} \ln R_{yt} = A_0 \sum_t \varphi_{xt} + A_1 \sum_t \ln R_{xt} \\ \quad + b_1 \sum_t (\ln^2 R_{xt} - \varphi_{xt}^2) + 2b_0 \sum_t \varphi_t \ln R_t. \end{cases} \tag{4.118}$$

By solving this system one can find the values of the unknown coefficients A_0,A_1, b_0, b_1. The result will be as follows:

$$y_{rt} + iy_{it} = e^{A_0 + iA_1} (x_{rt} + ix_{it})^{(b_0 + ib_1)}.$$

To obtain a model like (4.108) the proportionality coefficient should be taken out of exponential form and rewritten in arithmetic form. To do this we should find a_0 and a_1 according the available values of A_0 and A_1:

$$a_0 = e \cos A_1, \quad a_1 = e^{A_0} \sin A_1.$$

One of the subforms of power functions of complex variables is a power function of a complex argument. In the general case it may be represented as follows:

$$y_{rt} = (a_0 + ia_1)(x_{rt} + ix_{it})^{(b_0 + ib_1)}. \tag{4.119}$$

Taking the natural logarithms of the left- and right-hand sides of the equality we obtain the following linear model:

$$\ln y_{rt} = \ln(a_0 + ia_1) + (b_0 + ib_1) \ln(x_{rt} + ix_{it}). \tag{4.120}$$

Taking into account previously introduced designations we have

$$\begin{aligned} \ln y_{rt} &= (\ln R_a + i\varphi_a) + (b_0 + ib_1)(\ln R_x + i\varphi_x) \\ &= A_0 + iA_1 + (b_0 + ib_1)(\ln R_x + i\varphi_x). \end{aligned} \tag{4.121}$$

Compared to the model of complex variables, this one does not have an imaginary part in the left-hand side of the obtained equality. This is why the system of normal equations for a power function of complex argument may be represented as follows:

$$\begin{cases} \sum_t \ln y_{rt} = TA_0 + b_0 \sum_t \ln R_{xt} - b_1 \sum_t \varphi_{xt}, \\ 0 = TA_1 + b_1 \sum_t \ln R_{xt} + b_0 \sum_t \varphi_{xt}, \\ \sum_t \ln y_{rt} \ln R_{xt} = A_0 \sum_t \ln R_{xt} - A_1 \sum_t \varphi_{xt} + b_0 \sum_t (\ln^2 R_{xt} - \varphi_{xt}^2) \\ \quad - 2b_1 \sum_t \varphi_{xt} \ln R_{xt} \\ \sum_t \ln y_{rt} \varphi_{xt} = A_0 \sum_t \varphi_{xt} + A_1 \sum_t \ln R_{xt} \\ + b_1 \sum_t (\ln^2 R_{xt} - \varphi_{xt}^2) 2b_1 \sum_t \varphi_{xt} \ln R_{xt}. \end{cases} \quad (4.122)$$

Solving the system we find LSM assessments for the power function of a complex argument with complex coefficients.

Let us now consider the technique of assessing the parameters of a model of an exponential function of complex variables, which can be written in the following general form:

$$y_{rt} + iy_{it} = (a_0 + ia_1)e^{(b_0 + ib_1)(x_{rt} + ix_{it})}. \quad (4.123)$$

Again, let us make this linear by taking logarithms of the left- and right-hand sides of the equality:

$$\ln(y_{rt} + iy_{it}) = \ln(a_0 + ia_1) + (b_0 + ib_1)(x_{rt} + ix_{it}). \quad (4.124)$$

Taking into account the introduced designations (4.114) and (4.115):

$$\ln R_{yt} + i\varphi_{yt} = A_0 + iA_1 + (b_0 + ib_1)(x_{rt} + ix_{it}). \quad (4.125)$$

Again, assessing the model parameters is not hard since the system of normal equations has the following form:

$$\begin{cases} \sum_t \ln R_{yt} = TA_0 + b_0 \sum_t x_{rt} - b_1 \sum_t x_{it}, \\ \sum_t \varphi_{yt} = TA_1 + b_1 \sum_t x_{rt} + b_0 \sum_t x_{it}, \\ \sum_t \ln R_{yt} x_{rt} - \sum_t \varphi_{yt} x_{it} = A_0 \sum_t x_{rt} - A_1 \sum_t x_{it} + b_0 \sum_t (x_{rt}^2 - x_{it}^2) - 2b_1 \sum_t x_{rt} x_{it}, \\ \sum_t \ln R_{yt} x_{it} + \sum_t \varphi_{yt} x_{rt} = A_0 \sum_t x_{it} + A_1 \sum_t x_{rt} + b_1 \sum_t (x_{rt}^2 - x_{it}^2) + 2b_0 \sum_t x_{rt} x_{it}. \end{cases} \tag{4.126}$$

Solving this system we find the parameters of the original model.

A similar method may be applied to make LSM assessments of the parameters of a simpler model – an exponential function of a complex argument:

$$y_{rt} = (a_0 + ia_1)e^{(b_0 + ib_1)(x_{rt} + ix_{it})}. \tag{4.127}$$

Using logarithms and particular designations we obtain the following linear function:

$$\ln y_{rt} = A_0 + iA_1 + (b_0 + ib_1)(x_{rt} + ix_{it}). \tag{4.128}$$

Then the system of normal equations for a linear exponential function of a complex argument will look like this:

$$\begin{cases} \sum_t \ln y_{rt} = TA_0 + b_0 \sum_t x_{rt} - b_1 \sum_t x_{it}, \\ 0 = TA_1 + b_1 \sum_t x_{rt} + b_0 \sum_t x_{it}, \\ \sum_t \ln y_{rt} x_{rt} = A_0 \sum_t x_{rt} - A_1 \sum_t x_{it} + b_0 \sum_t (x_{rt}^2 - x_{it}^2) - 2b_1 \sum_t x_{rt} x_{it}, \\ \sum_t \ln y_{rt} x_{it} = A_0 \sum_t x_{it} + A_1 \sum_t x_{rt} + b_1 \sum_t (x_{rt}^2 - x_{it}^2) + 2b_0 \sum_t x_{rt} x_{it}. \end{cases} \tag{4.129}$$

A solution of this system would provide an economist with the necessary values of model coefficients.

It is just as easy to find the parameters of a model of a logarithmic function of complex variables with the general form

$$y_{rt} + iy_{it} = (a_0 + ia_1) + (b_0 + ib_1)\ln(x_{rt} + ix_{it}). \tag{4.130}$$

It should be noted with respect to this model that it is presented in additive form and need not be transformed into a linear model, in contrast to all the aforementioned nonlinear models. One should only find the principal value of a logarithm of a complex argument. This circumstance shows that the assessments of the complex coefficients of this model will not be shifted since the value to be minimized refers to the sum of squares of complex-valued deviations of model (4.130) from actual values but not its linear analog.

Taking into account designations (4.110), the system of normal equations for this model, which is relevant for logarithmic operations, will have the form

$$\begin{cases} \sum_t y_{rt} = Ta_0 + b_0 \sum_t \ln R_{xt} - b_1 \sum_t \varphi_{xt}, \\ \sum_t y_{it} = Ta_1 + b_1 \sum_t \ln R_{xt} + b_0 \sum_t \varphi_{xt}, \\ \sum_t y_{rt} \ln R_{xt} - \sum_t y_{it}\varphi_{xt} = a_0 \sum_t \ln R_{xt} - a_1 \sum_t \varphi_{xt} + b_0 \sum_t (\ln^2 R_{xt} - \varphi_{xt}^2) \\ \quad -2b_1 \sum_t \varphi_{xt} \ln R_{xt}, \\ \sum_t y_{rt}\varphi_{xt} + \sum_t y_{it} \ln R_{xt} = a_0 \sum_t \varphi_{xt} + a_1 \sum_t \ln R_{xt} + b_1 \sum_t (\ln^2 R_{xt} - \varphi_{xt}^2) \\ \quad +2b_0 \sum_t \varphi_{xt} \ln R_{xt} \end{cases} \tag{4.131}$$

The solution to this system will provide an economist with the needed values of assessments of model coefficients.

Thus, if we know the properties of the conformal mappings of simple functions of complex variables, then we can use them to solve various problems of modern econometrics by finding the values of the coefficients of econometric models using LSM, as proposed in the present section. Examples of building nonlinear econometric models will be given in subsequent chapters, along with solutions of applied economic problems.

4.9 Assessment of Confidence Limits of Selected Values of Complex-Valued Models

Since in this chapter we are looking at problems associated with building econometric models applicable exclusively to reversible processes, it is assumed a priori that researchers deal with selected values of random values. And since selected values are being assessed, it is necessary to determine the degree to which these selected values can be trusted, how close they are to their true values, the mathematical expectation. In mathematical statistics this problem has been successfully solved. Hence the natural wish to extrapolate the methods and approaches of mathematical statistics to complex-valued models.

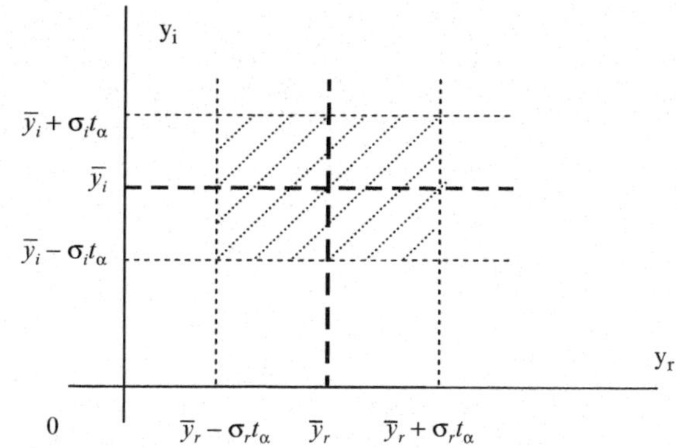

Fig. 4.1 Example of confidence domain of complex variables

It is clear that if a researcher aims at studying a simple stationary process that reveals itself in a sample of a random normally distributed value Y, then calculating the average

$$\bar{Y} = \frac{1}{n}\sum_{i=1}^{n} Y_i \tag{4.132}$$

and the variance of deviations of actual observations from this average σ^2, one can find the interval where the true value of Y lies:

$$\bar{Y} - \sigma t_\alpha \leq Y \leq \bar{Y} - \sigma t_\alpha. \tag{4.133}$$

If instead of the real case we consider a complex variable, the argument should not be violated at first glance –the averages are calculated separately for the real and imaginary parts, the confidence limits are defined for them, and then the confidence limits of the values of the two components of the complex value are found:

$$\begin{aligned} y_r - \sigma t_\alpha \leq y_r \leq y_r - \sigma t_\alpha, \\ y_i - \sigma t_\alpha \leq y_i \leq y_i - \sigma t_\alpha. \end{aligned} \tag{4.134}$$

Figure 4.1 graphically demonstrates this procedure. In a complex plane the confidence domain represents a rectangle with sides defined by the confident limits (4.134). The center of this confidence domain is a point in the complex plane with the coordinates $(\bar{y}_r, \bar{y}_i)$. However, we obtain a rectangle if the random variables are independent of each other, and in a complex-valued economy one of the principles is that of the dependence of the real and imaginary parts of a complex variable. They should show various sides of the same phenomenon or object. This is why the premise of the independence of y_r and y_i from each other that underlies the method for determining confidence limits is not true in our case.

By this logic, the confidence domain should represent a cloud of scattered possible and valid values that should not have sharp corners. In the general case this cloud should take the form of an ellipse.

The aforementioned standard procedure for finding confidence limits (4.134) is not the only one in mathematical statistics; it is just the most popular one. In multidimensional statistics the option of combining the distribution of random values has been studied, and this very method may be applied in a complex-valued economy. Since a complex variable is a two-dimensional value, it will be most suited for T^2 Hotelling statistics representing a locus of the points of an ellipsoid of a confidence domain for two random normally distributed values.

The confidence interval for expectation μ of a real random value is constructed by means of a t statistic that has a t-distribution with $n - 1$ degrees of freedom:

$$t = \frac{\bar{x} - \mu}{\hat{s}} \sqrt{n}. \tag{4.135}$$

This equality may be written in the following equivalent form:

$$t^2 = n(\bar{x} - \mu)\bar{C}^{-1}(\bar{x} - \mu), \tag{4.136}$$

where $\bar{C}$ is a matrix inverse to that of covariance evaluations.

To find a combined confidence domain for a two-dimensional random value using multidimensional statistical methods, namely Hotelling's distribution, we introduce an analog form of (4.136):

$$T^2 = \frac{ns_{rt}^2 s_{it}^2}{s_{rt}^2 s_{it}^2 - s_{rit}^2} \left(\frac{(y_{rt} - \bar{y}_r)^2}{s_{rt}^2} + \frac{(y_{it} - \bar{y}_i)^2}{s_{it}^2} - 2\frac{s_{rit}(y_{rt} - \bar{y}_r)(y_{it} - \bar{y}_i)}{s_{rt}^2 s_{it}^2} \right). \tag{4.137}$$

To calculate T^2, we first determine selected evaluations of the expectations of random values $\bar{y}_r$, $\bar{y}_i$ as the averages of the series y_{rt} and y_{it}, and then selected variances s_{rt}^2, s_{it}^2, s_{rit}^2. To simplify formula (4.137), it is presented in matrix form:

$$T^2 = n(\bar{Y} - \mu)^T C^{-1}(\bar{Y} - \mu), \tag{4.138}$$

where μ is a vector of expectations of two-dimensional random vector Y and $\bar{Y}$ is a vector of the average values (selected evaluations) of the expectations of two-dimensional random vector Y.

Hotelling connected T^2 with distribution F, which for the two-dimensional case has the following form:

$$T_\alpha^2 = \frac{2(n - 1)}{n - 2} F_\alpha. \tag{4.139}$$

The $F\alpha$ statistic has 2 and $n - 2$ degrees of freedom.

The confidence domain of a random complex-valued variable and its characteristics obtained by means of Hotelling's statistics can also be applied to the evaluation of confidence limits of selected evaluations of regression relationships between complex-valued variables. This approach was adapted for a complex-valued economy by A.F. Chanysheva.

Since any nonlinear elementary single-factor complex-values function can be transformed into a linear one by various methods (§4.8):

$$y_r + iy_i = (a_0 + ia_1) + (b_0 + ib_1)(x_r + ix_i), \tag{4.140}$$

it may be represented as an equation with one coefficient by centering the variables of the linear model with respect to their averages:

$$y'_r + iy'_i = (b_0 + ib_1)(x'_r + ix'_i). \tag{4.141}$$

As a result, the problem of determining the confidence domain of a regression model boils down to constructing a confidence domain for the two-dimensional random value $B = (b_0, b_1)$. LSM evaluations are the best way to evaluate complex proportionality coefficients for stochastic normally distributed values.

The possibility of finding a pair of coefficients for each observation is a specific feature of a simple linear complex-valued model without a free term:

$$b_{0t} = \frac{y_{rt}x_{rt} + y_{it}x_{it}}{x_{rt}{}^2 + x_{it}{}^2}, \tag{4.142}$$

$$b_{1t} = \frac{y_{it}x_{rt} - y_{rt}x_{it}}{x_{rt}{}^2 + x_{it}{}^2}. \tag{4.143}$$

Then, the method of finding a confidence domain of a complex proportionality coefficient proposed by Chanysheva is reduced to the following procedure:

1. Point estimates of the expectations of random values $\hat{b}_0$ and $\hat{b}_1$ are found using LSM: $\hat{B} = \hat{b}_0 + i\hat{b}_1$.
2. A random series of the complex variable $b_{0t} + ib_{1t}$ is formed using (4.142) and (4.143).
3. The two-dimensional covariance matrix $\bar{C}$ is calculated for the random vector $B = (b_0, b_1)$ according to

$$\bar{C} = \frac{1}{n-1}\sum_{t=1}^{n}(b_{kt} - \hat{b}_k)(b_{lt} - \hat{b}_l); k = 0, 1; l = 0, 1 \tag{4.144}$$

   and the inverse matrix $\bar{C}^{-1}$.
4. Using (4.135), the interval evaluation for the expectation of the random vector $B = (b_0, b_1)$ is calculated.

The confidence limits of the expectation of the two-dimensional random vector B with confidence probability P are described by the following equation:

$$(\bar{B} - \mu)^T \bar{C}^{-1} (\bar{B} - \mu) = \frac{2(n-1)}{n(n-2)} F_{1-P,2,n-2}. \tag{4.145}$$

The confidence domain for μ is defined by an inequality:

$$(\bar{B} - \mu)^T \bar{C}^{-1} (\bar{B} - \mu) < \frac{2(n-1)}{n(n-2)} F_{1-P,2,n-2}. \tag{4.146}$$

Equation (4.145) defines an ellipse with its center in random point $\bar{B}$. Its random sizes and the directions of its principal axes are defined by $\bar{C}^{-1}$ and the number $F_{1-P,2,n-2}$. The confidence domain represents a set of internal points of this random ellipse. According to the theory, this ellipse covers the unknown point μ with probability P. However, Chanysheva shows that in practice this is not the case. A considerable part of the points go beyond the ellipse and this domain does not appear to be a confidence domain. This is why she proposes a correction factor [9]:

$$H = \frac{(n-2)^2}{2(n-1),} \tag{4.147}$$

which should be taken into account to connect Hotelling's T^2 statistics with F statistics:

$$T^2 = H \times \frac{2(n-1)}{n(n-2)} F_{1-P,k,n-k} = \frac{n-2}{n} \times F_{1-P,2,n-2}. \tag{4.148}$$

Then the equation of an ellipse (4.145) will have the form

$$(\bar{B} - \mu)^T \bar{C}^{-1} (\bar{B} - \mu) = \frac{n-2}{n} F_{1-P,2,n-2}. \tag{4.149}$$

Let us demonstrate the essence of the procedure for calculating the confidence limits of a complex proportionality coefficient of a complex-valued regression model on some conditional data. The original data of the conditional example for random values are given in Table 4.2 and were selected to have similar dimensions and normal distributions. Therefore, they satisfy the initial assumptions for solving the problem of determining the confidence limits of LSM evaluations.

Before calculating the coefficients of this linear regression model, let us first center the original data with respect to their averages to represent the model as (4.141).

LSM applied to the centered data of this table made it possible to construct a linear regression model:

Table 4.2 Observations of random values

Number	x_r	x_i	y_r	y_i
1	50	20	25.6	78.4
2	49	20	25.4	81.4
3	42	21	22.6	75.8
4	56	23	27.3	79.2
5	47	21	29.5	65.4
6	51	22	29.0	71.4
7	50	24	22.7	67.8
8	56	27	31.0	64.6
9	48	23	26.6	74.2
10	50	25	30.1	71.2
11	46	26	30.0	63.4
12	57	25	23.7	74.0
13	54	23	30.4	80.2
14	51	28	22.6	72.9
15	53	27	25.0	81.4

$$\hat{y}_r + i\,\hat{y}_i = (0,\ 961457 + i0,\ 981518)\ (x_r + ix_i)$$

Now, using the fact that the real and imaginary parts of a complex proportionality coefficient are easy to calculate at any observation t using (4.142) and (4.143), we can obtain these values.

Comparing the dynamic series of the complex regression coefficient with an evaluation of it obtained on the whole set by means of LSM, we can see that they in fact represent evaluations of the expectations of this coefficient's values.

Now we have all the values to find a covariance matrix for this series, as well as an inverse one, using (4.144). Since later on we will use the inverse covariance matrix, its values are of utmost interest (Table 4.4)

Let us define the confidence probability at a level of 99 %. For $k = 2$ and $n = 15$ the tables give $F_\alpha = 5.8$. Substituting all the obtained characteristics into the modified Hotelling statistic (4.148), we obtain the equation of an ellipse (4.149) describing the confidence limit for the regression coefficient $b_0 + ib_1$:

$$157,55b_0^2 + 49,51b_1^2 - 255,614b_0 - 45,615b_1 - 56,06b_0b_1 + 144,4 \leq 0.$$

Knowing the confidence domain for the regression model coefficients we can easily construct the confidence domain for the calculated values of complex Y. For this the confidence limit equation for coefficients B should be multiplied by the complex variable $x_r + ix_i$:

$$(157,55b_0^2 + 49,51b_1^2 - 255,614b_0 - 45,615b_1 - 56,0.6b_0b_1 + 144,4)(x_r + ix_i) \leq 0.$$

It is true that this method of building confidence limits (which we called decomposition limits) is not quite correct since the complex argument is also a

Table 4.3 Evaluations of regression coefficient for any observation

Number	b_0	b_1
1	0.982069	1.175172
2	1.0253998	1.242229
3	1.152162	1.228133
4	0.9140683	1.038601
5	1.0415244	0.926152
6	0.9895694	0.9737
7	0.8989625	0.92503
8	0.9002417	0.719339
9	1.0525014	1.041949
10	1.0513292	0.898959
11	1.0847148	0.76597
12	0.8252822	0.935525
13	1.0113657	1.053599
14	0.9431685	0.912438
15	0.9963899	1.028254

Table 4.4 Matrix inverse to covariance one

157.55109	−28.0328
−28.03279	49.50967

Table 4.5 Matrix inverse to covariance one

0.0844132	−0.02444
−0.024437	0.039274

random value and it makes its own variance to the modeled result. We think it would be more correct to use the aforementioned procedure directly for the calculated values of the complex result $\hat{y}_r + i\hat{y}_i$. The logic of this approach is as follows.

1. LSM is used to evaluate the values $\hat{b}_0$ and $\hat{b}_1$: $\bar{B} = \hat{b}_0 + i\hat{b}_1$ and $\hat{y}_r + i\hat{y}_i$, respectively.
2. The random series of the complex variable $y_{rt} + iy_{it}$ is the original one and is at the researcher's disposal.
3. Thetwo-dimensional covariance matrix $\bar{C}$ is calculated using the formula

$$\bar{C} = \frac{1}{n-1}\sum_{t=1}^{n}(y_{kt} - \hat{y}_k)(y_{lt} - \hat{y}_l); k = r, i; l = r, i$$

   as well as the inverse matrix $\bar{C}^{-1}$.

   For the conditional example under consideration the inverse covariance matrix is shown in Table 4.5.

4. Formula (4.135) is used to calculate the interval evaluation for the expectation of the random vector $Y_r + iY_i$.

Using it, one can also determine the confidence limits for the modeled random complex variable:

$$186,9 - 0,475Y_r + 0,084{Y_r}^2 - 4,92Y_i - 0,048Y_rY_i + 0,039{Y_i}^2 = 0.$$

The confidence limits will look like an ellipse and, as studies have shown, the actual values fall within this confidence domain.

The approach proposed by Chanysheva can be extended to the problem of determining the confidence limits of other evaluations of sample value parameters.

4.10 Balancing Factor in Evaluating the Adequacy of Econometric Models

The TFCV provides economists with the possibility not only of describing more precisely particular complex socioeconomic processes but also of solving problems that in the domain of real variables are too huge or unsolvable.

In modeling various processes, modern researchers face the problem of evaluating the adequacy of the obtained model. Various coefficients or factors are usually calculated to determine how well the obtained model describes the original series of data. Based on these factors conclusions are drawn on the degree of the model's adequacy. The average approximation error is a coefficient characterizing the compliance of the modeled processes with the actual ones. It is recommended that it be calculated using one of the following formulae:

$$A_{first} = \frac{100\%}{\bar{y}} \sqrt{\frac{\sum\limits_{t=1}^{n} (y_t - {y'}_t)^2}{n}}, \tag{4.150}$$

$$A_{second} = \frac{100\%}{n} \cdot \sum_{t=1}^{n} \left| \frac{y_t - {y'}_t}{y_t} \right|. \tag{4.151}$$

Here, A_{first} and A_{second} are values of the average approximation error, $\bar{y}$ is the average of the original series of data, y_t is theactual value at observation t, ${y'}_t$ is the calculated value at observation t, and n is the number of observations.

In most cases these coefficients provide good results, and they are sufficient for evaluating model adequacy, though their values differ from each other. However, according to Ivan Svetunkov, there are a number of situations in practice in which none of the aforementioned formulae gives correct information on the properties of the constructed models and therefore misleads the researcher regarding model accuracy. This can happen in two situations:

1. Calculation of the average approximation error for a series whose average is close to zero;
2. Calculation of the average approximation error for a series of data where there are values close to zero.

Actually, it follows from formula (4.150) that if the average $\bar{y}$ of a series of data is close to zero, then the approximation error becomes very high and stops showing the real properties of the model however adequate it is to the real process. Researchers will face these situations when they, for example, center the original series with respect to its average.

On the other hand, it is clear from (4.151) that if in a series there are some values y_t close to zero, then the approximation error also becomes overinflated regardless of the adequacy of the constructed model as some errors represent a quotient where the denominator is close to zero. At the same time, if y'_t equals zero (or is close to it), then, as is clear from (4.151), the coefficient stops taking into account the difference between the actual and the calculated values because we have unity under the summation sign.

In these situations it is not possible to estimate the degree of adequacy of the modeled series by (4.150) and (4.151). This problem is especially serious when the original series contains both negative and positive values, and it should be reduced to dimensionless quantities. The results will be both positive and negative, as well as close to zero. It is difficult to find solutions to this problem in the domain of real variables. Ivan Svetunkov proposed considering actual y_t and calculated y'_t values of variables not in the form of independent series but as complex numbers [10]: $z_t = y_t + iy'_t$.

If we plot points of this series in a complex plane we will obtain a particular aggregate of it lying around the line issuing from the point of origin at a 45° angle, and the closer the calculated values are to the actual ones, the closer the series points are to this line.

Another picture is when these points are in the pseudo-Euclidean plane where we plot real numbers on the horizontal axis and imaginary ones on the vertical axis. The length of the vector $z_t = y_t + iy'_t$ in this plane can be found by the formula

$$|z_t| = \sqrt{y_t^2 + (iy'_t)^2} = \sqrt{y_t^2 - y_t'^2}. \tag{4.152}$$

The specifics of presenting complex variables in the pseudo-Euclidean plane were considered in the first chapter; here we will merely recall that in the Euclidean plane only a zero vector [with coordinates *(0;0)*] can have zero length, and in the pseudo-Euclidean plane, as we see from formula (4.152), nonzero vectors can also have zero length. For example, in the pseudo-Euclidean plane, the absolute value of the complex number 2 + 2*i* will be

$$R = \sqrt{2^2 + (2i)^2} = \sqrt{2^2 - (2)^2} = \sqrt{4 - 4} = 0.$$

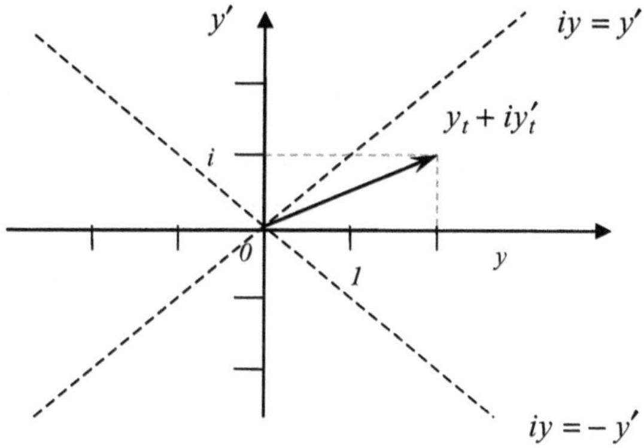

Fig. 4.2 Representation of complex number in pseudo-Euclidean plane

There many vectors with zero length in a plane, all of them satisfying the condition $|y_t| = |y'_t|$ more precisely, one of the following two conditions:

$$y_t = y'_t, \tag{4.153}$$

$$y_t = -y'_t. \tag{4.154}$$

Thus, if the coordinates of the vectors satisfy (4.153) or (4.154), then they lie on the respective lines of the pseudo-Euclidean plane and have zero lengths. These lines are called isotropic. In Fig. 4.2 isotropic lines are shown as dotted lines.

They divide the plane into four sectors:

$$|y_t| > |y'_t| \begin{cases} y_t > 0 - \text{right sector,} \\ y_t < 0 - \text{left sector,} \end{cases}$$

$$|y_t| < |y'_t| \begin{cases} y_t > 0 - \text{top sector,} \\ y_t < 0 - \text{bottom sector.} \end{cases}$$

It is clear from (4.152) that in the pseudo-Euclidean plane the vector length can also be:

- A real number if $|y_t| > |y'_t|$;
- An imaginary number if $|y_t| < |y'_t|$.

In a plane all vectors with real lengths will lie either in the right or left sector, while vectors with imaginary lengths will be either in the top or in the bottom sector.

For us these properties are interesting due to the opportunity it gives us to compare the actual values with the calculated ones and obtain information on the conformance of our model to reality:

1. If the absolute value of the actual data is greater than that of the calculated data, i.e., $|y_t|>|y'_t|$, then the absolute value of the complex number $z_t = y_t + iy'_t$ in the pseudo-Euclidean plane will be a real number.
2. If the absolute value of the actual data is lower than of the calculated data, i.e., $|y_t|<|y'_t|$, then the absolute value of the complex number $z_t = y_t + iy'_t$ in the pseudo-Euclidean plane will be an imaginary number.
3. If the actual and calculated data are equal, then the absolute value of the complex number $z_t = y_t + iy'_t$ will be equal to zero.

Taking into account these specific features of complex numbers in the pseudo-Euclidean plane, one can use them to solve the problem of the adequacy of the evaluation of an economic model. For that, let us use a correlation between two modules: in the pseudo-Euclidean plane R_2 and in the Euclidean plane R_1:

$$R_t^1 = \sqrt{y_t^2 + y_t'2},$$

$$R_t^2 = \sqrt{y_t^2 - y_t'2}.$$

It should be noted that:

1. R_t^2 can be both an imaginary and a real number, while R_t^1 can only be a real number.
2. $R_t^2 = 0$ if the absolute value of the actual values is equal to that of the calculated ones.
3. Always, $R_t^1 \geq 0$, where $R_t^1 = 0$ only when $y_t = y'_t = 0$.
4. If the actual values are equal to zero, then $R_t^1 = \sqrt{y_t'2} = y_t'$, $R_t^2 = \sqrt{-y_t'2} = iy_t'$:
5. If the calculated values are equal to zero, then $R_t^1 = \sqrt{y_t^2} = y_t$, $R_t^2 = \sqrt{y_t^2} = y_t$.

Using these properties, one can obtain the balancing factor of a series of actual data and a series of calculated ones proposed by Ivan Svetunkov:

$$B = \frac{\frac{\sum_{t=1}^{n} \sqrt{y_t^2 - y_t'2}}{n}}{\frac{\sum_{t=1}^{n} \sqrt{y_t^2 + y_t'2}}{n}} = \frac{\sum_{t=1}^{n} \sqrt{y_t^2 - y_t'2}}{\sum_{t=1}^{n} \sqrt{y_t^2 + y_t'2}}. \tag{4.155}$$

Since the numerator of formula (4.155) contain a radicand that may be negative for certain t, factor B will be complex in the general case.

Here, two extreme cases are possible.

The first occurs when the balancing factor is a real number. As we see from (4.155), this is possible if a model has systematic deviations so that $|y_t|>|y'_t|$. Then the radicand of the numerator will always be positive.

The second one is when the balancing factor is imaginary. It follows from (4.155) that this can occur when a model has systematic deviations from the real data $|y_t|<|y'_t|$.

If a model describes an actual series of data so that the approximation error is both positive and negative and the swing is almost the same, then the imaginary part of the complex balancing factor will be equal to its real part.

The proposed balancing factor possesses quite logical properties.

First of all, it follows from (4.155) that both the numerator and the denominator of the factor are always positive. Therefore, both the real and the imaginary parts of the balancing factor are nonnegative quantities.

Besides this evident property, the balancing factor has other characteristics. If a model ideally describes actual values, and the approximation error is zero, the numerator of the balancing factor will be equal to zero and the factor itself will be equal to zero, too. If a model describes a real series of values with some slight approximation error, then the numerator of formula (4.155) of the balancing factor will not be high, and after being divided by the denominator will be close to zero. Therefore, the closeness to zero of the real and imaginary parts of this factor testifies to the high degree of adequacy of the model.

If a model poorly describes the original series of data, the factor numerator will be far from zero, its value tending to the denominator, but it will never be greater than the latter. Thus, both the real and imaginary parts of the balancing factor will not exceed one.

Thus, if a balancing factor has the real and imaginary parts close to each other, this means that the model does not have a shift. If the values of the real and imaginary parts of the factor are close to zero, this indicates a good approximation of the model.

For example, if as a result of calculation $B = 0.5 + i0.1$, then this means that the actual values according to the absolute value are greater than the calculated values, and the model is shifted, and as the real part is much greater than zero, then this means that the model describes the original series of data poorly.

Let us consider now the practical applicability of the coefficient proposed by Ivan Svetunkov, namely, the production function of complex variables for the Diatom plant in the town of Inza, Russia [11], which described the dependence of the gross margin and production costs (complex production result) on capital and labor (complex production resources). Then three parameters are to be modeled:

- Gross profit,
- Production costs,
- Gross output.

Since the data on the production process at this plant constitute a commercial secret, all the original data were presented in nondimensional quantities, which are occasionally given in subsequent chapters of this study. Here we are not interested in these data. The plant operates in a tough competitive environment, which is why there are months without income and even losses. This means that the original data, when presented in nondimensional quantities, are not far from zero, taking both

Table 4.6 Actual and calculated data of gross margin of Diatom plant

t	Actual profit	Modeled profit
1	0.012308	−0.049609
2	−0.058620	−0.032465
3	−0.165870	−0.038947
4	−0.007310	−0.012894
5	0.007612	−0.008581
6	−0.000110	0.025115
7	0.063887	0.010122
8	0.023063	0.038876
9	0.050557	0.037702
10	0.026168	0.054776
11	0.062789	0.074427
12	0.140044	0.097013
13	0.085624	0.114934
14	0.123722	0.120919

positive and negative values, sometimes close to zero. Nondimensional original data on the gross profit and its calculated values obtained by our model are given in Table 4.6.

Let us first calculate approximation errors (4.150) and (4.151) as is done in econometrics today (14). We obtain

$$A_{first} = 173,7122\%, \ A_{second} = 1678,0838\%,$$

respectively.

As is clear, the approximation errors obtained are simply enormous, and if A_{first} is more than 100%, then A_{second} shows an error of more than 1,000 %. The formal conclusion is clear: the model should not be used.

However, it is clear that the original series corresponds to the second case where these coefficients give a distorted view – some values of the series are close to zero, and it is division by a value close to zero that distorts the picture.

If we calculate a coefficient of determination between the calculated and actual values, the picture will be somewhat different: $R^2 = 0,6309$. This means that the model is quite good at describing real values, not as bad as follows from the values of the average approximation errors in percentage terms.

Now let us calculate the balancing factor proposed by Ivan Svetunkov (4.155). It turned out to be equal to $B = 0.3749 + .02452i$.

This means that the model does not describe the original series of data in the best possible way because the real part of the balancing factor is not close to zero, and the imaginary part, though less than the real one, also cannot be considered as close to zero. However, the real and imaginary parts of the complex coefficient do not differ greatly from each other – only by 0.13. That is, there is a shift of the model with respect to the actual values, but it is not significant.

Values of the real and imaginary parts of the balancing factor also characterize to a certain extent the accuracy of approximation. It is interesting that the coefficient of determination indicates that 63 % of the changes in the real values are modeled by the model, but 37% cannot be explained. incidentally, the real part of the balancing factor estimated in percentage terms is 37.5 %. Clearly, this is not just a coincidence. The values of the real and imaginary parts of the balancing factor may really show how accurately the model describes real values.

Thus, the balancing factor is free from the aforementioned drawbacks of the average absolute approximation errors and provides the researcher with additional information on the adequacy of econometric models of real variables.

Ivan Svetunkov's balancing factor, which can be used to estimate the adequacy of econometric models of real variables, may be applied to complex-valued econometric models.

In fact, a complex-valued econometric model can be regarded as a system of two models of real variables – a separate equation for the real part and a separate one for the imaginary part. Then we can build a pseudo-Euclidean plane separately for the real part of the model and compare the real values with the modeled ones and do the same for the imaginary part of the complex variable and imaginary part of the complex-valued model.

To estimate the adequacy of a model's ability to describe real parts, we should calculate the their balancing factor:

$$B_r = \frac{\sum_{t=1}^{n} \sqrt{y_{rt}^2 - y_{rt}'^2}}{\sum_{t=1}^{n} \sqrt{y_{rt}^2 - y_{rt}'^2}}. \tag{4.156}$$

And to estimate the adequacy of a model's ability to describe the imaginary parts of complex variables, we should calculate the balancing factor of the imaginary part:

$$B_i = \frac{\sum_{t=1}^{n} \sqrt{y_{it}^2 - y_{it}'^2}}{\sum_{t=1}^{n} \sqrt{y_{it}^2 + y_{it}'^2}} \tag{4.157}$$

Both coefficients are independent and should therefore be considered in isolation, outside their relation. In other words, they should not be added to each other and on that basis make any general conclusions about the complex-valued model. Therefore, the general idea of the adequacy of complex-valued economic models consists of a notion of how well the real part is described and how well the imaginary part is described.

References

1. Arens R (1957) Complex processes for envelopes of normal noise. IRE Trans Inform Theory IT-3:204–207
2. Goodman NR (1963) Statistical analysis based on a certain multivariate complex Gaussian distribution. Ann Math Statist 34:152–176
3. Reed IS (1962) On a moment theorem for complex Gaussian processes. IRE Trans Inform Theory IT-8:194–195
4. Wooding RA (1956) The multivariate distribution of complex normal variables. Biometrika 43:212–215
5. Wentzel ES (2010) The probability theory. KnoRus, Moscow
6. Svetunkov SG (2007) Method least squares for econometric models of complex variables. Economic cybernetics: systems analysis in economics and management. SUEF, St. Petersburg, pp 86–91
7. Tavares GN, Tavares LM (2007) On the statistics of the sum of squared complex Gaussian randm variables. IEEE Trans Commun 55(10):1857–1862
8. Slutsky EE (2010) Economic and statistical works: selections. AST, Moscow
9. Chanysheva AF (2009) Finding the interval estimation of complex regression equations. In: Economic forecasting: models and methods: proceedings of the international scientific and practical conference, 5–6 April 2009, VSU, Voronezh
10. Svetunkov IS (2011) New coefficients of econometrics models quality estimation. Applied Econometrics 24(4):85–99, Publishing House "Market DS"
11. Svetunkov SG, Svetunkov IS (2008) Production functions of complex variables. Publishing House LKI, Moscow

Chapter 5
Production Functions of Complex Argument

The theory of production functions most often considers the dependence of the production result (output) on two production resources – capital and labor. This chapter considers the main models of production functions with a complex argument, when the real production result depends on the complex argument. Capital refers to the real part of the complex argument and labor to the imaginary one. Models of this type considerably expand the instrumental base of the theory of production functions. In addition, this chapter demonstrates an important property of linear complex models – their stability under conditions of multicollinearity.

We subsequently consider production functions of a complex argument in the form of linear, power, exponential, and logarithmic models, study their properties, and derive elasticity coefficients. Models of each type are built on the basis of methods of complex-valued econometrics with real data of production systems of various levels.

5.1 Fundamentals of Production Functions of a Complex Argument

In order to model the operation of an enterprise, region, branch of industry, or country in general, it is necessary to use the dependence of production results on the resources used without studying the very process of transformation of the resources into the production result. At each of the production hierarchy levels for modeling this relationship we proceed from a number of various assumptions and admissions that are of no any interest for the purposes of our research. We assume that we have some production structure where modeling is considered only as a process of transformation of production resources into a production result.

It is proposed that production be considered a "black box" where you have resources $x_1, x_2, x_3, \ldots, x_n$ coming in and some production result $y_1, y_2, y_3, \ldots y_m$ going out:

S. Svetunkov, *Complex-Valued Modeling in Economics and Finance*,
DOI 10.1007/978-1-4614-5876-0_5, © Springer Science+Business Media New York 2012

$$Y = f(X)\,. \tag{5.1}$$

In economics such a relationship is called the "production function." In the theory of production functions, which uses real variables, the production result is considered one real variable varying with respect to several production resources, i.e., we have a multifactor relationship:

$$y = f(x_i), \quad i = 1, 2, 3, \ldots n. \tag{5.2}$$

Various resources are used in manufacturing. They include labor resources, financial resources, capital resources (machines, equipment, etc.), intellectual resources, and others. In addition, each of the resources represents some aggregate value, for example, labor resources can be divided into production and nonproduction personnel. However, production personnel are also divided into workers and students, engineering and technical personnel, nonmanual workers, etc.

If you try to make a model closer to economic realities, it is necessary to include therein all these resources in as much detail as possible, which is why from the wide array of production resources we focus on two basic and highly aggregated ones – labor L and capital C. The reason for this is the following.

To produce some quantity of goods Q, besides capital and labor, one should expend a certain quantity of material and financial resources, heating, and electric power. The quantity of these resources used in production can be considered to a certain extent directly proportional to the output. Nevertheless, capital and labor, though varying with changes in output, do not vary according to a linear law; in addition, these two resources are interchangeable – increasing capital resources and implementing new technologies can result in increasing labor productivity and reduction in labor costs. This means that with higher capital resource and lower labor costs one can obtain the same output.

This interchangeability is inherent in any level of production hierarchy, for example, at the enterprise level one can achieve an increase in production with unchanged capital resources but increasing labor costs and organizing work in two or three shifts. The same production volume can be achieved if we increase the capital by implementing more productive equipment while making the number of employees unchanged. However, in the first case the prime cost, gross margin, and profitability will considerably differ from the same parameters in the second option, though these specifics are not discussed in the theory of production functions.

For the purposes of our scientific research it is important that from among the whole variety of resources production functions include two resources – labor and capital:

$$Q = f(K, L). \tag{5.3}$$

We will not review here parts of the theory of production functions since the reader must know the theory or, if necessary, can familiarize himself with it from various studies and textbooks. Here, we consider the possibility and specifics of applying complex-valued models to problems of the theory of production functions.

Since (5.3) represents a mapping of points of the resource plane to the real axis of production results, we could replace the plane of real variables by that of complex ones to preserve the meaning of the problem but change its form. The production function of a complex argument will look as follows:

$$Q = f(K + iL). \tag{5.4}$$

Association of labor resources with the imaginary part is not due to any economic reasons – only those of convenience. However, in Chap. 1, we introduced a rule according to which the real part of complex variables is associated with the active part of an economic indicator, and the imaginary part with a passive one. To some extent this rule is followed in the case under consideration; since labor resources in the production process are more subject to various changes compared to capital resources, they could be associated with the imaginary part.

In accordance with TFCV rules, the equality sign in (5.4) means that the production result represented in the form of a real variable can in reality be represented as a complex variable with the imaginary part equal to zero:

$$\begin{cases} Q + iQ_i = f(K + iL), \\ Q_i = 0. \end{cases} \tag{5.5}$$

Because in economics smooth functions exist only in the form of ideas of the most reckless idealists and are never encountered in practice, there is no precise functional compliance between resources and production result (5.4), only a regression relationship. Taking into account this fact, while constructing a production function of a complex argument, we could obtain the following relationship:

$$Q = f(K + iL) + \varepsilon_r + i\varepsilon_i. \tag{5.6}$$

Hence, it follows that the production functions of a complex argument (PFCA) represent a conformal mapping of a set of points in a complex plane of production resources to the complex plane of the production result, where along the real axis we plot Q, and along the imaginary one we plot the deviation e_i. In the general case, this mapping has the form of a linear regression coinciding with the axis of real numbers Q, with the deviation being random. This conformal mapping is shown in Fig. 5.1.

If we use the least-squares method (LSM) to find PFCA coefficients, then

$$\sum (\varepsilon_r + i\varepsilon_i) = 0. \tag{5.7}$$

The specifics of the complex approximation error and methods of measuring the degree of approximation were stated in the previous chapter and will not be discussed in this and subsequent chapters.

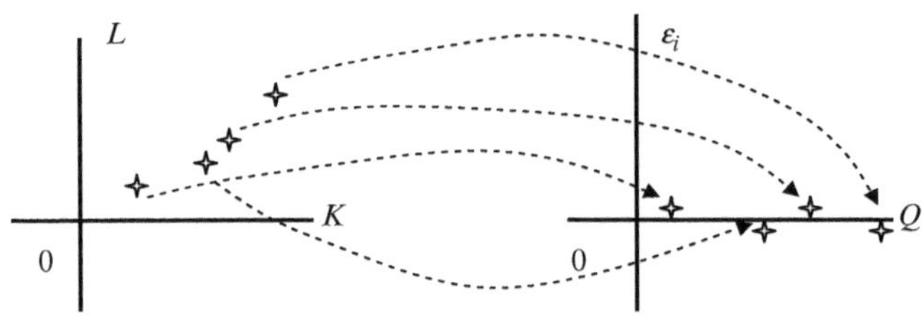

Fig. 5.1 Conformal mapping of PFCA of points in complex plane of production resources to complex plane of results

The formal mathematical basis for PFCA will be any elementary complex-valued function from among a set of known ones. The properties of the principal functions were studied in the Chap. 2.

In this chapter we will gradually consider PFCA on the basis of the general scientific principle of from the simple to the complex and therefore start with the linear production function of a complex argument. After obtaining particular results we will move on to more complex models. However, before that, we should pay attention to the problem domain – all the variables, both the production resources and the production result, and positive values.

5.2 Linear Complex-Valued Model of a Complex Argument and Multicollinearity

Before we begin examining the specific features of functions of a complex argument as production process models, let us devote some attention to a peculiarity of these functions. In Chap. 2 we studied conformal mappings of these models, which showed that they represent an equation of a line in three-dimensional space. This peculiarity of complex argument models may in certain cases be considered a considerable advantage compared to models of real variables, especially for a linear model of a complex argument. Let us show this.

Any complex-valued model is multifactor since a complex argument consists of two real variables consisting of the real and imaginary parts. This is why, before considering models of the production function of complex argument, it seems logical to carry out a comparative analysis of complex argument functions with two-factor models of real variables, which are widely used by economists in economic modeling.

While building multifactor models of socioeconomic dynamics (including two-factor ones), scientists practically always come up against the fact that many, and very often all, pair correlation coefficients between variables included in the models are close to one in their modulus. This phenomenon has been called multicollinearity. It takes place when model factors have similar monotonic, almost linear, trends in their dynamics.

There is one big problem, however. Attempts to use LSM for estimating the coefficients of multifactor models under conditions of multicollinearity results in a degeneration of the system of normal equations using LSM – coefficients of the equations of these systems appear to be close to each other.

This problem of building multifactor models under conditions of multicollinearity has occupied scientists' minds since the middle of the twentieth century, with scientists proposing various modeling methods under conditions of multicollinearity. There is even a special branch of mathematical statistics – robust statistics – aimed at building multifactor models under conditions of multicollinearity.

Let us show how to obtain satisfactory LSM estimations of a multifactor model of real variables under conditions of multicollinearity without complex mathematical procedures [1].

First of all let us write down an LSM system of a normal equation for a multifactor linear model:

$$Y_t = a_0 + a_1 x_{1t} + a_2 x_{2t} + \ldots + a_n x_{nt}, \tag{5.8}$$

omitting the summation limits to make the notation simpler, and taking into account the fact that it is valid for all t beginning from $t = T$:

$$\begin{cases} \sum Y_t = a_0 T + a_1 \sum x_{1t} + \ldots + a_n \sum x_{nt} \\ \sum Y_t x_{1t} = a_0 \sum x_{1t} + a_1 \sum x_{1t}^2 + \ldots + a_n \sum x_{nt} x_{1t} \\ \ldots \\ \sum Y_t x_{nt} = a_0 \sum x_{nt} + a_1 \sum x_{1t} x_{nt} + \ldots + a_n \sum x_{nt}^2. \end{cases} \tag{5.9}$$

If we present this system in the form of equations in segments, we obtain

$$\begin{cases} 1 = \dfrac{a_0}{\frac{\sum Y_t}{T}} + \dfrac{a_1}{\frac{\sum Y_t}{\sum x_{1t}}} + \ldots + \dfrac{a_n}{\frac{\sum Y_t}{\sum x_{nt}}} \\ 1 = \dfrac{a_0}{\frac{\sum Y_t x_{1t}}{\sum x_{1t}}} + \dfrac{a_1}{\frac{\sum Y_t x_{1t}}{\sum x_{1t}^2}} + \ldots + \dfrac{a_n}{\frac{\sum Y_t x_{1t}}{\sum x_{nt} x_{1t}}} \\ \ldots \\ 1 = \dfrac{a_0}{\frac{\sum Y_t x_{nt}}{\sum x_{nt}}} + \dfrac{a_1}{\frac{\sum Y_t x_{nt}}{\sum x_{1t} x_{nt}}} + \ldots + \dfrac{a_n}{\frac{\sum Y_t x_{nt}}{\sum x_{nt}^2}}. \end{cases} \tag{5.10}$$

Since every equation of this LSM system represents an equation of a hyperplane in a hyperspace of coefficients a_i, the solution will be a intersection of the hyperplanes in the hyperspace. However, if the coefficients of the equations of these hyperplanes are close to each other, the latter will be almost parallel, which takes place under conditions of multicollinearity . This means that at the slightest shift of at least one of the hyperplanes in the hyperspace the intersection of all the hyperplanes will significantly change its coordinates, which are the required LSM estimations. Hence, the solution of a system of normal equations rounded to the tenth decimal place will give certain values of the coefficients of the multifactor model, and rounded to the eighth place will give other values of the same multifactor linear model, sometimes opposite in sign.

Equations of the LSM system in (5.10) are represented as equations of hyperplanes in segments in the hyperspace of model coefficients. If in the single-factor case LSM estimations represent an intersection of two equations of LSM systems, since there are only two unknown parameters, a_0 and a_1, and the problem can be presented in a plane, then with two factors the number of model coefficients will be three – a_0, a_1, and a_2. This problem of estimating the parameters of a multifactor model should be considered not in a plane but in three-dimensional space. Actually, the number of unknown parameters becomes three and they can be shown as axes of a three-dimensional space $0a_0$, $0a_1$, and $0a_2$. In this case LSM conditions represent a system of three equations with three unknowns, each of them being an equation of a plane in space. The solution of an LSM system in this case will represent the intersection of three planes in this space of coefficients. The coordinates of this point will give the values of model coefficients.

If we present the problem of finding the coefficients of single-factor models under conditions of multicollinearity by LSM in the form of system (5.10), we see that the values of the segments obtained, cut off by LSM hyperplanes on each of the axes of the coefficients' hyperspaces, will almost always practically coincide, which means that the hyperplanes are almost parallel to each other. Apparently, this is why the solution of an LSM system, which represents an intersection of these practically parallel hyperplanes in a hyperspace, is quite unstable – the slightest rounding error may result in a new intersection due to a slight shift of the hyperplanes in the hyperspace, and this point will be considerably remote from the original one. Here, the solution of an LSM system as an intersection of hyperplanes changes in such a way that it is not only the absolute value of a multifactor model's coefficients that is distorted, but the sign itself of these coefficients, which is a universal consequence.

Therefore, the consequences of multicollinearity are due to the unacceptability of the existing algorithm for the estimation of multifactor models' parameters in this case. Then research should be aimed not at the struggle with objective reality – strong collinearity of virtually all indicators and factors of socioeconomic dynamics (which is confirmed by the false-correlation phenomenon), not at the improvement of mathematical algorithms of work with weakly justified matrices, but at the improvement of the involved coefficient estimation tool.

In order to increase the stability of estimations of parameters of multifactor models it is necessary to separate the hyperplanes of an LSM system of normal equations eliminating their practical parallelism. Therefore, we should create a situation where segments on the axes of the hyperspaces of model-parameter estimations differ from each other as much as possible but do not coincide, as always happens under conditions of multicollinearity. The best thing would be to propose a method of obtaining LSM estimations providing perpendicularity of the hyperplanes – this would make it possible to obtain stable estimations of model coefficients.

To solve this problem, recall the formula for finding the angle between two planes in a three-dimensional space. If we represent one plane by

$$Ax + By + Cz + D = 0, \tag{5.11}$$

and the second plane by

$$A'x + B'y + C'z + D' = 0, \tag{5.12}$$

the cosine of the angle between them can be found by the formula [2]

$$\cos\gamma = \frac{AA' + BB' + CC''}{\sqrt{A^2 + B^2 + C^2}\ \sqrt{A'^2 + B'^2 + C'^2}}. \tag{5.13}$$

If the planes are almost parallel, the angle between them is close to zero and the cosine of the angle is close to one. If the planes are perpendicular, the angle between them is 90° and the cosine is zero. This means that to obtain stable LSM estimations under conditions of multicollinearity, transformations are necessary such that the numerator is equal (or close) to zero or the denominator tends to infinity.

Using the formula for the cosine of the angle between planes (5.13), one can find the cosine of the angle between the planes of system (5.9) in the space of coefficients of the econometric model.

The cosine of the angle between the first and second planes described by the first and second equations of the system of normal equations, respectively, will be determined as follows:

$$\cos\gamma_{12} = \frac{T\sum x_{1t} + \sum x_{1t}\sum x_{1t}^2 + \ldots + \sum x_{nt}\sum x_{1t}x_{nt}}{\sqrt{T^2 + \left(\sum x_{1t}\right)^2 + \ldots + \left(\sum x_{nt}\right)^2}\sqrt{\left(\sum x_{1t}\right)^2 + \left(\sum x_{1t}^2\right)^2 + \ldots + \left(\sum x_{1t}x_{nt}\right)^2}}. \tag{5.14}$$

It has been stated that the cosine can be equal to zero in only two cases – where the numerator is equal to zero and where the denominator tends to infinity.

If the second condition – setting the denominator to infinity without increasing the numerator in the same direction – is impossible, the first one is quite feasible. To do that, the original variables should be centered around their average. This operation represents a transformation of the original series $\{x_t\}$ where the average of this series is subtracted from each value thereof:

$$x_t' = x_t - \bar{x}. \tag{5.15}$$

As follows from the elementary principles of mathematical statistics, the sum of this centered series will be equal to zero. This is why in formula (5.14) the following sums will be equal to zero:

$$\sum x_{1t} = 0, \ \sum x_{2t} = 0, \ \ldots, \sum x_{nt} = 0. \tag{5.16}$$

Since all of them are factors of each term (5.14), the numerator will be equal to zero. The denominator will always be greater than zero. This means that the cosine of the angle between the first and second hyperplanes is equal to zero, i.e., the angle between them is 90° and these two hyperplanes are perpendicular to each other. In turn, this testifies to the fact that the intersection of these hyperplanes will be defined quite stably regardless of rounding errors.

It is clear that with (5.15) for all the model coefficients, the angles between the first and the other hyperplanes of the hyperspace of the unknown coefficients that can be found by an LSM system of normal equations will be correct, and the first hyperplane will be perpendicular to the other hyperplanes. In centering with respect to the averages, the cosines of angles between other hyperplanes of system (5.9), for example the second and the third hyperplanes, will not be equal to zero, but they will be less than in the case where uncentered variables are used, which testifies to the fact that the hyperplanes intersect at a wider angle, and the intersection of all the hyperplanes will be more stable to the rounding errors and new information.

In many cases this simple procedure gives satisfactory results, which is why in conditions of multicollinearity all the original variables should always be centered in advance. However, in practice, there are situations where this approach does not give the needed results. An example is provided by M. Glushenkova and A. Zemlyanaya. Let us observe an socioeconomic process described by a functional two-factor linear model with a zero free term:

$$y_t = 7.3x_{1t} + 2x_{2t}. \tag{5.17}$$

Factors x_{1t} and x_{2t} in the period under consideration vary linearly, and this variation is described by a linear functional dependence of one factor on the other:

$$x_{2t} = 0.273972603x_{1t}. \tag{5.18}$$

It is evident that in this case, the pair correlation coefficient between them will be equal to one, which testifies to an extreme case of multicollinearity – a linear functional relationship between them. Table 5.1 shows the values of the resulting parameter y_t calculated according to formula (5.17) for linearly varying variables. It should again be noted that all the data in Table 5.1 result from the functional relationship (5.17) between all the variables.

Now let us pose another problem, namely, to build a multifactor linear model on the data from the table, without knowing the coefficients (5.17), which means calculating the model coefficients according to the table data:

$$y_t = a_1x_{1t} + a_2x_{2t}.$$

Table 5.1 Data of conditional example

t	x_{1t}	x_{2t}	Calculated value Y_t, obtained by (5.17)
1	2	0.547945205	15.69589
2	4	1.095890411	31.39178
3	6	1.643835616	47.08767
4	8	2.191780822	62.78356
5	10	2.739726027	78.47945
6	12	3.287671233	94.17534
7	14	3.835616438	109.8712
8	16	4.383561644	125.5671
9	18	4.931506849	141.263
10	20	5.479452055	156.9589
11	22	6.02739726	172.6548
12	24	6.575342466	188.3507
13	26	7.123287671	204.0466
14	28	7.671232877	219.7425
15	30	8.219178082	235.4384
16	32	8.767123288	251.1342
17	34	9.315068493	266.8301
18	36	9.863013699	282.526
19	38	10.4109589	298.2219
20	40	10.95890411	313.9178

Since there is a strict functional relationship between all the variables of the problem, one might think that it would be easy to solve it since there is no random error or uncertainty; the problem is completely determined. Moreover, one should expect accuracy as a result of applying LSM. Let us use it to find estimations of the original model from the table data, taking into account that the free term is equal to zero. The system of normal equations for this model will have the form

$$\begin{cases} \sum Y_t x_{1t} = a_1 \sum x_{1t}^2 + a_2 \sum x_{2t} x_{1t} \\ \sum Y_t x_{2t} = a_1 \sum x_{1t} x_{2t} + a_2 \sum x_{2t}^2 \end{cases}$$

The equations that may help to calculate the coefficients calculated for Table 5.1 data will look like the following system:

$$\begin{cases} 90094.411 = a_1 11480 + a_2 3145.206 \\ 24683.403 = a_1 3145,206 + a_2 861.700 \end{cases}. \tag{5.19}$$

The next step is to solve the system and obtain the unknown coefficient values. But before that, let us present this system of equations as a system of equations in segments, as is done in (5.10):

$$\begin{cases} 1 = \dfrac{a_1}{7.8479452} + \dfrac{a_2}{28.645} \\ 1 = \dfrac{a_1}{7.8479452} + \dfrac{a_2}{28.645} \end{cases}. \tag{5.20}$$

We see that this system of equations is degenerated and does not have solutions. The planes of LSM equations are not only parallel to each other, they fully coincide, i.e., they do not intersect. Therefore, multifactor LSM does not allow for solving the set problem.

Now let us center the original data given in Table 5.1 around their averages hoping for an improvement in the situation. Then the system of normal equations will look like this:

$$\begin{cases} 20875.53 = a_1 2660 + a_2 728.77 \\ 5719.32 = a_1 728.77 + a_2 199.66 \end{cases}. \tag{5.21}$$

Clearly, the values of this system differ from the uncentered ones (5.19). However, will it be possible to solve the problem? Let us again present the system obtained in the form of a segment system:

$$\begin{cases} 1 = \dfrac{a_1}{7.8479452} + \dfrac{a_2}{28.645} \\ 1 = \dfrac{a_1}{7.8479452} + \dfrac{a_2}{28.645} \end{cases}$$

It is easy to see that this system coincides with (5.20), which tells us that the system cannot be solved.

Between all the variables there is a strict linear relationship that, in the three-dimensional space of the original variables, represents points lying on one line, and since these points lie on one line, there is an infinite number of planes that this line will belong to. System (5.20) demonstrates this fact – it makes it possible to infinitely combine pairs of values of planar model coefficients in three-dimensional space, i.e., it allows one to obtain an infinite set of equations of planes intersecting and containing a line. One such plane is that described by (5.17). However, it is not possible to single this plane out of the infinite number of others using LSM.

We have considered an extreme case that does not occur in actual economic practice, but situations close to the it exist everywhere – where pair correlation coefficients between factors and modeled results are close to unity in the modulus, for example they are equal to 0.97. In these cases our task is to estimate the coefficients of a multifactor linear model that does not have a satisfactory solution in the domain of real variables.

Linear models of a complex argument make it possible to solve this problem. In the general case, linear models of a complex argument have the form

$$y_r = (a_0 + ia_1)(x_r + ix_i) + (b_0 + ib_1). \tag{5.22}$$

After centering the original variables with respect to their averages, this model will look much easier:

$$y_r = (a_0 + ia_1)(x_r + ix_i). \tag{5.23}$$

Taking out the real and imaginary parts of this equality we obtain a system of two equations:

$$\begin{cases} y_r = a_0 x_r - a_1 x_i, \\ 0 = a_1 x_r + a_0 x_i \end{cases} \tag{5.24}$$

where

$$\begin{cases} y_r = a_0 x_r - a_1 x_i, \\ x_r = -\dfrac{a_0}{a_1} x_i \end{cases} \tag{5.25}$$

The first equation of the obtained system tells us that the result linearly depends on the factor variables, and the second one testifies to the fact that the linear functions of a complex argument are characterized by a linear functional relationship between the factor variables. In other words, the linear functions of a complex argument may exist only when a linear relationship exists between the model factors, i.e., under the extreme condition of multicollinearity – a linear functional dependence between factors! Scientists struggle to eliminate the consequences of multicollinearity, and a function of a complex argument introduced in scientific use should not be used in other cases!

Let us apply LSM to the estimation of coefficients of complex-valued model (5.23), as was shown in Chap. 4. There, we derived a formula for assessing linear models of a complex argument:

$$\begin{cases} \sum_t x_{rt} y_t = a_0 \sum_t (x_{rt}^2 - x_{it}^2) - 2a_1 \sum_t x_{rt} x_{it}, \\ \sum_t x_{it} y_t = a_1 \sum_t (x_{rt}^2 - x_{it}^2) + 2a_0 \sum_t x_{rt} x_{it}. \end{cases}$$

With reference to the data of Table 5.1 this system has the form

$$\begin{cases} 90094.41 = a_0 10618.29 - a_1 6290.41, \\ 24683.4 = a_1 10618,29 + a_0 6290.41, \end{cases} \tag{5.26}$$

or, in segments,

$$\begin{cases} 1 = \dfrac{a_0}{8.4848} - \dfrac{a_1}{14.3225}, \\ 1 = \dfrac{a_1}{2.3246} + \dfrac{a_0}{3.9239}. \end{cases} \quad (5.27)$$

It is clear that we have obtained a stable system of normal equations, which, if solved, gives us the following complex-argument model:

$$y_r = (7.3 - i2)(x_1 + ix_2) \quad (5.28)$$

If we open the brackets and group the real and imaginary parts of the obtained equality, it will be transformed into the following one:

$$y_r = 7.3x_1 + 2x_2 + i(-2x_1 + 7.3x_2). \quad (5.29)$$

The real part of the equality fully corresponds to the original function (5.17), and the imaginary part can easily be transformed into the following equality:

$$0 = i(-2x_{1t} + 7.3x_{2t}) \rightarrow x_{2t} = \frac{2}{7.3}x_{1t} = 0.273972603x_{1t}. \quad (5.30)$$

This means that the original dependence between the factors of (5.18) is identified with absolute precision. M. Glushenkova and N. Zemlyanaya made numerous calculations for this and other conditional examples setting the variance in factors, thereby decreasing the correlation between them, as well as introducing variance to the equation of the dependence, transferring it from a functional to a regression dependence. Each time, estimations of the linear function of a complex variable appeared stable to these variances.

To conclude this section, it should be noted that there is a fundamental difference between a linear two-factor model of real variables such as

$$y_t = a_0x_{1t} + a_1x_{2t} \quad (5.31)$$

and a linear model of a complex argument:

$$y_t = (a_0 + ia_1)(x_{1t} + ix_{2t}) \quad (5.32)$$

Although both models (5.31) and (5.32) have two variables and two coefficients, they describe different figures. The model of real variables (5.31) represents an equation of a plane in the space of original variables. The model of complex argument (5.32) is an equation of a line in this space.

This means that in a situation where there is no linear relationship between variables x_{1t} and x_{2t} the complex-argument model will provide poor results, which is why a model of real variables should be used. In this case the observed points are located around some plane in space with some deviation from the line; the smaller the pair correlation coefficient, the larger the deviation of the points in the plane described by (5.31) from the line. However, if the points in three-dimensional space are on the same straight line, they should be described by a linear complex-argument model.

Once again, it follows that models of complex variables are different from those of real variables and usage thereof in economic modeling extends the instrumental base of the economy, facilitating the solution of the complicated problems associated with modeling socioeconomic processes.

In the cases, when the system of equations (5.25) is violated than the suggested approach becomes non effective.

5.3 Linear Production Function of a Complex Argument

Economic dynamics are complex and diverse. The version of its development described by the linear character of variations in its indicators cannot be excluded. In this case a linear PFCA can be used for various purposes of economic analysis. It is the properties of linear PFCA specified in early 2005 that served as the basis for the formation of a complex-valued economy [3]. Its main features are well studied and described in the study published in 2008 [4], which is why here we will focus only on its most important properties. To build this simple function, let us use three variables describing the production process: production volume Q_t, labor costs L_t, and capital expenditures K_t. Let us present the production resources in the form of a complex variable.

Let us examine the linear function of a complex argument without a free term, bearing in mind that, if necessary, it is easy to get rid of it by centering the original variables around their averages. The model of the linear PFCA will have the form

$$Q_t = (a_0 + ia_1)(K_t + iL_t). \tag{5.33}$$

Multiplying the two factors on the right-hand side of equality (5.33) and grouping the real and imaginary parts separately, we obtain

$$Q_t = (K_t a_0 - L_t a_1) + i(L_t a_0 + K_t a_1). \tag{5.34}$$

Thus, the production function (5.34) can be represented in the form of a system of two equations:

$$Q_t = K_t a_0 - L_t a_1. \tag{5.35}$$

and

$$0 = L_t a_0 + K_t a_1. \tag{5.36}$$

Since it is clear from the discussion in the previous section that (5.33) represents an equation of a line in three-dimensional space, in the resource plane it will describe the following kind of linear projection:

$$K_t = -\frac{a_0}{a_1} L_t. \tag{5.37}$$

Since labor resources and capital resources are positive by definition, (5.37) indicates that one of the proportionality coefficients should be negative. Since most often, growth of resources leads to growth of production results, to meet this condition, coefficient a_1 in (5.35) should be negative.

This means that if the coefficients and resources are positive, we should use the following linear PFCA:

$$Q_t = (a_0 - ia_1)(K_t + iL_t). \tag{5.38}$$

Now it is quite simple to find the coefficients of this model and their economic meaning. To do that, let us find the complex proportionality coefficient from the production volume and production resources:

$$a_0 - ia_1 = \frac{Q_t}{K_t + iL_t} = \frac{Q_t(K_t - iL_t)}{K_t^2 + L_t^2}. \tag{5.39}$$

This equality, as follows from the properties of complex numbers, holds only if the real and imaginary parts of the complex numbers of the left- and right-hand sides of the equality are equal to each other (5.39). This property makes it possible to obtain the formulae for the calculation of each coefficient – by opening the brackets and grouping the real and imaginary parts separately. Then we obtain a formula for the calculation of each of the coefficients. For the real part of the proportionality coefficient:

$$a_0 = \frac{Q_t K_t}{K_t^2 + L_t^2} \tag{5.40}$$

and for its imaginary part:

$$a_1 = \frac{Q_t L_t}{K_t^2 + L_t^2}. \tag{5.41}$$

It is evident from the given formulae that a pair of coefficient values is calculated when there is at least one observation of both resource and production result values. This property distinguishes the proposed function from its analogs in the domain of real numbers, where in order to find two unknown coefficients you should have at least two observations. This property of production function (5.38) is easy to explain: the function has only one unknown complex coefficient, which is why its values can be easily determined by one observation. If we wanted to find the coefficients of a more complicated linear model of a complex argument, for example,

$$Q_t = (b_0 + ib_1) + (a_0 + ia_1)(K_t + iL_t),$$

then we would have a model with two complex coefficients, and we would need two observations to find the their values. However, since we are studying a simple function without a free term, this means that for its practical use one should first center the original variables with respect to their averages.

The obtained formulae (5.40) and (5.41) not only make it possible to find the numerical values of the coefficients on the basis of the known values of costs and expenditures, but they also give an economic interpretation to coefficients a_0 and a_1 if the production result is well modeled by this linear function.

The denominators of the formulae are the same and characterize the magnitude of the resources involved. These formulae differ only in their numerators, which makes it possible to understand the meaning of each coefficient.

The numerator of the real coefficient a_0 characterizes the use of capital resources, the numerator of the imaginary part characterizes that of labor resources.

This is why it makes sense to name them as follows: coefficient a_0 is the coefficient of use of capital resources and coefficient a_1 that of the use of labor resources.

Let us build a linear PFCA according to the statistical values of the produced national income, the value of the fixed production assets, and the mean average number of industrial and production workers in the Soviet Union from 1972 till 1989. These data, expressed in relative values, as well as those of resource usage coefficients, and calculated according to (5.40) and (5.41), are given in Table 5.2.

To calculate the coefficients of model (5.33) all the original table data are oriented relative to their averages.

According to variations in the complex proportionality coefficient in time, one can see that for observations from 1981, 1982, and 1983 the complex proportionality coefficient differs significantly from the values of the whole series. This is easy to explain. The original data were centered around their arithmetic averages, and therefore, in the points where the values assigned to observations of resources approach the arithmetic averages, their difference is close to zero, and division by a small value provided by formulae (5.40) and (5.41), yields considerable coefficient values.

Actually, for example, for 1981, the centered relative value of capital resources of the Soviet Union was 0.015389, though for 1984 it was 0.432389. If we ignore the observations of the complex coefficient of these years we notice that the real part of the complex proportionality coefficient varies around 0.53422, and its imaginary part around 0.04278, with a slightly decreasing trend for the whole observation period.

Table 5.2 Calculation of coefficients of resource use for Soviet economy from 1972 to 1989 [5]

Year	National income, Q_t	Fixed production assets, K_t	Annual average number of industrial and production workers L_t	Resource use coefficient	
				a_0	a_1
1972	1	1	1	0.553187	0.066240
1973	1.079	1.091	1.013	0.521700	0.061203
1974	1.130	1.193	1.029	0.526128	0.058947
1975	1.159	1.292	1.049	0.562921	0.055536
1976	1.232	1.393	1.073	0.537856	0.039692
1977	1.295	1.496	1.091	0.523597	0.026845
1978	1.361	1.612	1.109	0.510070	0.008363
1979	1.399	1.724	1.124	0.591209	−0.02362
1980	1.476	1.846	1.136	0.530158	−0.10133
1981	1.554	1.973	1.147	0.200023	0.420264
1982	1.671	2.107	1.159	0.822334	0.244040
1983	1.750	2.247	1.165	0.713303	0.124065
1984	1.819	2.390	1.169	0.641294	0.080584
1985	1.847	2.518	1.174	0.546466	0.057859
1986	1.875	2.649	1.178	0.484325	0.044366
1987	1.914	2.778	1.175	0.456662	0.033584
1988	2.014	2.904	1.151	0.502928	0.019308
1989	2.097	3.024	1.122	0.524799	0.003609

Table 5.3 Original data for building production functions and calculated data of resource use coefficients [5]

Year	Gross domestic product, Q_t		Investments in fixed capital, K_t		Number of people participating in economy, L_t		Resource use coefficient	
	Absolute values (billions of rubles)	Relative values	Absolute values (billions of rubles)	Relative values	Absolute values (millions of rubles)	Relative values	a_0	a_1
1998	2630	1	407.1	1	63.6	1	0.89868	0.00634
1999	4823	1.834	670.4	1.651	62.7	0.986	0.78225	0.01232
2000	7306	2.778	1165.2	2.860	64.2	1.009	0.78447	0.00901
2001	8944	3.401	1504.7	3.702	64.5	1.014	1.19895	0.07636
2002	10834	4.119	1762.4	4.331	66.2	1.041	1.12936	0.04457
2003	13285	5.051	2186.2	5.371	65.8	1.035	0.97630	0.00944
2004	16779	6.380	3105.1	7.627	66.9	1.053	0.74852	0.00649

The latter means that in the Soviet Union the national income was provided by a small number of workers, meaning a slight growth of labor productivity.

Let us see now how the linear PFCA (5.38) behaves if we consider Russia as an example, and if we can use it for these purposes. Table 5.3 shows the corresponding statistical data for 1998–2004. Gross domestic product is taken as the result of production function Q_t, the number of people involved in the economy as labor costs L_t, and investments in the fixed capital as capital K_t. The last two columns of this table give the results of calculation of the resource usage coefficient.

Again, we are looking at the effect of the scale of centered values on model coefficients in the middle of the segment under consideration. For 2001–2002 the coefficient is greatly affected by centering results. It is evident that it is difficult to draw a conclusion on the nature of the production process on the basis of the data obtained – either we need longer series or we should get rid of the free complex coefficient in a different way, avoiding centering with respect to the averages.

Thus, the simplest model of complex argument (5.38), despite its coefficients' having a simple economic meaning, is not universal and, therefore, of little interest and use for describing real economic situations.

Linear PFCAs with a free term are more complicated:

$$Q_t = (b_0 + ib_1) + (a_0 + ia_1)(K_t + iL_t). \tag{5.42}$$

This free term removes stringent requirements from the original data and initial points. This is why it shows real processes more clearly. However, economic processes are never linear or nonlinear according to some set form. All the processes in an economy change, as do the resulting proportions and quantitative laws.

Since there could be situations that describe models like (5.42), we should consider its properties and methods of finding coefficients.

For that let us consider model (5.42), not only at moment t but also in the following observation:

$$Q_{t+1} = (b_0 + ib_1) + (a_0 + ia_1)(K_{t+1} + iL_{t+1}). \tag{5.43}$$

If we delete the left-hand side of equality (5.42) from the left-hand side of equality (5.43) and the right-hand side of equality (5.42) from the right-hand side of equality (5.43), we obtain

$$\Delta Q_t = (a_0 + ia_1)(\Delta K_t + i\Delta L_t). \tag{5.44}$$

Hence, we can derive a formula for calculating the complex proportionality coefficient.

For the real part of the proportionality coefficient this will have the form

$$a_0 = \frac{\Delta Q_t \Delta K_t}{\Delta K_t^2 + \Delta L_t^2}, \tag{5.45}$$

and for its imaginary part:

$$a_1 = \frac{\Delta Q_t \Delta L_t}{\Delta K_t^2 + \Delta L_t^2}. \tag{5.46}$$

Table 5.4 Calculation of resource extensiveness coefficients for Soviet Union from 1972 to 1989

Year	Increment of national income ΔQ_t	Increment of fixed production assets, ΔK_t	Increment of annual average number of industrial and production workers ΔL_t	Resource use coefficients a_0	a_1
1972					
1973	0.079	0.091	0.013	0.850769	0.121538
1974	0.051	0.102	0.016	0.487992	0.076548
1975	0.029	0.099	0.020	0.281443	0.056857
1976	0.073	0.101	0.024	0.684142	0.162568
1977	0.063	0.103	0.018	0.593524	0.103723
1978	0.066	0.116	0.018	0.555588	0.086212
1979	0.038	0.112	0.015	0.333307	0.044639
1980	0.077	0.122	0.012	0.625100	0.061485
1981	0.078	0.127	0.011	0.609600	0.052800
1982	0.117	0.134	0.012	0.866188	0.077569
1983	0.079	0.140	0.006	0.563251	0.024139
1984	0.069	0.143	0.004	0.48214	0.013486
1985	0.028	0.128	0.005	0.218417	0.008532
1986	0.028	0.131	0.004	0.213541	0.006520
1987	0.039	0.129	−0.003	0.302162	−0.00703
1988	0.100	0.126	−0.024	0.765864	−0.14588
1989	0.083	0.120	−0.029	0.65350	−0.15793

Since in the denominator of each formula we have the same value characterizing the change in the return of the resource scale, the meaning of the coefficients is determined by their numerators.

The first coefficient comprising the real part of the complex proportionality coefficient shows the increment of capital resources, and the second one that of labor resources.

Let us estimate the values of the formulae for calculating the coefficients for the above-mentioned examples.

Table 5.4 give the results of calculations for the economy of the Soviet Union.

The real part of the complex proportionality coefficient varies around some average equal to 0.535. These variations have a wide range, however, the coefficient does not show explicit trends in the changes in values.

The imaginary part of the proportionality coefficient showing to some extent the intensity of use of the labor force demonstrates a decrease in its values with time. If we try to interpret this trend, we could say that year over year, in the Soviet Union the labor forces were used less and less extensively, i.e., more and more intensively compared to the previous year. This means a growth in labor productivity. This conclusion was drawn using model (5.33). In the first approximation, this change of trend in the labor force extensiveness coefficient can be described by a linear trend model:

$$a_{1t} = 0.1575 - 0.0137t$$

Table 5.5 Original data for building of a production function and calculated values of resource use coefficients

Year	GDP increment, ΔQ_t	Increase in investments in fixed capital, ΔK_t	Increase in number of people participating in the economy, ΔL_t	Resource extensiveness coefficients a_0	a_1
1998					
1999	0.834	0.651	−0.014	−38.7810	−0.018330
2000	0.944	1.209	0.023	49.62157	0.017623
2001	0.623	0.842	0.005	104.9132	0.003678
2002	0.718	0.629	0.027	16.72674	0.029552
2003	0.932	1.04	−0.006	−161.5470	−0.005410
2004	1.329	2.256	0.018	166.5680	0.010520

Then, with the help of linear PFCA, the production process in the former USSR can be more or less successfully described by the model

$$\Delta Q_t = (0.535 + i(0.1575 - 0.0137t)(\Delta K_t + i\Delta L_t)$$

Where $t = T\text{-}1972$, T – current year.

It should be noted here that formulae (5.45) and (5.46) will tend to infinity if capital and labor resource increments are minor and close to zero.

Let us see what this model of linear PFCA could provide for the example of Russian economy. Since we are interested in the possibilities of this model we could consider the data given in Table 5.3.

Table 5.5 shows the results of calculation of capital and labor resource extensity coefficients. From the analysis of variations of these coefficients in time we can see that neither the first, nor the second coefficient demonstrate any variation trend. Therefore, linear PFCA cannot be applied for modeling of the economy of Russia with the data used.

Naturally, it is not necessary to calculate coefficient values for each observation but estimate their values on the whole set of observations using LSM.

Thus, for centered original data (i.e. for model (5.33)), to calculate LSM estimations of model of complex argument one should solve the following complex equation:

$$a_0 + ia_1 = \frac{\sum Q_t(K_t + iL_t)}{\sum (K_t + iL_t)^2} \tag{5.47}$$

With reference to the former USSR, LSM estimations of the complex proportionality coefficient appeared to be as follows:

$$a_0 + ia_1 = 0.52936 - i0.03976$$

Calculation of coefficient values for each observation and calculation of their average values given earlier provided the following values of the complex proportionality coefficient:

$$a_0 + ia_1 = 0.53422 + i0.04278$$

The values of the real part of the coefficient are close to each other, and those of the imaginary part are different. This is why, taking into account the fact that LSM estimations are free of the shortcomings of the calculation procedure at each observation, economists who decide to use a model of a production function in the form of a linear complex argument function should use LSM estimations.

Since we are considering a new production function, we should determine a very important characteristic of the production function in general – coefficients of elasticity of production results for resources. The linear function without a free term

$$Q_t = (a_0 + ia_1)(K_t + iL_t)$$

is of interest.

An elasticity coefficient represents the following value for discrete values:

$$\varepsilon_{yx} = \frac{\frac{\Delta y}{y}}{\frac{\Delta x}{x}}.$$

For a continuous function, the elasticity coefficient may be written as follows:

$$\varepsilon_{yx} = \frac{\frac{dy}{y}}{\frac{\partial x}{x}} = \frac{dy}{dx}\frac{x}{y}. \tag{5.48}$$

The elasticity coefficient for capital for the function under consideration will look like this:

$$\varepsilon_K = \frac{\partial Q}{\partial K}\frac{K}{Q} = \frac{\partial Q}{\partial K}\frac{K}{(a_0 + ia_1)(K + iL)}. \tag{5.49}$$

A partial derivative of the linear PFCA for capital will be equal to

$$\frac{\partial Q}{\partial K} = (a_0 + ia_1). \tag{5.50}$$

Substituting this value of the partial derivative into (5.49) we obtain the formula for the coefficient of the production elasticity of capital:

$$\varepsilon_K = \frac{K}{K + iL}. \tag{5.51}$$

This means that the coefficient of production elasticity of capital is a complex value and varies with variations in the number of observations of a complex resource.

Similarly, one can determine the coefficient of production elasticity of labor:

$$\varepsilon_L = \frac{\partial Q}{\partial L}\frac{L}{Q} = \frac{\partial Q}{\partial L}\frac{L}{(a_0 + ia_1)(K + iL)}. \tag{5.52}$$

Since the partial derivative of the output volume for labor is

$$\frac{\partial Q}{\partial L} = i(a_0 + ia_1), \tag{5.53}$$

the necessary coefficient of elasticity of the output volume of labor will also be complex:

$$\varepsilon_L = \frac{iL}{K + iL}. \tag{5.54}$$

It is easy to see that

$$\varepsilon_K + \varepsilon_L = 1. \tag{5.55}$$

Since the elasticity coefficient of a production function shows the change in the production result in percentage terms with a 1% change in the resource, let us determine the essence of the coefficients obtained. Each of them, (5.51) and (5.52), shows the change in the production result with changes in the resource under the influence of the other resource.

If the complex argument changes by 1%, the production result, as follows from (5.55), will also change by one.

Isoquantum is an important characteristic of production functions. It represents an aggregate of points in the resource plane, each of which is associated with the same value of the production result, i.e., $Q = Q_c = const.$

The isoquantum equation for the model of production function without a free term will have the form

$$K = \frac{Q_c - a_1 L}{a_0}. \tag{5.56}$$

The obtained equation shows that isoquantum represents a set of parallel lines shifting upward to the right with an increase in the production volume.

5.4 Power Production Function

A two-factor linear relationship in the domain of real variables represents an equation of a plane in three-dimensional space. The linear complex argument model, as was discovered previously, represents an equation of a straight line.

Nonlinear two-factor models of real variables represent an equation of nonlinear surfaces in three-dimensional space, and nonlinear complex argument models, accordingly, represent an equation of some curved line in three-dimensional space. The specifics of these lines were considered in Chap. 2. Let us now consider nonlinear models of a PFCA.

We start with the power model, which is traditional for the theory of production functions. From the whole variety of complex-valued power functions we first take the functions with real coefficients. It has the form

$$Q_t = a(K_t + iL_t)^b. \tag{5.57}$$

Its linear form looks as follows:

$$\ln Q_t = \ln a + b\ln(K_t + iL_t). \tag{5.58}$$

For the real and imaginary parts of the linear function, using the principal logarithm value, we obtain the following equalities:

$$\begin{cases} \ln Q_t = \ln a + b\ln\sqrt{L_t^2 + K_t^2}. \\ 0 = barctg\dfrac{L_t}{K_t} \end{cases} \tag{5.59}$$

From the second equality we have an obligatory condition: $L_t = 0$, which means that it is not possible to use this model in modeling production processes.

After the introduction of a complex proportionality coefficient into the model,

$$Q_t = (a_0 + ia_1)(K_t + iL_t)^b. \tag{5.60}$$

Taking the logarithms of the left- and right hand sides we get

$$\ln Q_t = \ln\sqrt{a_0^2 + a_1^2} + iarctg\frac{a_1}{a_0} + b\ln(K_t + iL_t). \tag{5.61}$$

Then, taking separately the real and imaginary parts of the obtained equalities, we can represent the model as a system of two equalities:

$$\begin{cases} \ln Q_t = \ln \sqrt{a_0^2 + a_1^2} + b \ln \sqrt{L_t^2 + K_t^2}, \\ arctg \dfrac{a_1}{a_0} = -barctg \dfrac{L_t}{K_t}. \end{cases} \tag{5.62}$$

It follows from the system above that the power PFCA (5.60) can be used when there is a linear relationship between the resources with a constant angle between them, i.e., the line goes through the zero point.

Incidentally, the relationship between the production resources and production result is complicated and can be determined from the exponential form of model (5.60):

$$Q_t = \sqrt{a_0^2 + a_1^2}(L_t^2 + K_t^2)^b. \tag{5.63}$$

Let us now find the values of the elasticity coefficient of this function (5.60) for capital and labor.

The coefficient of production elasticity for capital,

$$\varepsilon_K = \frac{\partial Q}{\partial K}\frac{K}{Q} = \frac{\partial Q}{\partial K}\frac{K}{(a_0 + ia_1)(K + iL)^b}, \tag{5.64}$$

can be determined if we know the partial derivative of the function of a complex argument of capital:

$$\frac{\partial Q}{\partial K} = b(a_0 + ia_1)(K + iL)^{b-1}. \tag{5.65}$$

Then the coefficient of production elasticity of capital will have the form

$$\varepsilon_K = \frac{bK}{K + iL} \tag{5.66}$$

Similarly, we can determine the elasticity coefficient of function (5.60) for labor:

$$\varepsilon_L = \frac{ibL}{K + iL}. \tag{5.67}$$

If not one but the entire complex resource varies by 1%, the production result will change by

$$\varepsilon = \varepsilon_K + \varepsilon_L = b. \tag{5.68}$$

Thus, the exponent of the function under consideration is the coefficient of general elasticity of production for a complex argument.

Now let us define the isoquantum equation. If $Q = Q_c = const$, the isoquantum equation for model (5.60) will have the form

$$L_t^2 + K_t^2 = \left(\frac{Q_c}{\sqrt{a_0^2 + a_1^2}} \right)^{\frac{1}{b}}, \tag{5.69}$$

which means that isoquanta of model (5.60) represent circumferences in the production resource plane with various diameters determined by production result Q_c: if it grows, the circumference radius grows too.

Let us now consider a model of PFCA with a complex proportionality coefficient and imaginary exponent:

$$Q_t = (a_0 + ia_1)(L_t + iK_t)^{ib}. \tag{5.70}$$

Taking the logarithm and singling out the real and imaginary parts we obtain

$$\begin{cases} \ln Q_t = \ln \sqrt{a_0^2 + a_1^2} - barctg \dfrac{K_t}{L_t}, \\ arctg \dfrac{a_1}{a_0} = b \ln \sqrt{L_t^2 + K_t^2}. \end{cases} \tag{5.71}$$

From the second equality we see that the relationship between the resources represents a circumference. In three-dimensional space this means a model of one-fourth of a cylinder perpendicular to the resource plane.

There is a complex nonlinear relationship between the production resources and the production result:

$$Q_t = \sqrt{a_0^2 + a_1^2} e^{-barctg \frac{K_t}{L_t}}. \tag{5.72}$$

Additional information on the function and its properties is obtained from the elasticity coefficients. The elasticity coefficients of the production result for a complex argument applicable to function (5.70) will be calculated as follows:

$$\varepsilon_K = \frac{dQ}{d(K + iL)} \frac{(K + iL)}{Q}. \tag{5.73}$$

To find it, we should calculate a derivative of this function for the complex argument:

$$\frac{dQ}{d(K + iL)} = \frac{(a_0 + ia_1)(K + iL)^{ib}}{d(K + iL)}.$$

To do that, let us represent function (5.70) in exponential form:

$$Q_t = \sqrt{a_0^2 + a_1^2} e^{iarctg\frac{a_1}{a_0}} \left[\sqrt{L^2 + K^2} e^{iarctg\frac{L}{K}}\right]^{ib}.$$
$$= \sqrt{a_0^2 + a_1^2} e^{iarctg\frac{a_1}{a_0}} \left(\sqrt{L^2 + K^2}\right)^{ib} e^{-barctg\frac{L}{K}}$$

According to the d'Alembert-Euler (Riemann-Cauchy) condition, we can find the derivative of a complex-valued function by calculating the derivative of the real part for each of the constituents of the complex argument. The real part of the function will then have the form

$$Q_t = R\cos\theta = ae^{-barctg\frac{L}{K}} \cos(\alpha + b\ln\sqrt{L^2 + K^2}), a = \sqrt{a_0^2 + a_1^2}, \alpha = arctg\frac{a_1}{a_0}. \tag{5.75}$$

The first derivative will be

$$\frac{dQ}{d(K + iL)} = \frac{\partial Q}{\partial K} - i\frac{\partial Q}{\partial L}. \tag{5.76}$$

The first partial derivative of the volume for capital,

$$\frac{\partial Q}{\partial K} = \frac{\partial\left(ae^{-barctg\frac{L}{K}} \cos\left(\alpha + b\ln\sqrt{L^2 + K^2}\right)\right)}{\partial K}, \tag{5.77}$$

can be defined as the derivative of a complex function:

$$\frac{\partial Q}{\partial K} = \frac{\partial(ae^{-barctg\frac{L}{K}})}{\partial K} \cos\left(\alpha + b\ln\sqrt{L^2 + K^2}\right)$$
$$+ \frac{\partial(\cos(\alpha + b\ln\sqrt{L^2 + K^2}))}{\partial K} ae^{-barctg\frac{L}{K}}. \tag{5.78}$$

Its first part will be equal to

$$\frac{\partial(ae^{-barctg\frac{L}{K}})}{\partial K} \cos\left(\alpha + b\ln\sqrt{L^2 + K^2}\right)$$
$$= \frac{\partial(-barctg\frac{L}{K})}{\partial K} ae^{-barctg\frac{L}{K}} \cos\left(\alpha + b\ln\sqrt{L^2 + K^2}\right). \tag{5.79}$$

Since

$$\frac{\partial(-barctg\frac{L}{K})}{\partial K} = -b\frac{\frac{\partial\frac{L}{K}}{\partial K}}{1 + (\frac{L}{K})^2} = \frac{b\frac{L}{K^2}}{1 + (\frac{L}{K})^2}, \tag{5.80}$$

the first part of derivative (5.78) will have the form

$$\frac{b\frac{L}{K^2}}{1+\left(\frac{L}{K}\right)^2}ae^{-barctg\frac{L}{K}}\cos(\alpha+b\ln\sqrt{L^2+K^2})=\frac{b\frac{L}{K^2}}{1+\left(\frac{L}{K}\right)^2}R\cos\theta=b\frac{bL}{K^2+L^2}R\cos\theta. \tag{5.81}$$

Now let us consider the second part of the derivative of complex function (5.78):

$$\frac{\partial\left(\cos\left(\alpha+b\ln\sqrt{L^2+K^2}\right)\right)}{\partial K}ae^{-barctg\frac{L}{K}}=-R\sin\theta\frac{bK}{L^2+K^2}. \tag{5.82}$$

Taking into account (5.81) and (5.82) the first partial derivative of the volume for capital will be

$$\frac{\partial Q}{\partial K}=\frac{bL}{K^2+L^2}R\cos\theta-\frac{bK}{K^2+L^2}R\sin\theta=\frac{bR}{K^2+L^2}(L\cos\theta-K\sin\theta). \tag{5.83}$$

Let us now find the first partial derivative for labor since it constitutes the second term of the unknown derivative:

$$\frac{\partial Q}{\partial L}=\frac{\partial\left(ae^{-barctg\frac{L}{K}}\cos\left(\alpha+b\ln\sqrt{L^2+K^2}\right)\right)}{\partial L}. \tag{5.84}$$

Let us consider it as a derivative of a complex function:

$$\frac{\partial Q}{\partial L}=\frac{\partial(ae^{-barctg\frac{L}{K}})}{\partial L}\cos(\alpha+b\ln\sqrt{L^2+K^2}) \\ +\frac{\partial\left(\cos\left(\alpha+b\ln\sqrt{L^2+K^2}\right)\right)}{\partial L}ae^{-barctg\frac{L}{K}}. \tag{5.85}$$

The first part of this sum will be equal to

$$\frac{\partial(ae^{-barctg\frac{L}{K}})}{\partial L}\cos\theta=\frac{\partial(-barctg\frac{L}{K})}{\partial L}ae^{-barctg\frac{L}{K}}\cos\theta. \tag{5.86}$$

Since

$$\frac{\partial(-barctg\frac{L}{K})}{\partial L}=-b\frac{\frac{\partial\frac{L}{K}}{\partial L}}{1+\left(\frac{L}{K}\right)^2}=-\frac{\frac{b}{K}}{1+\left(\frac{L}{K}\right)^2}=-\frac{bK}{K^2+L^2}, \tag{5.87}$$

for the first term (5.85) we have

$$-\frac{bK}{K^2+L^2}R\cos\theta. \tag{5.88}$$

The second part of the derivative of a complex function is

$$\frac{\partial\left(\cos\left(\alpha + b\ln\sqrt{L^2+K^2}\right)\right)}{\partial L}R = -R\sin\theta\frac{bL}{L^2+K^2}. \tag{5.89}$$

Then the first derivative of the function of volume for labor under consideration can be written taking into account (5.88) and (5.89) as follows:

$$\frac{\partial Q}{\partial L} = -\frac{bK}{K^2+L^2}R\cos\theta - R\sin\theta\frac{bL}{L^2+K^2} = -\frac{bR}{K^2+L^2}(K\cos\theta + L\sin\theta). \tag{5.90}$$

Now we can obtain a general formula for the first derivative of a complex function for a complex argument by substituting the values of the first, (5.83), and second, (5.90), terms into (5.76):

$$\frac{\partial Q}{\partial(K+iL)} = \frac{bR}{K^2+L^2}(L\cos\theta - K\sin\theta) - i\left(-\frac{bR}{K^2+L^2}(K\cos\theta + L\sin\theta)\right). \tag{5.91}$$

Grouping will give us

$$\frac{\partial Q}{\partial(K+iL)} = \frac{bR}{K^2+L^2}[L(\cos\theta + i\sin\theta) - iK(\cos\theta + i\sin\theta)], \tag{5.92}$$

which is easily transformed into

$$\frac{\partial Q}{\partial(K+iL)} = \frac{(L-iK)}{K^2+L^2} = \frac{bRe^{i\theta}}{L+iK}. \tag{5.93}$$

Now it is possible to find the coefficient of elasticity of the required function for a complex argument:

$$\varepsilon_{K+iL} = \frac{dQ}{d(K+iL)}\Big/\frac{Q}{K+iL} = \frac{bRe^{i\theta}}{L+iK}\frac{K+iL}{Re^{i\theta}} = ib. \tag{5.94}$$

The elasticity coefficient of the function under consideration is an imaginary value!

The elasticity coefficient of the function under consideration for capital can be calculated using the formula obtained for the first partial derivative of the function for capital (5.83):

$$\varepsilon_K = \frac{\partial Q}{\partial K}\Big/\frac{Q}{K} = \frac{bR}{K^2+L^2}(L\cos\theta - K\sin\theta)\Big/\frac{Re^{i\theta}}{K} = \frac{bK(L\cos\theta - K\sin\theta)}{(K^2+L^2)(\cos\theta + i\sin\theta)}. \tag{5.95}$$

Using the previously derived values of the partial derivative of the function for labor (5.90) it is also easy to find the elasticity of the volume for this resource:

$$\varepsilon_L = \frac{\partial Q}{\partial L}/\frac{Q}{L} = -\frac{bR}{K^2+L^2}(K\cos\theta + L\sin\theta)/\frac{Re^{i\theta}}{L} = -\frac{bL(K\cos\theta + L\sin\theta)}{(K^2+L^2)(\cos\theta + i\sin\theta)}. \tag{5.96}$$

If we assume that the production result is a constant value, we can find an equation of the isoquantum of this production function.

From (5.72), on the assumption of a constant result, we will have an obvious equality:

$$-\,barctg\frac{K_t}{L_t} = \text{const.} \tag{5.97}$$

This means that the isoquantum represents a straight line in the resource plane issuing from the point of origin.

It is easy to notice that a change in the exponent from the real to the imaginary symmetrically changed the properties of the respective parts.

Therefore, with a restriction on the form of resource change (cylinder), the dependence of the production result on the *n*th resources shown in the space represents a nonlinear curve located on the cylinder surface. The form of this model is much more complicated than that of the model with a real exponent.

The power form of the PFCA can have a more complicated form if a complex exponent is used:

$$Q_t = a_0(K_t + iL_t)^{(b_0+ib_1)}. \tag{5.98}$$

Taking the logarithms of the left- and right-hand side of the function we obtain

$$\begin{aligned}\ln Q_t &= \ln a_0 + (b_0 + ib_1)\ln(K_t + iL_t)\\ &= \ln a_0 + (b_0 + ib_1)(\ln\sqrt{L_t^2 + K_t^2} + iarctg\frac{L_t}{K_t}).\end{aligned} \tag{5.99}$$

Then, singling out the real and imaginary parts yields

$$\begin{cases}\ln Q_t = \ln a_0 + b_0\ln\sqrt{L_t^2+K_t^2} - b_1 arctg\dfrac{L_t}{K_t},\\ 0 = b_0 arctg\dfrac{L_t}{K_t} + b_1\ln\sqrt{L_t^2+K_t^2}.\end{cases} \tag{5.100}$$

It follows from the second equality of this system that model (5.98) is suitable for modeling production processes characterized by a complex nonlinear relationship between the resources, which are not easy to represent in an explicit form. Since for various values of coefficients b_0 and b_1 this relationship takes various forms, it is more applicable in practice than previous models of power PFCAs.

Power PFCAs with complex coefficients should be considered universal:

$$Q_t = (a_0 + ia_1)(K_t + iL_t)^{(b_0+ib_1)} \tag{5.101}$$

since when we take logarithms of the left- and right-hand sides of this equation and single out the real and imaginary parts, we obtain

$$\begin{cases} \ln Q_t = \ln\sqrt{a_0^2 + a_1^2} + b_0 \ln\sqrt{L_t^2 + K_t^2} - b_1 arctg\dfrac{L_t}{K_t}, \\ 0 = arctg\dfrac{a_1}{a_0} + b_0 arctg\dfrac{L_t}{K_t} + b_1 \ln\sqrt{L_t^2 + K_t^2}. \end{cases} \tag{5.102}$$

Various combinations of this model make it possible to model most varied nonlinear dependences of the production result from resources, which vary from linear (for $b_1 = 0$) to complex nonlinear ones. Coefficients of these models should be found using LSM, as was pointed out in Chap. 4.

However, use of a complex exponent makes it possible to find intermediate values of this indicator. To do that, let us take the ratios of the left- and right-hand sides of the model at close times. We obtain

$$\frac{Q_t}{Q_{t-1}} = \left(\frac{K_t + iL_t}{K_{t-1} + iL_{t-1}}\right)^{b_0+ib_1}. \tag{5.103}$$

From here it is easy to find a complex exponent:

$$b_{t0} + ib_{t1} = \ln\frac{Q_t}{Q_{t-1}} / \ln\left(\frac{K_t + iL_t}{K_{t-1} + iL_{t-1}}\right).$$

Let us demonstrate the change in the exponent for this model for the Russian economy, taking a longer series of data from 1995 to 2009. The results of calculation of the complex exponent are given in Table 5.6.

During the entire calculation period the complex exponent varies in value. This is especially evident for the default situation of 1998–1999.

The calculated value of the coefficient has changed significantly. If we consider the complex exponent for the following period, we see that its real and imaginary parts did not vary so dramatically. This means that for this period model (5.98) describes production more or less satisfactorily.

There is no need to carry out complex calculations and prove that the elasticity of the function of the complex argument under consideration (5.98) will be equal to

$$\varepsilon_{K+iL} = b_0 + ib_1 \tag{5.104}$$

Table 5.6 Original data (in dimensionless quantities) for building a power PFCA and calculated values of complex exponent

Year	GDP, Q_t	Fixed assets, K_t	Economically active population, L_t	Coefficients b_0	b_1
1995	1	1	1		
1996	1.406	2.523	0.983	0.373	0.238
1997	1.640	2.564	0.961	5.833	6.783
1998	1.841	2.726	0.950	1.836	0.801
1999	3.376	2.749	1.019	15.272	−19.114
2000	5.114	3.204	1.021	2.715	0.919
2001	6.261	3.906	1.008	1.016	0.312
2002	7.574	4.714	1.022	1.016	0.222
2003	9.246	5.853	1.028	0.924	0.176
2004	11.919	6.280	1.029	3.609	0.603
2005	15.127	7.404	1.042	1.450	0.204
2006	18.843	8.457	1.046	1.653	0.210
2007	23.274	10.469	1.059	0.990	0.105
2008	29.001	12.457	1.071	1.266	0.111
2009	27.371	14.371	1.056	−0.404	−0.035

since this function can also be represented as follows:

$$Q_t = (a_0 + ia_1)(K_t + iL_t)^{b_0}(K_t + iL_t)^{ib_1}, \tag{5.105}$$

which will definitely result in (5.104).

An indisputable advantage of PFCA is that its application is much broader than that of linear models since for $b_0 = 1$ and $b_1 = 0$ model (5.98) is transformed into a model of a linear PFCA, which testifies to the fact that the linear model of PFCA is a special case of a power model. Incidentally, for $b_0 = 1$ and $b_1 = 0$ an inversely proportional complex-valued relationship is modeled, which simulates stagnation – an increase in production resources in this model results in a reduction of the production result.

5.5 Exponential Production Function of Complex Argument

From among the family of possible models of exponential PFCAs let us consider an exponential function with a natural base, taking into account that other bases are decimal. Binary and other bases can be applied similarly.

Studies of this function start according to the principle of from the simple to the complex. The model of an exponential function with real coefficients is the simplest in this family:

$$Q_t = ae^{b(K_t + iL_t)}. \tag{5.106}$$

This function may easily be transformed as follows:

$$Q_t = ae^{bK_t}e^{ibL_t}. \tag{5.107}$$

Due to the fact that one can speak about an equality of complex variables only if their real and imaginary parts are equal to each other, one can see that this model means a system of two equalities:

$$\begin{cases} Q_t = ae^{bK_t}, \\ 2\pi k = bL_t, \end{cases} \tag{5.108}$$

where $k = 0,1,2,3,\ldots$

Of course, it is more convenient to consider $k = 0$. In any case, the second equation of the system (5.108) testifies to one thing – labor resources are considered here as a constant value. From the first equality it is clear that this model represents a single-factor power dependence of the production volume on capital resources. Both the first and second equations of the system show that this model describes the effect of capital on production volume with a constant labor force, which influences neither the result nor the capital.

An exponential function with an imaginary exponent has a similar meaning, though it is symmetrical with reference to production resources:

$$Q_t = ae^{ib(K_t+iL_t)}. \tag{5.109}$$

This function can easily be transformed into

$$Q_t = ae^{ibK_t}e^{-bL_t}, \tag{5.110}$$

where

$$\begin{cases} Q_t = ae^{-bL_t}, \\ 2\pi k = bK_t, \end{cases} \tag{5.111}$$

which again shows that it is the single-factor dependence of the production result on labor, with constant capital resources, that is being modeled.

If now we use not the real but the imaginary proportionality coefficient, the practical value of the model will not change:

$$Q_t = iae^{b(K_t+iL_t)}. \tag{5.112}$$

This function can also be transformed into a convenient form:

$$Q_t = ae^{i(\pi/2)}e^{bK_t}e^{ibL_t}. \tag{5.113}$$

The argument of the imaginary proportionality coefficient is determined up to one period. Now it is easy to obtain a system of two equations characterizing the real and imaginary parts of model (5.112):

$$\begin{cases} Q_t = ae^{bK_t}, \\ \dfrac{\pi}{2} = -bL_t. \end{cases} \tag{5.114}$$

This means that model (5.112) implies that the capital resource is a priori constant.

Let us make the model more complicated due to a complex proportionality coefficient with a real proportionality coefficient of the exponent:

$$Q_t = (a_0 + ia_1)e^{b(K_t+iL_t)} \tag{5.115}$$

This complex-argument model can be represented as equalities of the modulus and the argument, which comprises the following system of equations:

$$\begin{cases} Q_t = \sqrt{a_0^2 + a_1^2}e^{bK_t}, \\ 2\pi k = arctg\dfrac{a_1}{a_0} + bL_t. \end{cases} \tag{5.116}$$

We see again that this model implies the a priori constant nature of the labor resource [the second equation of system (5.116)] and the single-factor dependence of the production result on capital resources.

Thus, in contrast to the model of a power PFCA, the model of an exponential PFCA is not diversified and for practical purposes can be presented only with a complex exponent:

$$Q_t = (a_0 + ia_1)e^{(b_0+ib_1)(K_t+iL_t)}. \tag{5.117}$$

Singling out its modulus and argument, we obtain the following system of equations:

$$\begin{cases} Q_t = \sqrt{a_0^2 + a_1^2}\dfrac{e^{b_0K_t}}{e^{b_1L_t}}, \\ 2\pi k = arctg\dfrac{a_1}{a_0} + b_0L_t + b_1K_t. \end{cases} \tag{5.118}$$

It follows from the second equation that the model of an exponential power function of a complex argument implies an a priori linear relationship between the production resources.

The first equation shows that the exponential model is an analog of complex-value function (5.117) in the domain of real variables (for a linear variation of resources):

$$Q_t = \sqrt{a_0^2 + a_1^2}\frac{e^{b_0 L_t}}{e^{b_1 K_t}} = ae^{b_0 L_t - b_1 K_t}.$$

Its logarithm will look as follows:

$$\ln Q_t = A + b_0 L_t - b_1 K_t. \tag{5.119}$$

Graphically, the model will represent an exponent located in a plane perpendicular to the resource axis, where all the points satisfy the second equality of system (5.118).

It is not difficult to find coefficients of this model using LSM.

However, it is possible to discover if the model is suitable for describing real economic production situations without appealing to LSM.

For that, let us divide the left- and right-hand sides of equality (5.117) at time t by the the left- and right-hand sides of the same equality at the previous moment:

$$\frac{Q_t}{Q_{t-1}} = e^{(b_0 + ib_1)(\Delta K_t + i\Delta L_t)}. \tag{5.120}$$

Then we can determine the complex exponent basing on two values of the original variables:

$$b_0 + ib_1 = \frac{\ln \frac{Q_t}{Q_{t-1}}}{\Delta K_t + i\Delta L_t}. \tag{5.121}$$

Let us demonstrate this possibility for the Russian economy. The original data are given in Table 5.7.

It is seen from the table that the complex exponent of an exponential PFCA varies, especially its real part. This testifies to the fact that this model is not suitable for modeling the current production process.

We see that the power model of a complex argument considered in Sect. 5.4 possesses much more interesting properties than the exponential model, which is applicable for practical purposes. There might also be cases where the exponential functions of a complex argument will better describe some production process than the power one, but they will be few in number. This is why in this section we do not study elasticity coefficients of this production function.

5.6 Logarithmic Production Function of Complex Argument

A logarithmic function of a complex variable is a periodical function and, as we agreed at the beginning of the book, we will use only the principal value of the logarithm. In this case the logarithmic PFCA will have the form

$$Q_t = (a_0 + ia_1) + (b_0 + ib_1)\ln(K_t + iL_t). \tag{5.122}$$

Table 5.7 Exponential PFCA and calculated values of complex exponent

Year	GDP, Q_t	Fixed assets, K_t	Economically active population, L_t	Coefficients	
				b_0	b_1
1995	1	1	1		
1996	1.406	2.523	0.983	2.891	0.002
1997	1.640	2.564	0.961	0.711	1.562
1998	1.841	2.726	0.950	2.717	0.046
1999	3.376	2.749	1.019	0.912	−7.959
2000	5.114	3.204	1.021	0.288	−0.004
2001	6.261	3.906	1.008	0.235	0.005
2002	7.574	4.714	1.022	0.175	−0.004
2003	9.246	5.853	1.028	0.595	−0.001
2004	11.919	6.280	1.029	0.212	−0.001
2005	15.127	7.404	1.042	0.209	−0.002
2006	18.843	8.457	1.046	0.105	−0.001
2007	23.274	10.469	1.059	0.111	−0.001
2008	29.001	12.457	1.071	−0.030	−0.001
2009	27.371	14.371	1.056	0.000	0.000

In this function, the effect of the free term is evident – the real part of this coefficient a_0 characterizes a shift in the production result for initial values of the variables, and the imaginary part of coefficient a_1 characterizes the correction of the imaginary part of the equality. Since there are no other interpretations or influences on the results of modeling production, we can neglect this coefficient to consider the model properties. Then the model of the logarithmic PFCA can be presented in the following general form:

$$Q_t = (b_0 + ib_1)\ln(K_t + iL_t). \tag{5.123}$$

This function can easily be transformed into a form convenient for investigation, namely, a system of two equalities, real and imaginary parts:

$$\begin{cases} Q_t = b_0 \ln \sqrt{L_t^2 + K_t^2} - b_1 arctg \dfrac{L_t}{K_t}, \\ 2\pi k = b_1 \ln \sqrt{L_t^2 + K_t^2} + b_0 arctg \dfrac{L_t}{K_t}. \end{cases} \tag{5.124}$$

The second equality of this system shows the relationship between production resources, and the first one represents an analog of a complex-valued model on a set of real numbers.

As we can see from the second equality, this model implies the most varied forms of relationship between the production resources. These forms are determined first of all by the values of coefficients b_0 and b_1. Thus, for example, if $b_0 = 0$, the relationship between the resources should be described by the equation

Table 5.8 Complex proportionality coefficient of logarithmic PFCA

Year	Coefficients	
	b_0	b_1
1995	0.470	−1.066
1996	1.239	−0.462
1997	1.445	−0.514
1998	1.578	−0.499
1999	2.831	−0.934
2000	3.961	−1.007
2001	4.346	−0.787
2002	4.726	−0.641
2003	5.139	−0.501
2004	6.392	−0.561
2005	7.483	−0.520
2006	8.766	−0.504
2007	9.871	−0.423
2008	11.468	−0.389
2009	10.252	−0.282

of a circle. If $b_1 = 0$, the relationship between the resources is a straight line. If these coefficients are not equal to zero, the relationship between the production resources can take the most diverse forms, for example isoquanta of "neoclassical production functions." From this point of view, the model of the logarithmic PFCA is universal. The analytical properties of the logarithmic function of a complex argument were studied in Chap. 2; therefore, it would be unnecessary to focus on those results here.

Since the first equality of the system modeling the relationship between the production resources and the production result also represents a combination of the equation of a circle and the equation of an arctangent, it can, depending on the values and signs of the coefficients, describe rather varied combinations of a straight line and circle. A curve in space that results from the intersection of these two nonlinear surfaces is complicated itself, and the model is able to describe complicated trajectories of the development of production systems.

The obvious fact that (5.126) represents a system of two equations with two unknowns is an important advantage, which is why the model coefficients can be found by means of only one observation of the production process. That is, the practical application of this model appears simple.

In fact, from (5.123) we derive a formula for the complex proportionality coefficient:

$$b_0 + ib_1 = \frac{Q_t}{\ln(K_t + iL_t)}. \tag{5.125}$$

Again, let us apply this formula to the data on the Russian economy over the last years. Table 5.8 gives only the results of calculation of the complex proportionality coefficient.

Table 5.9 Complex proportionality coefficient of logarithmic PFCA with the free term

Year	Coefficients b_0	b_1
1995	0.444122	0.282994
1996	8.863934	10.30833
1997	3.191134	1.391758
1998	38.66129	−48.387
1999	11.36293	3.844329
2000	5.759372	1.769688
2001	7.008043	1.528732
2002	7.741272	1.471488
2003	37.98628	6.349217
2004	19.5134	2.739132
2005	27.95692	3.559588
2006	20.77293	2.194427
2007	32.95931	2.880673
2008	−11.3934	−0.99884
2009	0.444122	0.282994

The results of the calculations of the coefficients of the logarithmic model show that both parts of the complex coefficient increase in time. Beginning in 2001 they do it almost linearly. This is why the logarithmic function can be applied in practice using trends to describe variation in the coefficients over time. Then the model describing properly the original production series in Russia will have the following form:

$$Q_t = [(0.9281\mathrm{t} + 2.9644) + i(0.0498\mathrm{t} - 0.7611)] \ln(K_t + iL_t).$$

Obviously, the logarithmic model with a free term (5.122) will be more adequate for real production processes:

$$Q_t = (a_0 + ia_1) + (b_0 + ib_1) \ln(K_t + iL_t).$$

To verify the possibility of modeling macro-level production in Russia, model coefficients should be calculated. Let us use the differences between the left- and right-hand sides of the exponents that differ from each other by a unit of time:

$$Q_t - Q_{t-1} = (b_0 + ib_1) \ln \frac{K_t + iL_t}{K_{t-1} + iL_{t-1}}. \tag{5.126}$$

Hence it is easy to calculate the complex proportionality coefficient for the logarithmic PFCA with a complex free term (5.122):

$$b_0 + ib_1 = \frac{Q_t - Q_{t-1}}{\ln \frac{K_t + iL_t}{K_{t-1} + iL_{t-1}}}. \tag{5.127}$$

Substituting these coefficients into (5.122), it is easy to calculate the value of the free complex term. Let us show that this procedure is even possible for the same example with respect to Russia. Calculations of the complex proportionality coefficient are given in Table 5.9.

Over time, both coefficients in the table vary in their values and signs. This variation is virtually unpredictable and is characterized by a high variance; therefore, the conclusion will be as follows. The logarithmic PFCA with a free term cannot be used for modeling this process.

It is again evident that the power PFCA is more interesting for researchers and practicing economists than the logarithmic function of a complex argument.

In conclusion it should be noted that the logarithmic function is a periodic function and one should always take that into account when using it in practice.

5.7 Summary

The theory of functions of a complex variable does not consider models of a complex argument when the real variable depends on the complex one. This may have been done in specific chapters of TFCV but they are not accessible for analysis. Since a complex argument represents a two-factor relationship, one could speak of an analogy between two-factor production functions of real variables and models of a complex argument.

This chapter showed that models of a complex argument can operate as models of production functions, that they can model various production processes. The examples given in this chapter demonstrated that in certain cases particular models are quite good at describing real production processes at the macro level, though other models are not applicable. But the scope of production processes is not limited to the examples with real data given in this chapter. There are many more. Real production processes are so diverse that they cannot be fit into the Procrustean bed of neoclassical functions as most economists do. These models often distort reality.

This is why an extension of the tools of the theory of production functions is quite timely. PFCA turns out to be rather helpful here. The research whose results are provided in Chap. 5 shows that each of the considered models of a complex argument will be the best at describing some particular original production process. Calculation of the trajectories of a production process for each of the considered production functions of a complex argument on conditional examples shows that they generate the most diverse forms – convex, concave, increasing, decreasing, with asymptote, and without it... Readers can obtain these trajectories independently, substituting their own series into the model formulae and setting various coefficient values. This is why it was senseless to provide the entire range in this chapter – it was sufficient to show the main way to use complex-argument models in economics. Its use in modeling production processes enriches the economist's arsenal of tools.

The possibility of building a linear two-factor model under conditions of multicollinearity turned out to be an extremely important result that revealed in this chapter new properties of the mathematical apparatus of the TFCV referring to solution of economic problems. It was shown in this chapter that the method of real variables does not provide a satisfactory solution of this problem, while TFCV demonstrated quite promising results. The two-factor linear model of real variables represents an equation of a plane in three-dimensional space. This is why in a situation where points lying in this space are on a straight line (multicollinearity) LSM demonstrates an inability to find an equation of this plane. It is this straight line in three-dimensional space that is described by a linear function of a complex argument. This is why LSM applied to this function provides stable estimations of a two-factor model under conditions of multicollinearity.

Since nonlinear functions of a complex argument often mean a linear dependence between the resources, the problem of the influence of multicollinearity on the results of building two-factor nonlinear models can also arise in this case. Nonlinear functions of a complex argument will provide stable values of model coefficients since this is the only case where they may exist.

We should also point out an important feature that distinguishes models of a complex argument from two-factor models of real variables. Models of real variables describe a plane (in the linear case) or nonlinear surfaces in three-dimensional space, and complex-argument models describe a straight line lying in a plane (in the linear case) or a curve lying on the surfaces perpendicular to the plane of production resources.

This fact shows that PFCAs has quite a narrow sphere of application compared to real-variable models, but they describe those very processes that are poorly described by models of real variables.

All the power of the TFCV, the significant advantages of models involving TFCV tools are demonstrated in the building of model dependences of one complex variable on another. Here, complex-argument models appear to be an "introduction" to the new apparatus of the theory of production functions, namely, the apparatus of production functions of complex variables.

References

1. Svetunkov SG (1994) Modeling under conditions of multicolinearity. Indus Power Eng N 6:28–32
2. Corn G, Corn T (1984) Reference book on mathematics for scientists and engineers. Nauka, Moscow
3. Svetunkov SG, Svetunkov IS (2005) Study of properties of production function of complex argument (preprint). Publishing House SPbGUEF, St. Petersburg
4. Svetunkov SG, Svetunkov IS (2008) Production functions of complex variables. LKI, Moscow
5. Short-term economic parameters of the Russian federation (Goscomstat of Russia) http://www.cir.ru. Accessed Oct 2004

Chapter 6
Production Functions of Complex Variables

The whole power of the apparatus of TFCV is demonstrated with reference to economic modeling on production functions of complex variables (PFCVs) – when the complex production result depends on the complex production resource. In this chapter the complex production result is represented as a variable with the gross margin as its real part and production costs as its imaginary part. Capital refers to the real part of the complex resource and labor to the imaginarypart. The main properties of linear, power, logarithmic, and exponential models of complex variables are considered consistently. Numerous complex coefficients of effectiveness are derived and their economic interpretation is given. Here we show how complex-valued econometrics can be used to model real production systems.

6.1 General Provisions of the Theory of Production Functions with Complex Variables

Complex-argument functions represent a certain "truncation" of properties of functions with complex variables – they described the dependence of a real variable on a complex one that served as as the complex argument of a function. Mere formulation of the problem with reference to one of the branches of economics – the theory of production functions – has already yielded new scientific results. Therefore, even more diverse and exciting results are naturally expected from the use of functions with complex variables, i.e., the dependence of one complex variable on another. Since a complex variable inherently represents some two-factor model, the focus will be on the dependence of one pair of economic indicators on another. With reference to economic problems it is natural to assume that one pair – the complex argument – can represent production indicators. Such a dependence that connects production resources with the production result will be a production function.

S. Svetunkov, *Complex-Valued Modeling in Economics and Finance*,
DOI 10.1007/978-1-4614-5876-0_6,

Formally, production functions represent some sort of mathematical dependence of the production result on the production resources given a number of assumptions. Of the majority of production resources only two are used to build production functions – production capital K in its various forms and labor L. These two resources are interchangeable to a certain extent, which justifies their use.

In the previous chapter we used these very resources as one complex variable. Material will be presented in this chapter that says there is not much difference in terms of which variable we assign to the real and which one to the imaginary part of the complex variable of a complex production resource (argument). Properties of models of complex-argument production functions most often varied symmetrically without changing their essence.

In models of PFCVs, the procedure for assigning both the complex argument and the complex result to the real or imaginary parts has a pronounced economic sense.

This is why we will stick to a particular rule of creation of a complex production argument – capital will constitute the real part and labor the imaginary part. Then for these functions, the complex production result will have the form

$$K_t + iL_t \tag{6.1}$$

Since all the examples to be considered in this chapter refer only to socioeconomic dynamics, all the variables have a regulation index t. If there is a need to build certain production functions on other than a time set, the index will be easily replaced by another.

Production performance can be demonstrated by the most diverse technical and economic indicators. In the theory of production functions we use mainly one of them – the volume of produced and sold products Q_t. Clearly, in this case assumptions arise regarding production functions: that the demand is unsaturated, that the price is unchanged, etc.

However, in real economic practice nobody judges production performance only by the production volume (output); one should have an idea of successful economic activity, which is confirmed by various indicators of economic efficiency, primarily gross margin G_t, production costs C_t, and, based on those, profitability R_t.

Since the gross margin, production costs, and gross output are connected with each other by a simple relationship

$$Q_t = G_t + C_t$$

then, copying any pair of these three, one can calculate the third parameter.

These indicators are also helpful to calculate another indicator of economic activity – profitability:

$$R_t = \frac{G_t}{C_t} = \frac{Q_t - C_t}{C_t}.$$

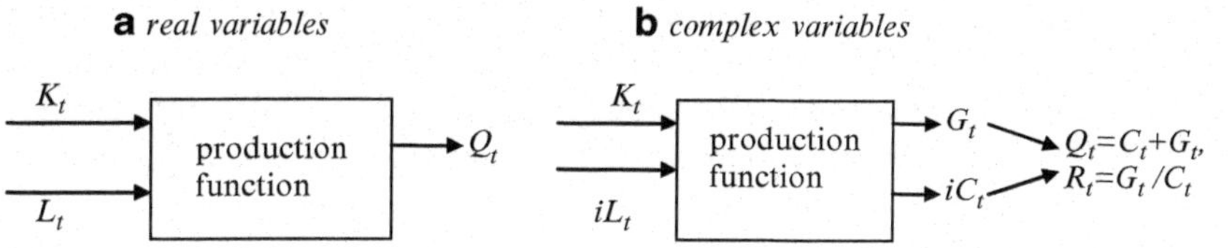

Fig. 6.1 Structural scheme of production functions of real variables (**a**) and complex variables (**b**)

To create a complex variable of the production result we need a pair of variables showing various sides of one process and having the same dimension and scale. Because various combinations of production resources lead to various combinations of production costs and gross margin, and therefore various gross output and profitability, it is gross margin G_t and production costs C_t variables that should be parts of the complex variable of the production result.

It is proposed to represent the complex variable of the production result including gross margin G_t and production costs C_t in the following form:

$$G_t + iC_t. \tag{6.2}$$

We deliberately assign the gross margin to the real part and the costs to the imaginary part of the complex variable of the production resources. This procedure is defined by the formation of the complex variable of the production resources (6.1) whose meaning will become clear in the course of study of the corresponding production functions. In addition, Chap. 1 introduced a rule – we will assign the indicator reflecting the active part of the socioeconomic process to the real part and the passive one to the imaginary part. Comparing the gross margin and production costs with each other one could probably say that economists pay more attention to the former; therefore, from this point of view it seems to be quite in line with the rule introduced to assign the gross margin to the real part of the complex variable.

Figure 6.1 gives two structural schemes that can help us obtain a visual idea of the difference between production functions of real variables and those of complex variables.

Production functions of real variables model the influence of production resources on the gross output and PFCVs first model the influence of production resources on the gross margin and production costs and then, on the basis of this information, their influence on the gross output. The correlation between the gross margin and costs characterize profitability.

It is evident now that PFCVs provide a more detailed description of the production process than those of real variables, which is why one expects more reliability and accuracy from a description of production processes from complex-valued models.

Figure 6.1b helps us to understand that in the general form PFCVs can be represented as follows:

$$G_t + iC_t = F(K_t + iL_t). \tag{6.3}$$

Many functions can us help associate, by dependence (6.3), two complex variables $G_t + iCt$ and $K_t + iL_t$. We will use only those that were considered in Chap. 3, except for Zhukovsky's function and trigonometric functions because it is very difficult to imagine a production process described by these models.

Since production processes differ from each other:

- By hierarchy (enterprise, group of enterprises, regional manufacturing, world production, etc.),
- By the specificity of the production (agriculture, machine-building, light industry, oil refinery, electric power engineering, etc.), and
- By national geographic features (surplus laboring or scarce labor population; availability of sources of raw materials and infrastructure; warm, hot, or cold climate, etc.),

there can be no uniform standard production function of complex variables that would best describe all these diverse production processes changing only the coefficient values depending on the situation. In each case economists should choose the best indicators from the many available ones. That is why this chapter will focus on PFCVs of various forms from among those elementary functions whose conformal mapping was studied in Chap. 3.

It follows from (6.3) that using complex-valued functions one can immediately model two economic indicators – gross margin and production costs; however, this was said to be a model of three production performance. After all, the sum of the gross margin and production costs simply represents gross output:

$$G_t + C_t = Q_t \tag{6.4}$$

Function (6.4) can also be presented in a different form, representing a complex result in exponential form:

$$G_t + iC_t = R_t e^{i\theta_t}, \tag{6.5}$$

where we obtain

$$\begin{cases} G_t = R_t \cos\theta_t, \\ C_t = R_t \sin\theta_t, \\ Q_t = R_t(\cos\theta_t + \sin\theta_t). \end{cases} \tag{6.6}$$

Here the polar angle can be found in the following way:

$$\theta_t = arctg\frac{C_t}{G_t} = arctg\frac{1}{R_t}.$$

That is, it shows the profitability of production – the higher the polar angle, the less efficient is the production. If a company is profitless but not running at loss, i.e., with zero profitability, the polar angle tends to plus infinity.

More often, none of model forms (6.6) is used in modern economic and mathematical modeling. Moreover, nor is the entire whole system of dependence of a pair of production results on a pair of production resources used. One should note the fact that finding estimations of model coefficients (6.6), for example using the least-squares method (LSM), is most often an extremely complicated problem. For this, one should resort to numerical methods of solving systems of nonlinear equations. For modern scientists armed with computer equipment and software this is not an obstacle, but for an economist whose knowledge of mathematics is at the level of a university course this task is insurmountable. Coefficients of complex-valued functions can be found quite easily. In Chap. 4 we showed how to use LSM for these purposes.

New economic indicators appear in PFCVs that have not been encountered in the theory of production functions, which is based on real variables. These are moduli of complex variables and their polar angles. If polar angles are more or less understood – they characterize resources from the point of view of the capital-to-labor ratio (the tangent of a polar angle represents the labor-to-capital ratio) and the production result from the point of view of profitability at cost – the interpretation of the moduli of these complex variables is quite hard. In fact, the moduli of these variables are $R_{GC} = \sqrt{G^2 + C^2}, R_{KL} = \sqrt{K^2 + L^2}$. What economic meaning do they have? They show the production scale and the resource scale, respectively.

PFCVs possess another unique property. An inverse relationship follows from (6.3):

$$K_t + iL_t = f(G_t + iC_t). \tag{6.7}$$

This means that if we describe some production process using a PFCV, we can build an inverse function (6.7), and using this we could solve a problem that has not arisen in the modern theory of production functions –achieving a desired gross margin, production costs, or production volume. What labor and capital resources would have to be used to obtain a given profitability level? Function (6.7) makes it rather easy to answer these questions, though functions of real variables will be more complicated and will be represented by a system of equations.

6.2 Linear Production Function of Complex Variables

The simplest – linear – function of complex variables will have the form

$$G_t + iC_t = (b_0 + ib_1)(K_t + iL_t). \tag{6.8}$$

Since this function has only one complex coefficient, namely, the proportionality coefficient (b_0 + ib_1), it will be the subject of our study of this function. From equality (6.8) we easily obtain

$$b_0 + ib_1 = \frac{G_t + iC_t}{K_t + iL_t}.$$

Multiplying the numerator and denominator of the right-hand side of the equality by the conjugate term we obtain

$$b_0 + ib_1 = \frac{G_t + iC_t}{K_t + iL_t} \times \frac{K_t - iL_t}{K_t - iL_t} = \frac{G_tK_t + C_tL_t + i(C_tK_t - G_tL_t)}{L_t^2 + K_t^2}.$$

After grouping the real and imaginary parts it is easy to derive each of the coefficients b_0 and b_1:

$$b_0 = \frac{G_tK_t + C_tL_t}{L_t^2 + K_t^2}, \tag{6.9}$$

$$b_1 = \frac{C_tK_t - G_tL_t}{L_t^2 + K_t^2}. \tag{6.10}$$

In contrast to the coefficients of the production functions of a complex argument, it is not easy to give an economic interpretation of each of the coefficients of the function of complex variables (6.8). These coefficients have the same denominator, which shows the resource scale; however, the numerators (6.9) and (6.10) greatly differ and do not have clear economic parallels.

The coefficient b_0 will grow linearly with both the growth of the production volume $(G + C)$ the growth of the gross margin and production costs, given constant overhead costs. If both resources and performance grow according to the direct proportionality law, this coefficient will stay constant. In all other cases, its dynamics are more complex.

As for b_1, it will grow with the growth of the prime cost and, to a certain extent, with the growth of the number of people employed in production. The latter relationship is nonlinear. Growth of the gross margin will be demonstrated by a reduction in the values of this coefficient.

If we take the first observation as a starting point, making all the other values relative, b_0 in the first observation will be equal to one and b_1 to zero. However, not only the initial value but any other, for example, the last one, can be taken as the starting point. Then, for this year's observation of the production process, the coefficient b_0 will be equal to one and b_1 to zero. Since in the real economy, situations where the linear dependence of some complex indicator on a complex factor is modeled with a zero value of the free term are quite rare, the model can be transformed into this form by preliminary centering of the original variables around their averages.

Opening the brackets of equality (6.8) and grouping the real and imaginary parts of the equality we have

$$G_t + iC_t = (b_0 + ib_1)(K_t + iL_t) \leftrightarrow G_t + iC_t = b_0K_t - b_1L_t + i(b_0L_t + b_1K_t),$$

where for the real part of the equality

$$G_t = b_0 K_t - b_1 L_t \tag{6.11}$$

and for the imaginary part

$$C_t = b_0 L_t + b_1 K_t. \tag{6.12}$$

The expressions obtained have a simple economic meaning and determine in what cases of modeling production can a linear PFCV be used (if we consider the coefficients to be positive). Equation (6.11) uniquely specifies that with the growth of the capital resource, the gross margin increases, and with the growth of labor resources it decreases. It follows from (6.12) that production costs increase. This testifies to the fact that the modeled process is characterized by a permanent return on the capital resource and a decrease in the return on labor resources.

However, if coefficient b_1 is negative, the increase in labor resources leads to an increase in profits and costs, and growth of the capital resources leads in this case to the modeling of a situation where the gross margin increases and the costs decrease. This means that there is a permanent return on the labor resource and – regardless of the correlation between b_0 and b_1 – increasing, permanent, or decreasing return on the capital resource.

This means that linear a PFCV possesses certain analytical properties that can facilitate study of the core significance of the production processes of various hierarchy levels. Apparently, this can be done only if the linear PFCV provides a good description of the modeled production.

Since all the variables are made dimensionless by division by their first values (G_t/G_1, C_t/C_1, K_t/K_1, L_t/L_1), there are several important points to pay attention to. The values obtained make it possible to determine some analog of the gross margin:

$$G_t + C_t = (b_0 K_t - b_1 L_t) + (b_0 L_t + b_1 K_t) = (b_0 + b_1) K_t + (b_0 - b_1) L_t. \tag{6.13}$$

We call this sum not the gross revenue but its analog for the following reasons. All the input variables are scaled. This means that the gross profit Gt is divided by the initial value of profit in the first year - G_1 - and production costs Ct are divided by the costs of the first year - C_1. Their sum is not equal to the quotient of gross revenue Q_t to the value of gross revenue in the first year - Q_1,

$$\frac{G_t}{G_1} + \frac{C_t}{C_1} = \frac{G_t C_1 + C_t G_1}{G_1 C_1} \neq \frac{G_t + C_t}{G_1 + C_1} = \frac{Q_t}{Q_1}.$$

This is why (6.13) is just an analog of the revenue. For (6.13) to have the necessary meaning, the gross margin and the prime cost should be dimensionless relative to the gross revenue:

$$\frac{G_t}{Q_1}, \frac{C_t}{Q_1}. \tag{6.14}$$

In this case (6.13) means the gross revenue in relative values.

In the same way, the ratio

$$\frac{G_t}{C_t} = \frac{b_0 K_t - b_1 L_t}{b_0 L_t + b_1 K_t} \tag{6.15}$$

is an analog of profitability, as here we calculate not the profitability itself. In dimensionless values,

$$\frac{G_t/G_0}{C_t/C_0} = \frac{R_t}{R_0}.$$

However, if the variables of the production result are transformed into dimensionless values by presenting them as the original value of the gross revenue, then (6.15) will characterize profitability by the cost.

This is why for practical purposes all the original values of this and other models of PFCVs should be transformed into dimensionless values by dividing their values by the gross output at the initial time Q_0:

$$\frac{G_t}{Q_1}, \frac{C_t}{Q_1}, \frac{K_t}{Q_1}, \frac{L_t}{Q_1}.$$

Formulae (6.9) and (6.10) make it possible to find the corresponding coefficients of the linear PFCV (6.8) for each observation. To do that one should only substitute the available statistical data therein. Let us take an actually operating enterprise as an example to calculate these coefficients.

The board of Inza Diatom plant (Ulyanovsk oblast of Russia) kindly supplied us with the necessary statistical data concerning their enterprise. They were preprocessed and systematized by I.E. Nikiforova. The absolute production values concerning operation of the plant are given in Table 6.1.

A cursory analysis of the data supplied shows a high variance with respect to the trends, and this is not surprising for a real production process – the results of production activity are affected by numerous factors most of which remain unidentifiable.

There is an interesting fact that a change in the capital resource takes place by stages – July is characterized by the development of new production capacities, and according to the table the return comes gradually reaching its maximum by December.

All these circumstances convince us that no model can describe a production process with an error lower than 10–20%. However, first of all, we are interested in the mere opportunity to build complex-valued production functions and, secondly, model general relationships between resources and production result, albeit with some error.

Using the values of revenue, production costs, payroll, and fixed assets one can build a production function like (6.8). For that we should first transform all the

Table 6.1 Production activity of Diatom production enterprise by months

Month	Profit (thousands of rubles)	Costs (thousands of rubles)	Payroll (thousands of rubles)	Work (man-hours)	Number of people	Fixed assets (thousands of rubles)
February	59	2,604	213.5	52,100	354	4,263
March	72	3,178	231.3	51,347	357	4,263
April	26	1,146	289.1	57,095	364	4,263
May	47	2,059	246.1	62,898	401	4,263
June	21	897	266.6	62,742	400	4,263
July	73	3,202	294.1	57,005	404	5,684
August	47	2,045	396.4	61,662	437	5,684
September	49	2,152	310.2	64,484	457	5,684
October	41	1,804	402.4	63,071	454	5,684
November	60	2,615	511.9	64,599	465	5,684
December	107	4,736	439.4	63,905	460	5,684

values into dimensionless relative ones. The data for February for each of the economic parameters will serve as the base. At this stage we will not transform the variables into a uniform scale by dividing them by the output for the first month. We will simply study dimensionless variables whose values for February are considered to be one.

Since labor is characterized by three parameters – payroll, work, and number of people involved – we should take the one that shows resource costs more accurately. To transform labor resource values into one dimension and scale, it is payroll that will go with capital, because it is what characterizes labor costs in monetary units.

Since it is necessary to center the original variables in order not to use the free factor by using model (6.8) directly, Table 6.2 gives dimensionless variables centered with respect to the averages. Since formulae (6.9) and (6.10) make it possible to find the coefficients for each observation, let us use them to obtain two series of coefficients with the dynamics shown in Table 6.3.

It is evident that the function coefficients do not remain constant, which is expected with these initial data. But to answer the question as to whether this model is suitable for studying production processes at a production plant, one should determine if there is a trend in the changes of the coefficients. If there is a trend in the change of at least one coefficient, it will demonstrate that the model is shifted and is a bad one for describing production relationships. In this case one should either change the model or modify it significantly.

The model coefficients vary chaotically, the spread of values is quite high, and no trend of coefficient change is identified. This means that the model of a linear production function of a complex argument can be used to describe the production process under consideration but the approximation accuracy will be very slight.

To provide a more correct economic interpretation of the calculated economic variables, the gross margin and production costs should be assigned dimensionless

Table 6.2 Dimensionless centered data for Diatom production enterprise

Month	Profit	Costs	Payroll	Fixed assets
February	0.072419	0.077014	−0.53332	−0.18182
March	0.292758	0.297444	−0.44995	−0.18182
April	−0.4869	−0.48289	−0.17922	−0.18182
May	−0.13097	−0.13228	−0.38063	−0.18182
June	−0.57165	−0.57852	−0.28461	−0.18182
July	0.309707	0.306661	−0.1558	0.151515
August	−0.13097	−0.13766	0.323355	0.151515
September	−0.09707	−0.09656	−0.08039	0.151515
October	−0.23267	−0.23021	0.351458	0.151515
November	0.089368	0.081239	0.864339	0.151515
December	0.885978	0.895755	0.52476	0.151515

Table 6.3 Coefficients of production functions (6.8) according to data of Diatom production enterprise

Month	Coefficient b_0	Coefficient b_1
February	−0.17084	0.077546
March	−0.79429	0.329688
April	2.686068	0.008219
May	0.416793	−0.145
June	2.354795	−0.50422
July	−0.01806	2.005393
August	−0.50469	0.168554
September	−0.23606	−0.76258
October	−0.79301	0.320131
November	0.108772	−0.08433
December	2.025593	−1.10349

values. This can be done by dividing their values by the gross revenue, as was stated previously (6.14). Let us compare the result obtained with the previous one.

The dimensionless values of the variables expressed as a function of gross revenue values are again centered around their averages. These data are given in Table 6.4.

With the changes in the original variables, the model coefficient values also changed (6.8). To show this more clearly Table 6.5 provides two series of values of the complex coefficient – those from Table 6.3 and those calculated according to the data of Table 6.4.

The change in the scale of the variables caused a change in the values of the model coefficients (6.8) but not a change in trends since both the first and second cases involve a linear model but with different scaling rules. New coefficients vary as chaotically as previously, almost repeating the dynamics of change in the coefficients calculated according to Table 6.2.

The fact that a linear PFCV should not be used for this production is also confirmed by the calculation of the complex pair correlation coefficient (4.84), which for the applied data of the Diatom plant is

$$r_{xy} = 0.301931 + i0.055403.$$

Table 6.4 Dimensionless data for Diatom production enterprise. The profit and costs are divided by gross revenue

Month	Profit	Costs	Payroll	Fixed assets
February	0.001604	0.075308	−0.53332	−0.18182
March	0.006486	0.290854	−0.44995	−0.18182
April	−0.01079	−0.47219	−0.17922	−0.18182
May	−0.0029	−0.12935	−0.38063	−0.18182
June	−0.01267	−0.5657	−0.28461	−0.18182
July	0.006862	0.299867	−0.1558	0.151515
August	−0.0029	−0.13461	0.323355	0.151515
September	−0.00215	−0.09443	−0.08039	0.151515
October	−0.00515	−0.2251	0.351458	0.151515
November	0.00198	0.079439	0.864339	0.151515
December	0.019629	0.875909	0.52476	0.151515

Table 6.5 Coefficients of production function (6.8) according to the production data of Diatom production enterprise

	According to Table 6.2		According to Table 6.4	
Month	Coefficient b_0	Coefficient b_1	Coefficient b_0	Coefficient b_1
February	−0.17084	0.077546	−0.12742	−0.04043
March	−0.79429	0.329688	−0.56069	−0.21215
April	2.686068	0.008219	1.328494	1.287556
May	0.416793	−0.145	0.279659	0.125965
June	2.354795	−0.50422	1.431754	0.87016
July	−0.01806	2.005393	−0.96716	0.984596
August	−0.50469	0.168554	−0.34478	−0.15258
September	−0.23606	−0.76258	0.246949	−0.49218
October	−0.79301	0.320131	−0.54544	−0.22047
November	0.108772	−0.08433	0.089557	0.013408
December	2.025593	−1.10349	1.550685	0.410326

Since the real part of the complex pair correlation coefficient is small, it is evident that a linear complex-valued relationship should not be used to model production at the Diatom plant. The proximity of the imaginary part to zero testifies to the existence of a certain nonlinear dependence between the two complex variables, but one can judge its character only after additional studies.

Let us see now whether this function can be used with the Russian economy. Table 6.6 shows data on the cost of fixed assets in Russia, the level of active participation in the economy of the people, and the final product, which is divided into final consumption and gross. Final consumption represents gross margin at the macro level, which is why we will use this indicator as a real part of the complex production result.

Before building a model, let us calculate complex pair correlation coefficient (4.84) between two complex variables – the complex resource (fixed assets and active population) and the complex result (final consumption and gross margin):

$$r_{xy} = 0.991963 - i, 0.005319$$

Table 6.6 Overall development of Russian economy

Year	Fixed assets (K_t)		Economically active population (L_t)		Final consumption (G_t)		Gross (C_t)	
	Absolute values (millions of rubles)	Relative values	Absolute values (thousands of people)	Relative values	Absolute values (millions of rubles)	Relative values	Absolute values (millions of rubles)	Relative values
1995	5182040	1,000000	70740	1,000000	1095820,9	1,000000	391588,4	1,000000
1996	13072378	2,522632	69660	0,984733	1544658,8	1,409591	528694,9	1,350129
1997	13286272	2,563908	68079	0,962383	1891846,7	1,72642	564244,2	1,440911
1998	14125670	2,72589	67339	0,951923	2100663,3	1,916977	443978,2	1,133788
1999	14246427	2,749193	72176	1,020300	3303947,9	3,015044	729214,5	1,862196
2000	16605251	3,204385	72332	1,022505	4476851,0	4,085386	1365734,0	3,487677
2001	20241428	3,906073	71411	1,009485	5886861,0	5,372101	1963110,0	5,013198
2002	24430544	4,714465	72421	1,023763	7443199,0	6,79235	2169314,0	5,539781
2003	30329106	5,852735	72835	1,029615	9024756,0	8,235612	2755048,0	7,035571
2004	32541444	6,279659	72909	1,030662	11401444,0	10,40448	3558952,0	9,088502
2005	38366273	7,403701	73811	1,043412	14318964,0	13,06688	4338731,0	11,07983
2006	43822840	8,456677	74156	1,048290	17629743,0	16,08816	5748727,0	14,68053
2007	54251541	10,46915	75060	1,061069	21785787,0	19,88079	8031682,0	20,51052
2008	64552706	12,45701	75892	1,072830	27237356,0	24,85566	10642560,0	27,17792

Table 6.7 Coefficients of production function (6.8) according to data of Table 6.6

Year	Complex proportionality coefficient	
	Coefficient (b_0)	Coefficient (b_1)
1995	1.163592	−1.43889
1996	2.230591	−2.70000
1997	2.270221	−2.50794
1998	2.87264	−2.64283
1999	2.15053	−1.90593
2000	1.224434	−1.60172
2001	140.9579	34.74048
2002	1.484029	−1.94932
2003	1.386099	−1.54839
2004	2.007539	−2.17582
2005	1.940288	−2.23109
2006	2.285952	−2.37343
2007	2.476524	−2.21761
2008	2.683951	−2.27884
2009	1.163592	−1.43889

Since the real part of the complex pair correlation coefficient is close to one and the imaginary one is close to zero, the linear complex-valued function will be quite acceptable for modeling the dependence under consideration.

Let us find the coefficients of linear model (6.8) according to the data of this table as was done previously, i.e., complex coefficients of model (6.11) and (6.12) that change over time. The calculation results are given in Table 6.7.

It is seen from the table that the coefficients vary with respect to some average values, except for those calculated for 2001. They differ considerably from the whole series.

This is due not to some particular features from this year for Russia but to the fact that to calculate the complex proportionality coefficients all the original data were centered with respect to the averages. The averages of the original data of Table 6.6 fall exactly in the year 2001 and deviations from the average for this year are close to zero, which means that the centered values of the variables for the year 2001 are close to zero. And since in (6.11) and in (6.12) division is used, then with the denominator close to zero, the quotient will be high, which we observe from the table.

If a researcher wishes to build a complex-valued model not for each observation but for the whole series, it is necessary to find model coefficients using LSM. In Chap. 4, certain operations were performed that could provide such a complex-valued model of the dependence of the final product of Russia on the labor resources and fixed assets:

$$G_t + iC_t = (-1.4095 - i5.3207) + (2.2105 + i1.9415)(K_t + iL_t). \quad (6.16)$$

This model describes very well the process under consideration and can be used, for example, for purposes of multivariant forecasting of economic development in Russia. For that, it is necessary to find, in addition to the obtained coefficient values, the confidence interval of changes in the obtained point values. However, the process under consideration can by no means be associated with a stationary one and methods of mathematical statistics should not be used in this case formally in their full scope. Moreover, this study does not consider the problem of applying complex-valued models to irreversible processes of socioeconomic dynamics. The aim of this study is to elaborate the principles of the application of methods and models of the TFCV to economic problems.

6.3 Model of Power Production Function of Complex Variables with Real Coefficients

In economics linear relationships are known to be quite rare, occurring mainly in certain, rather short periods of time. This refers mainly to dynamic processes. In most cases nonlinear dependences prevail, which, as a matter of fact, also operate in a relatively small period of time, since one nonlinear trend replaces another. A change in trends of the development of economic projects that we observe can be explained by the evolution of those economic projects when they change their structure, composition of the elements, relationships between the objects, and others. In the same way, the production process develops according to a complex cyclical trajectory and in particular periods of time can be described by various nonlinear models. This is why economists should most often work with nonlinear complex-valued models.

From among a great variety of possible complex-valued models, power PFCVs occupy a special place. As in the theory of production functions of real variables where power functions prevail due to their remarkable properties, in the theory of complex-valued economies power functions play an important role. They are universal, easy to use, and are very good at describing particular real production situations. The previous chapter, which considered production functions of a complex argument, showed that the power function possesses the most interesting properties of all the functions already studied.

The general form of power PFCVs can have the form

$$G_t + iC_t = (a_0 + ia_1)(K_t + iL_t)^{b_0 + ib_1}. \tag{6.17}$$

Let us first study the simplest case – where the imaginary parts of complex coefficients of this function are equal to zero and function (6.17) is a power production function with real coefficients:

$$G + iC = a(K + iL)^b. \tag{6.18}$$

The properties of the model of power PFCVs with real coefficients has been studied in detail by Ivan Svetunkov. Here it is appropriate to specify only the most important properties of this model.

With reference to our problem of mapping a complex variable of production resources onto a complex plane of production results, there are restrictions due to the core economic meaning of the variables. Naturally, these restrictions influence linear complex-valued production functions, but in the above-mentioned example going beyond these limits is hardly possible, though – with respect to nonlinear models – quite probable.

Thus, the first group of restrictions is due to the fact that complex variables of production resources lie in the first quadrant since $K > 0$ and $L > 0$, i.e., argument ц of the complex variable of resources varies within a range of 0 to $\frac{\pi}{2}$. If it is equal to zero, this means that no unit of labor resources is involved in production. If it becomes equal to $\frac{\pi}{2}$, this means that only labor resources are involved in production with capital resources being equal to zero. It is evident that in practice these cases do not exist and we should exclude coordinate axes from the problem domain.

According to their economic meaning, complex variables of production performance cannot be defined on the entire complex plane. Though they lie in that plane within wider ranges defined by the polar angle falling within the range from 0 to $\frac{3}{4}\pi$, they cannot go beyond the limits. Thus production performance is defined in the first and, partially, the second quadrants of the complex plane.

If the polar angle θ of a complex variable of production performance is equal to zero, it means that the production costs are equal to zero and the gross margin is at its maximum. Such situations in actual economic practice are exceedingly rare, which is why the limit in this part should be shown as a strict inequality. Since in the second quadrant of the complex plane of production performance the gross margin plotted on the axis of real numbers becomes negative, the enterprise is working at a loss – the negative gross margin is numerically equal to gross losses at the company. In its economic meaning, the negative gross margin (loss) cannot be higher than the production costs: $-G \leq C$. If no units of the goods produced are sold, then the gross margin G is numerically equal to the sum of incurred production costs C, its sign being negative. It is in this case where the polar angle of production performance becomes $\frac{3}{4}\pi$. The case where $-G = C$ is a rare but possible occurrence in actual practice.

This is why any model of PFCVs, including the power function, should be supplemented by the conditions imposed on the polar angles by complex variables:

$$0 < \varphi < \frac{\pi}{2},\ 0 < \theta \leq \frac{3\pi}{4}. \tag{6.19}$$

However, due to the periodic character of polar angles and taking into account the TFCV, this condition should have the form

$$2\pi k < \varphi < 2\pi k + \frac{\pi}{2},\ 2\pi k < \theta \leq 2\pi k + \frac{3\pi}{4}.$$

From among the whole set of k due to the economic meaning of the variables, we will use only $k = 0$. Incidentally, the power production function of complex variables should be monosemantic; otherwise, the model stops reflecting the real economic situation. This means that exponent b should be limited so that the extreme allowable value of the polar angle of production resources φ is associated with the extreme allowable value of the polar angle of production performance θ. Since for the power function under consideration $\theta = b\varphi$, the exponent should satisfy the condition

$$0 < b\varphi \leq \frac{3\pi}{4}. \tag{6.20}$$

If the exponent is negative, $b < 0$, then any increase in production resources will definitely lead to a reduction in production performance and vice versa – a reduction in labor and capital resources will cause an increase in production performance. Here the polar angle of production performance becomes negative, which means negative production costs – an impossible situation in economics. This is why we consider only functions with positive exponents.

For a power function with real coefficients used as a model of production processes, the increase in the radius and polar angle of the complex variable of production resources (which means a greater growth in labor than in capital resources) will mean an increase in the production performance with growth in costs outpacing the gross margin. If we consider an inverse economic process – capital rather than labor resource growth (which in the complex plane of production performance means a decrease in the polar angle with simultaneous growth in the variable radius), – we will have an increase in the production performance with growth in the gross margin outpacing production costs.

Since in most real production processes investments in fixed capital lead to improvements in production technology and productivity growth and a reduction in scrap and production wastes, this means a reduction in costs with a simultaneous increase in the gross margin. It is this process that is modeled by the production complex-valued power function with real coefficients. If in this production it is necessary to achieve a fast increase in the production volume, additional workers will be hired with the same capital resources, and the enterprise will start to work in two or three shifts. It is clear that allowances should be made for work in the second and third shifts, which increasing costs per labor unit leading to growth in total production costs. The gross margin starts decreasing. This process is modeled by means of the production function under consideration.

Therefore, according to its properties, the production complex-valued power function with real coefficients corresponds to real production processes. In periods of stagnation, this model can also be used, but this will be shown subsequently. Now let us consider the model properties.

Production function (6.18) can be presented in trigonometric form:

$$G + iC = a\left(\sqrt{K^2 + L^2}\right)^b \left(\cos\left(barctg\frac{L}{K}\right) + i\sin\left(barctg\frac{L}{K}\right)\right). \tag{6.21}$$

This allows us to derive two simple formulae for calculating gross margin G and production costs C:

$$G = a\left(\sqrt{K^2 + L^2}\right)^b \cos\left(b arctg \frac{L}{K}\right), \tag{6.22}$$

$$C = a\left(\sqrt{K^2 + L^2}\right)^b \sin\left(b arctg \frac{L}{K}\right). \tag{6.23}$$

These formulae show how the profit and costs will be modeled with various combinations of production resources. If the production technologies do not change and the resources used increase, the proportions between the resources will not change, nor will the polar angle in the complex plane of production resources [$arg\ (K + iL) = const$]. The modulus of the complex variable will increase. For this case, as follows from (6.3.6) and (6.3.7), the gross margin and gross costs grow at the same rate at increases in resources. The degree of this growth is determined by the real coefficients of the model.

The power production function of complex variables with real coefficients has a remarkable property. To find the coefficient values of function (6.18) it is sufficient to have only one observation of the production process, since (6.13) and (6.23) represent a system of two equations with two unknowns a and b, which can be found using the formulae

$$b = \frac{arctg \frac{C}{G}}{arctg \frac{L}{K}}, \tag{6.24}$$

$$a = \exp\left(\ln\left(\sqrt{G^2 + C^2}\right) - \frac{arctg \frac{C}{G}}{arctg \frac{L}{K}} \ln\left(\sqrt{K^2 + L^2}\right)\right). \tag{6.25}$$

One can see from (6.24) that coefficient b characterizes a relationship of two well-known economic indicators – profitability at cost $\frac{G}{C}$ and capital-labor ratio $\frac{K}{L}$. This circumstance makes it possible to consider the model exponent as one of the analytical features of the proposed model.

Let us show the extremely positive value of coefficient b to satisfy condition (6.20) as b_{NQ}:

$$b_{NQ} = \frac{3\pi}{4 arctg \frac{L}{K}}. \tag{6.26}$$

If $b = b_{NQ}$, the company revenue becomes 0 ($Q = G + C = 0$) due to the fact that in this case the gross margin is negative, $G < 0$, i.e., a loss that equals the production costs $|G| = C$. This is possible in a situation where no manufactured unit is sold.

Looking at exponent b as a variable lying within the range $0 < b < b_{NG}$, one can examine the influence of this variable on the production resources with fixed costs of production resources, for example, determine the conditions for reaching the maximum of real parameters with changing real variables. Calculating the first derivative of function (6.18) for variable b one can also find some other conditions [1].

Gross margin G reaches its maximum if the following condition is valid:

$$b = b_G = \frac{\operatorname{arctg}\left(\frac{\ln\sqrt{K^2+L^2}}{arctg\frac{L}{K}}\right)}{arctg\frac{L}{K}}. \tag{6.27}$$

Gross margin G is equal to costs C (i.e., profitability at cost is 100%) when the exponent becomes

$$b = b_{prof} = \frac{\pi}{4arctg\frac{L}{K}}. \tag{6.28}$$

Gross margin G is equal to zero, i.e., production is neither unprofitable nor profitable, when

$$b = b_{NG} = \frac{\pi}{2arctg\frac{L}{K}}. \tag{6.29}$$

Revenue from production Q becomes maximal if the exponent is

$$b = b_Q = \frac{\frac{3\pi}{4} - \operatorname{arctg}\left(\frac{arctg\frac{L}{K}}{\ln\sqrt{K^2+L^2}}\right) - \pi l}{arctg\frac{L}{K}}, \tag{6.30}$$

where $l = 1$ if $\sqrt{K^2 + L^2} < 1$ and $l = 0$ in all other cases.

Production costs C become maximal when

$$b = b_C = \frac{\pi m - \operatorname{arctg}\left(\frac{arctg\frac{L}{K}}{\ln\sqrt{K^2+L^2}}\right)}{arctg\frac{L}{K}}, \tag{6.31}$$

where $m = 0$ if $\sqrt{K^2 + L^2} < 1$ and $m = 1$ in all other cases.

Now, if we know these characteristic points, we can distinguish 13 areas and points of change in exponent b that characterize the various variations in production efficiency (Fig. 6.2). Along coefficient b's axis we plot the numerical values calculated for the conditional example for resource values equal to one. Let us consider the characteristic points and the model with an increasing exponent beginning from zero and in compliance with Fig. 6.2.

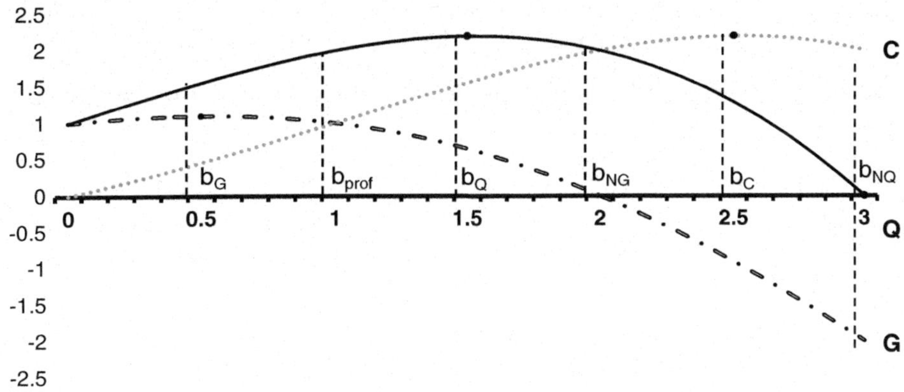

Fig. 6.2 Coefficient b values

$b \in [0; b_G)$ is an area of highly profitable production. With growth in coefficient b from zero to b_G, the growth in production costs C is accompanied by even higher growth in profit G. Profitability increases, and revenue Q increasing accordingly.

$b = b_G$ is the maximal gross margin. Here gross margin G reaches its highest value.

$b \in (b_G; b_{prof})$ means production is efficient but profit G decreases and costs C grow. Revenue Q continues increasing. Profitability at cost G/C decreases.

The point $b = 1$ is interesting because in this case K does not influence C and L does not influence G, as we see from formula (6.18), $G = aK, C = aL$. And it is clear that one can derive the revenue using the formula $Q = a(K + L)$.

$b = b_{prof}$ is the point where profitability at cost is equal to 100%.

$b = b_{prof}$ is where the profit G decreases but the company revenue Q is still growing due to faster growth in production costs. Profitability is lower than 100% and continues to fall.

$b = b_Q$ is the point of maximal revenue of the company.

$b \in (b_Q; b_{NG})$ indicates that production is still efficient despite the continued fall in profit G with increasing costs. Revenue Q in this interval starts decreasing.

$b = b_{NG}$ is the point of profitless production (known in the economic theory as "critical point"). Here, $G = 0$ and revenue Q is equal to costs C.

$b \in (b_{NG}; b_C)$ denotes inefficient unprofitable production. Profit is negative (loss) but lower than costs in absolute value. In real production this may reflect the situation where one has to sell goods at a lower price than the prime cost. Further increase in the exponent means that costs C grow and revenue Q decreases.

$b = b_C$ is the point of maximum costs. This is an extreme point where costs are at their highest, gross margin is negative and lower than the costs in absolute value, i.e., fewer and fewer products are sold; therefore, loss is still lower than the production costs but is increasing rapidly.

$b \in (b_C; b_{NG})$ denotes extremely inefficient production. Costs, gross margin, and revenue decrease, while loss grows.

$b = b_{NG}$ indicates an absence of revenue and is the point of production termination since the gross margin G is equal to costs C, the revenue Q being equal to zero. This means that the whole production volume is unprofitable, and no product unit is sold.

Since the economy is diverse, real practice admits the most diverse alternatives besides those mentioned previously, namely:

- An alternative is possible when production resources take the following value:
-

$$\sqrt{K^2 + L^2} = 1.$$

- In this case $b_G = 0$;
- if the production resources yield $\sqrt{K^2 + L^2} < 1$,
- b_G becomes negative; under the given conditions it is impossible to attain maximum profit;
- when the equality

$$arctg \frac{L}{K} = \ln \sqrt{K^2 + L^2}$$

holds for the production resources, the values of the points
$b_{NG} = b_Q$, $b_C = b_{NQ}$ become equal to each other;

$$arctg \frac{L}{K} < \ln \sqrt{K^2 + L^2},$$

when
the points take each other's places:
b_{NG} of b_Q and b_C of b_{NQ}.
This means that additional revenue is gained at a loss.

This additional information allows the researcher to understand better the character of production if the resources take one of these values.

The point of maximum revenue b_G makes it possible to obtain the analytical character of the proposed power function of complex variables, namely, to determine the production efficiency level using the distance of the actual value b to point b_G. Ivan Svetunkov suggests using a parameter that shows this level and that can be found by means of the formula

Table 6.8 Results of calculation of the Russian economy efficiency level

Year	1998 (%)	1999 (%)	2000 (%)	2001 (%)	2002 (%)	2003 (%)
S	8.3	20.7	23.4	20.4	17.8	18.8
G/C	12.7	25.5	24.7	18.5	14.4	13.5

$$S = 1 - \frac{b - b_G}{b_{NG} - b_G}. \tag{6.32}$$

It is evident that S is positive when $b < b_{NG}$, i.e., when company profits are positive. The coefficient is close to zero if b is close to b_{NG}, i.e., when the profit is close to zero. $S = 0$ only in the case of profitless production when $b = b_{NG}$. S is equal to one when exponent b coincides with b_G. In the zone of "highly profitable production" [when b falls within $(0; b_G)$] S becomes higher than one. If $b > b_{NG}$, the proposed coefficient becomes negative, which shows the unprofitability of production. For convenience S can also be represented in percentage terms just by multiplying its value by 100%.

Studies of numerous conditional and real examples have shown that this coefficient correlates with the dynamics of profitability at cost, and the higher the profitability, the closer the parameter values to unity and vice versa. Table 6.8 gives an example of the calculation of coefficient S for the Russian economy according to data from the Russian Federal State Statistics Service from 1998 to 2003 and G/C ratio as a reflection of the average profitability.

As we see from the table, coefficient S can really be used as one of the characteristics for estimating the production efficiency level since to a certain extent it shows the average profitability of the production (the pair correlation coefficient between S and G/C on this set of values is 0.71).

Since the proposed model has a clear economic meaning and shows real production processes in progress, its analytical properties should be more carefully studied for other production processes as well.

6.4 Power Production Complex-Valued Functions with Real Coefficients of the Diatom Plant and Russian Industry

The remarkable properties of the power PFCV mentioned in the previous chapter require confirmation by real economic examples. Since the data of the Diatom plant seem to be quite representative and in compliance with the above-mentioned procedure, Ivan Svetunkov calculated a, b, b_G, b_{NG}, b_Q, and S values for each observation. These values are given in Table 6.9. Unfortunately, the requirements of commercial secrecy do not allow us to furnish here the original data for this plant for the period under consideration from 2004 to 2007, which is why we provide only the results of the calculations.

Table 6.9 Characteristics of power production function of complex variables with real coefficients for Diatom production works

Quarter	a	b	b_G	b_Q	b_{NG}	S (%)	G/C (%)
Q1 2004	0.939	1.241	0.032	0.658	1.251	0.80	1.23
Q2 2004	1.928	1.294	0.009	0.644	1.270	−1.92	−2.99
Q3 2004	1.643	1.332	0.053	0.683	1.260	−6.02	−9.08
Q4 2004	1.558	1.288	0.066	0.708	1.284	−0.28	−0.41
Q1 2005	1.127	1.283	0.069	0.713	1.288	0.40	0.59
Q2 2005	1.344	1.332	0.016	0.682	1.332	−0.01	−0.01
Q3 2005	1.335	1.273	0.060	0.715	1.309	2.86	4.29
Q4 2005	1.728	1.329	0.086	0.756	1.339	0.78	1.14
Q1 2006	1.272	1.308	0.092	0.760	1.336	2.30	3.36
Q2 2006	1.476	1.348	0.061	0.742	1.362	1.06	1.59
Q3 2006	1.403	1.349	0.091	0.782	1.383	2.60	3.82
Q4 2006	1.218	1.326	0.092	0.799	1.415	6.75	9.94
Q1 2007	1.383	1.397	0.062	0.787	1.449	3.72	5.60
Q2 2007	1.502	1.385	0.105	0.829	1.449	4.77	6.96

Based on the interpretation of the values of the given model characteristics and proposed coefficient S, one can draw a conclusion on the low efficiency of the plant. Exponent b is in zone 8 of the 13 previously considered and is close to the outer value of b_{NG}, which characterizes profitless activity. The low S value also testifies to the fact. Since our task was to verify the analytical properties of the production power function with real coefficients, the table gives the profitability indicator G/C.

It is easy to see that parameter S correlates with profitability, confirming the suitability of the complex-valued model for economic analysis.

It is evident from the table that the proportionality coefficient and exponent vary in time, but this variation is not crucial. Obviously, any model of socioeconomic dynamics, however well it described a series of observations, at some point will stop doing it satisfactorily since socioeconomic dynamic processes are subject to evolution. This is why in forecasting it is necessary to adapt the model, i.e., to correct the model coefficients in the event of changes in trends. The complex-valued production function with real coefficients for which coefficients are calculated for each observation shows this in its full scope.

Besides the fact that the model under study makes it possible to estimate production efficiency, it can help to obtain recommendations on how to increase production efficiency. On the basis of the last observation for the Diatom plant and especially the calculated values of the coefficients of the power PFCV, which looks has the form

$$G + iC = 1.502(K + iL)^{1.385}, \tag{6.33}$$

we can provide for the company's leadership recommendations that follow from the model.

First of all let us determine what level of profits and costs the plant can attain if the production is improved to the maximum degree. Certainly, this proposal is idealized since we do not take into consideration all the possible factors and conditions. This is why we consider the extreme case. Thus, if we improve production to this unattainable extreme case, the model shows this when $S = 100\%$ or

$$b = b_G = 0.105.$$

Let us take the same K and L resource values for Q2 of 2007 and calculate G and C values for the exponent $b = bG = 0.105$. We will have the following in relative values:

$$G^* = 1.512, \quad C^* = 0.172. \tag{6.34}$$

We see that production costs for the case under consideration drop considerably since they relate to their original value for Q1 of 2004 and have been increasing for the whole considered period from 1 onward. In (6.34) we see a more than fivefold cost reduction! The type of production that provides such profits and costs is ideal, the obtained calculated values being marginal. These unattainable values show the direction for possible company development, in particular, they show that with existing production technology and more rational usage of the available resources, including labor resources, the Diatom plant could earn much bigger profits and incur lower costs. But what should be done to attain this?

To answer this question, let us determine how the enterprise can obtain the ultimate combination of profits and production costs, equal to $G* = 1.512;$ $C* = 0.172$; if it does not improve organizational and economic mechanisms, what resources should be used? For that let us substitute these extreme values of the gross margin and costs into formula (6.33), which shows the state of affairs at the company. On this basis we can calculate the necessary sizes of capital and labor to attain the necessary values of profits and costs without changing anything else at the company. In absolute values they will be equal to

$$K=5.55 \text{ million rubles}, L=0.46 \text{ million rubles} \tag{6.35}$$

For comparison, it should be noted that in Q2 of 2007, the cost of fixed production assets at the Diatom plant was 2.92 million rubles, and payroll was 5.52 million rubles.

What does this comparison mean? To make the production process at the plant more efficient, one should increase the fixed production assets there by $5.55/2.92 = 1.9$ times, and the payroll should be reduced by $5.52/0.46 = 12$ times. This can be done by reducing excessive labor resources and increasing productivity at the expense of investments in fixed capital.

Now we can give an economic interpretation to the obtained results.

The calculated values of labor and capital make it possible to draw the conclusion that the company's resources are used inefficiently: to increase production efficiency, we should increase investments in fixed production assets and reduce labor costs. The given proportions do not specify that the resource changes should be made on such a scale. Any model represents the result of abstraction, and many real factors are not taken into account, which is why we called the values of the gross margin and production costs calculated according to the ideal model "extreme" values. The proportions obtained make it possible to determine the main direction of actions to take. Since to attain efficient production it is necessary to increase capital by 1.9 times and reduce labor resources by 12 times, it is evident that labor resources are the main direction of improvement in company operations. It is this indicator that is used at the plant inefficiently, which is why rationalization of labor is an important factor in the improvement of production efficiency. It is seen from the calculations that investments should not be ignored by economic analysis, but the main focus of the company management, in terms of increasing efficiency, is on labor resources. Company management should examine the ratio between industrial, production, and other company personnel and the organization of labor and wages since it is here where they would be able to find redundancies for increasing production efficiency.

It should be noted that the criterion of maximum gross margin is the main one for company operations; however, sometimes competitive market conditions are such that companies are forced at all costs to be the leader in sales. This means that the main criterion of company operations is maximum production volume. Understanding that production and sales volumes are related but nonetheless different concepts, all the same we emphasize their closeness, not their differences, to simplify the problem. Therefore, let us assume that the maximum production volume criterion is associated with maximum revenue. Then we can make similar calculations for the Diatom plant, where we use bQ instead of bG, i.e., when

$$b = b_Q = 0.829$$

we get the following results.

Maximum company revenue will be $Q = 61.75$ million rubles, but profit will be $G = 27.36$ million rubles, costs $C = 34.39$ million rubles. This level can be reached with an increase in fixed production assets K of up to 4.73 million rubles and a reduction in labor costs L down to 3.6 million rubles. This means that to reach maximum revenue we should increase the cost of the fixed production assets by only $4.73/2.92 = 1.6$ times and reduce labor costs by $5.52/3.6 = 1.53$ times. This means that the Diatom company works at almost maximum production volumes with the given technology and production organization. If we recall that this company is the only supplier in Russia of diatom insulating bricks for nonferrous metallurgy, but that it has competitors from China, then it is understandable that in striving not to allow its competitors enter the market the company ensures, to the maximum degree, that it will have customers for its products.

Now it is interesting to compare the results and recommendations obtained by means of the power PFCV having real coefficients with those provided by the production function of Cobb-Douglas. Estimating the parameters of the power production function of complex variables with real coefficients and the production function of Cobb-Douglas with the data of the Diatom company using the LSM we obtain the following models:

$$G + iC = 1.398(K + iL)^{1,319}, \tag{6.36}$$

$$Q = 2.348K^{0.457}L^{0.543}. \tag{6.37}$$

The revenue approximation error for the first model (6.36) is 15.3% and for the second one (6.37) it is 14.8%. This means that these two models approximate the original data of the production under consideration with a practically similar approximation accuracy.

Recommendations that can be obtained using the production function of Cobb-Douglas mean that to increase the revenue, the Diatom plant should increase investments in fixed production assets and increase labor resources (number of workers), with an increase in the number of workers being more desirable since the elasticity coefficient (6.37) for labor is higher than for capital. The exponent value of the Cobb-Douglas function, calculated using LSM, with labor resources L equal to 0.543 shows that increasing the number of employees by 1%, one obtains a revenue increase of 0.543%.

Thus, if we leave the value of the fixed production assets for the last year of observation unchanged at an amount of 2.92 million rubles and increase the number of employees, it is possible to double the gross output, as follows from Cobb-Douglas function (6.37) by increasing the labor resources to 2,187 people (leaving current wages unchanged). If we substitute these values of capital and labor resources into to our function (6.36), then we will model another result, that is, gross margin will be negative and equal to −60.5 million rubles, and production costs will be 191million rubles.

Gross output will be 130.55 million rubles, which means that the number of employees should not be increased and, conversely, this number should be reduced by optimizing the organization of labor.

Since the two production functions (6.42) and (6.43) possess an accuracy of approximation that is almost equal to that of previous values of modeled production (15.3% and 14.8%, respectively) but propose diametrically opposite recommendations with respect to production development, there is a natural alternative: choose the model that actually describes the situation without distorting it.

To determine which recommendations are closer to the real situation at the plant and which model adequately describes the production situation, we appealed to the management of the Diatom plant. The general director, E.A. Nikiforov, who holds a Ph.D. in economics, explained that the number of employees was excessive. This was due to the fact that the company was a city-forming enterprise and to decrease the level of social tension and reduce unemployment in the city the board decided to provide employment to the maximum possible number of Inza residents using labor resources in a not very efficient manner. In the opinion of the company director, the labor resources at the enterprise are excessive. The situation with the company

Table 6.10 Original statistical data for Russian industry

Year	Production volume (billions of rubles)	Annual average number of production workers (thousands of workers)	Investments in fixed capital in industry producing goods, million rubles	Profitabiliy level of goods (works, services) sold
1998	1,707	13,173	165,092	0.127
1999	3,150	13,077	297,278	0.255
2000	4,763	13,294	527,544	0.247
2001	5,881	13,282	699,366	0.185
2002	6,868	12,886	817,504	0.144
2003	8,498	12,384	980,188	0.135
2004	11,209	11,977	1,179,744	0.179

capital resources is as follows. After the crisis of the 1990s, when the factory almost shut down, its material and technical base was in decline and was partly ruined (the unemployed Inza residents started removing the company's fixed assets for sale in the form of metal scrap). Only since the late 1990s did the plant start to recover, primarily by investment in fixed capital. That is why speedy growth of investments in fixed capital with an unchanged number of employees (due to socioeconomic reasons) is a strategic direction of the development of the Diatom plant, which, as we see from the above-mentioned example, is fully in line with the conclusions and recommendations of the power PFCV with real coefficients (6.42) and does not agree with the recommendations of the Cobb-Douglas production function (6.43).

From the foregoing example it does not mean that our function is always better than the Cobb-Douglas function. But since in the example in question it adequately described the production process but the Cobb-Douglas function did not, we can conclude that there could be other production situations where the complex-value function will be more efficient than the production function of real variables.

Let us verify if it is possible to apply power PFCVs with real variables at the macro level. Let us take statistical data for Russian industry for the period 1998–2004 given in Table 6.10.

To build a power PFCV, we need data for gross margin and production costs, which cannot be found for Russia in general in the statistical reference books of the TF State Committee for Statistics. However, the necessary calculated values of profit and costs can be found on the basis of production profitability values given in the table. Instead of the fixed production assets, we will use investments in fixed capital.

Transforming these data into relative values (1998) we can, using LSM, calculate the model coefficient values:

$$G + iC = 0.001(K + iL)^{3.457} \tag{6.38}$$

A high exponent value should be frightening, as for the period under consideration the polar angle of the complex resource decreased almost eightfold! The polar angle of the complex production result did not change. This is why the model exponent (6.38) appeared to be so high. According to model recommendations, an increase in investments in fixed capital would boost production efficiency.

The calculation of control points for this model is senseless since instead of fixed assets the model uses investments.

6.5 Coefficients of Elasticity of the Complex Exponential Production Function with Real Coefficients

Elasticity is one of the most important indices used in simulating socioeconomic processes. By definition, elasticity is calculated as follows:

$$\varepsilon_{\overset{/\sim}{\infty}} = \frac{\frac{\Delta y}{y}}{\frac{\Delta x}{x}}. \tag{6.39}$$

Elasticity (6.39) has a clear interpretation showing in terms of percentage the change in the result with a 1% factor change. For example, in economic theory, characterizing demand behavior often involves the price elasticity of the demand volume, with elasticity showing consumer reaction (change of the consumption volume) at slight changes of the price per product unit.

Moving from discrete values to constants, we can represent the elasticity as follows:

$$\varepsilon_{\overset{/\sim}{\infty}} = \frac{\frac{dy}{y}}{\frac{dx}{x}} = \frac{dy}{dx}\frac{x}{y}. \tag{6.40}$$

This form of notation allows for calculating the elasticity for various differentiable functions including those of production. Since we are discussing the properties of PFCVs, then, after becoming acquainted with the peculiarities of the behavior of an exponential complex function with real coefficients, we would like to focus on the resource elasticity for this function. For that, as follows from (6.40), it is necessary to calculate the first derivatives of this function with respect to its variables – labor and capital.

First, to calculate the derivatives, let us represent the model of the production function (6.18) in exponential form:

$$G + iC = a(K + iL)^b = a\left(Re^{i\theta}\right)^b, \tag{6.41}$$

where

$$R = \sqrt{K^2 + L^2}, \quad \theta = arctg\frac{L}{K}.$$

Now it is easy to group the model modulus and its polar angle:

$$G + iC = aR^b e^{ib\theta}. \tag{6.42}$$

With this in mind, the model of the exponential complex production function may be represented in trigonometric form:

$$G + iC = aR^b[\cos(b\theta) + i\sin(b\theta)]. \tag{6.43}$$

This makes it possible to calculate the first and second (if necessary) partial derivatives of the complex function with respect to resources – labor and capital. According to the d'Alembert-Euler (Riemann-Cauchy) condition for finding complex function derivatives, it will be enough to take derivatives of its real and imaginary parts. The real part of the model (6.43) may be represented in the following form:

$$\mathrm{Re}(G + iC) = aR^b\cos(b\theta) = U. \tag{6.44}$$

This is why the derivative of the function (6.18) with respect to resources may be calculated as follows:

$$\frac{\partial(G + iC)}{\partial(K + iL)} = \frac{\partial U}{\partial K} - i\frac{\partial U}{\partial L} = \frac{\partial(aR^b\cos(b\theta))}{\partial K} - i\frac{\partial(aR^b\cos(b\theta))}{\partial L}. \tag{6.45}$$

Let us calculate the first component of derivative (6.45), namely, $\frac{\partial(aR^b\cos(b\theta))}{\partial K}$, as a derivative of a complex function:

$$\frac{\partial(aR^b\cos(b\theta))}{\partial K} = \frac{\partial(aR^b)}{\partial K}\cos b\theta + \frac{\partial\cos(b\theta)}{\partial K}(aR^b). \tag{6.46}$$

The first part of (6.46) takes the form

$$\frac{\partial(aR^b)}{\partial K}\cos(b\theta) = abR^{b-1}\frac{K}{\sqrt{K^2 + L^2}}\cos(b\theta) \tag{6.47}$$

or

$$\frac{\partial(aR^b)}{\partial K}\cos(b\theta) = abR^{b-2}K\cos(b\theta). \tag{6.48}$$

The second form of (6.46) is a derivative of the cosine of the argument in K. It may also be calculated as follows:

$$\frac{\partial \cos(b\theta)}{\partial K}(aR^b) = -aR^b \sin(b\theta)\frac{\partial(b\theta)}{\partial K} = -aR^b \sin(b\theta)\frac{\partial(arctgb\frac{L}{K})}{\partial K}.$$

After applying the formula for calculating the derivative of the arctangent we have

$$\begin{aligned}\frac{\partial \cos(b\theta)}{\partial K}(aR^b) &= -abR^b \sin(b\theta)\frac{-L}{K^2(1+\frac{L^2}{K^2})} = abR^b \sin(b\theta)\frac{L}{R^2}\\ &= abR^{b-2}L\sin(b\theta).\end{aligned} \tag{6.49}$$

Then the partial derivative of the exponential complex production function with real coefficients (6.46) for capital will be written as follows:

$$\begin{aligned}\frac{\partial(aR^b\cos(b\theta))}{\partial K} &= abR^{b-2}K\cos(b\theta) + abR^{b-2}L\sin(b\theta)\\ &= abR^{b-2}(K\cos(b\theta) + L\sin(b\theta)).\end{aligned} \tag{6.50}$$

The same procedure can be applied to the partial derivative of the exponential production function (6.18) for labor: $\frac{\partial(aR^b\cos(b\theta))}{\partial L}$.

Since this is a complex function derivative, it should be calculated as follows:

$$\frac{\partial(aR^b\cos(b\theta))}{\partial L} = \frac{\partial(aR^b)}{\partial L}\cos(b\theta) + \frac{\partial\cos(b\theta)}{\partial L}(aR^b). \tag{6.51}$$

The summand (6.51) taking into account the derivative of the modulus may be represented as follows:

$$\frac{\partial(aR^b)}{\partial L}\cos(b\theta) = abR^{b-1}\frac{L}{\sqrt{K^2+L^2}}\cos(b\theta) = abR^{b-2}L\cos(b\theta) \tag{6.52}$$

The addend (6.51) includes the derivative of the cosine of the argument for labor. This derivative may also be determined as follows:

$$\frac{\partial\cos(b\theta)}{\partial L}(aR^b) = -abR^b \sin\theta\frac{\partial(arctg\frac{K}{L})}{\partial L}.$$

After calculating the derivative of the arctangent we finally obtain for this component

$$\frac{\partial \cos(b\theta)}{\partial L}(aR^b) = -abR^b \sin(b\theta)\frac{K}{K^2(1+\frac{L^2}{K^2})} = -abR^{b-2}K\sin(b\theta). \tag{6.53}$$

Then the derivative with respect to labor will have the form

$$\begin{aligned}\frac{\partial(aR^b\cos(b\theta))}{\partial L} &= abR^{b-2}L\cos(b\theta) - abR^{b-2}K\sin(b\theta)\\ &= abR^{b-2}(L\cos(b\theta) - K\sin(b\theta)).\end{aligned} \tag{6.54}$$

Now we can obtain the required formula of the first derivative of the production function in question for the complex resource (6.45):

$$\frac{\partial(G+iC)}{\partial(K+iL)} = abR^{b-2}(K\cos(b\theta)+L\sin(b\theta)) - iabR^{b-2}(L\cos(b\theta)-K\sin(b\theta)).$$

This huge expression can easily be simplified:

$$\begin{aligned}\frac{\partial(G+iC)}{\partial(K+iL)} &= abR^{b-2}[(K\cos(b\theta)+L\sin(b\theta) - iL\cos(b\theta)+iK\sin(b\theta)]\\ &= abR^{b-2}[(K-iL)(\cos(b\theta)+i\sin(b\theta)].\end{aligned} \tag{6.55}$$

Since $\cos(b\theta)+i\sin(b\theta) = e^{ib\theta}$ and $R^{-2}(K-iL) = \frac{K-iL}{K^2+L^2} = \frac{1}{K+iL}$, (6.55) can be much simplified:

$$\frac{\partial(G+iC)}{\partial(K+iL)} = abR^b e^{ib\theta}(K+iL)^{-1} = b(K+iL)^{-1}[aR^b e^{ib\theta}]. \tag{6.56}$$

If we know the value of the first derivative of the exponential complex function with real coefficients for a complex argument, we can determine the elasticity of this function. It will have the following form:

$$\varepsilon = \frac{\partial(G+iC)}{\partial(K+iL)}\frac{K+iL}{G+iC} = b(K+iL)^{-1}[aR^b e^{ib\theta}]\frac{K+iL}{G+iC} = b\frac{aR^b e^{ib\theta}}{G+iC}\frac{K+iL}{K+iL} = b. \tag{6.57}$$

Thus, the elasticity of the exponential complex function with real variables for a complex resource is equal to the function exponent. This means that the model under study possesses the following property: with a simultaneous 1% increase in production resources, the production result will increase by b percent.

Remarkable properties of models of complex variables revealing their additional advantages compared to models of real variables are demonstrated in the fact that using an exponential complex production function we can determine the contribution of each of the components of a complex resource per each component of the production result – the gross margin G and the production costs C. Since a real economy allows for an increase in, say, only labor resources with capital resources unchanged, it is important to determine how the gross margin and the production costs will change. To answer this question let us calculate the elasticity coefficients of the complex result for each of the resources –labor and capital.

First let us calculate the first partial derivative of the complex function for capital, using the same symbols:

$$\frac{\partial(G+iC)}{\partial K}=\frac{\partial(aR^b\cos(b\theta))}{\partial K}+i\frac{\partial(aR^b\sin(b\theta))}{\partial K}. \tag{6.58}$$

For the first summand the following formula has been obtained in (6.50):

$$\frac{\partial(aR^b\cos(b\theta))}{\partial K}=abR^{b-2}(K\cos(b\theta)+L\sin(b\theta)).$$

Omitting huge calculations like those given previously by (6.58), we have the following for the addend:

$$\frac{\partial(aR^b\cos(b\theta))}{\partial K}=abR^{b-2}(K\sin(b\theta)-L\cos(b\theta)). \tag{6.59}$$

Substituting the obtained results into (6.58) we find the form of the first partial derivative of the required function for capital:

$$\frac{\partial(G+iC)}{\partial K}=abR^{b-2}[(K\cos(b\theta)+L\sin(b\theta))+i(K\sin(b\theta)-L\cos(b\theta))]. \tag{6.60}$$

The expression in square brackets may be easily transformed into

$$\begin{aligned}&K\cos(b\theta)+iK\sin(b\theta)-i(L\sin(b\theta)+L\cos(b\theta))\\&\quad=(K-iL)(\cos(b\theta)+i\sin(b\theta))=(K-iL)e^{ib\theta}.\end{aligned}$$

The factor before the square brackets in (6.60) may be presented as follows:

$$abR^{b-2}=ab\frac{R^b}{K^2+L^2}.$$

Taking the previous expressions into account, it is easy to obtain a convenient formula of the first partial derivative of the complex production result for capital:

$$\frac{\partial(G+iC)}{\partial K}=ab\frac{R^b}{K^2+L^2}(K-iL)e^{ib\theta}=b(aR^be^{ib\theta})\frac{K-iL}{K^2+L^2}. \tag{6.61}$$

Hence we have the elasticity coefficient of the exponential PFCV for capital:

$$\varepsilon_K = \frac{\partial(G+iC)}{\partial K}\frac{K}{G+iC} = b(aR^b e^{ib\theta})\frac{K-iL}{K^2+L^2}\frac{K}{aR^b e^{ib\theta}}$$
$$= b(\frac{K^2}{K^2+L^2} - i\frac{LK}{K^2+L^2}). \quad (6.62)$$

Since this elasticity coefficient is complex, its real part characterizing the effect of capital on the gross margin and its imaginary part the effect of capital on production costs, the former and the latter should be considered separately:

$$\varepsilon_K = \varepsilon_{GK} + i\varepsilon_{CK}. \quad (6.63)$$

The real part of the complex elasticity coefficient εGK characterizes the value by which the gross margin will change with a 1% increase in capital:

$$\varepsilon_{GK} = b\frac{K^2}{K^2+L^2}. \quad (6.64)$$

In the denominator of the real part of the complex elasticity coefficient (6.63) there is a value that was previously called the scale of production resources. It is evident that a fraction of an expression (6.64) is always lower than one, which is why the real part of the elasticity for capital showing growth in the gross margin is determined by the value of exponent b. Since exponent b for this function characterizes the overall elasticity of the complex production result for a complex production resource (6.57), (6.64) may be written as follows:

$$\varepsilon_{GK} = \varepsilon\frac{K^2}{K^2+L^2} \rightarrow \varepsilon = \varepsilon_{GK}(1+\frac{L^2}{K^2}). \quad (6.65)$$

This means that the gross margin elasticity coefficient for capital of the production function in question is always lower than the elasticity.

It follows from (6.64) that an increase in capital resources will always, in this model, result in an increase in the gross margin – more or less depending on the value of exponent b.

Let us consider now the imaginary component of the complex elasticity of the production function for capital. It shows the effect of capital on production costs and is equal to

$$\varepsilon_{CK} = -b\frac{LK}{K^2+L^2} = -\varepsilon\frac{LK}{K^2+L^2}. \quad (6.66)$$

The fraction of this expression is always positive, like coefficient b, which is why the imaginary component of the complex elasticity for capital is always negative. This means that the model under consideration describes production processes under which any increase in capital results in a reduction in production costs, i.e., to a reduction in the unit cost. Most real production processes do behave in this way as capital increases lead mainly to an increase in labor productivity, which results in cost reductions. It is evident from (6.66) that the degree of reduction in production costs is determined by exponent b of the production function and labor and capital values.

Since the elasticity of the complex production result for capital is a complex number, it may be presented in exponential form. For that, let us calculate the modulus of the complex number and its polar angle.

The modulus of the complex elasticity coefficient (6.62) is equal to

$$R_{\varepsilon K} = bK, \tag{6.67}$$

and its polar angle is

$$\varphi_{\varepsilon K} = arctg\frac{L}{K}. \tag{6.68}$$

Since the sum of the gross margin and costs represents the sales volume, when both the gross margin and production costs are adjusted to the gross output as a result of preliminary scaling, the sum of the real and imaginary parts of the elasticity coefficient for capital will characterize the elasticity of the output for capital in the following way:

$$\varepsilon_{QK} = \varepsilon_{rk} + \varepsilon_{ik} = b\frac{K^2 - LK}{K^2 + L^2}. \tag{6.69}$$

It then follows that the elasticity of the output for capital may also be negative if the numerator (6.69) is negative – for labor-intensive processes. However, with an increasing capital resource and stable labor resource, the elasticity of the output for capital becomes positive.

Similarly, we can calculate the first derivative of the exponential production function for labor and determine on its basis the formula for calculating the elasticity of the complex production result for this resource. If we omit the calculations similar to those above, the following formula will result:

$$\varepsilon_L = \frac{\partial(G + iC)}{\partial L}\frac{L}{G + iC} = b(\frac{L^2}{K^2 + L^2} + i\frac{LK}{K^2 + L^2}). \tag{6.70}$$

The complex elasticity coefficient may be given an economic interpretation. Since it is complex, it is more convenient to consider its real and imaginary parts:

$$\varepsilon_L = \varepsilon_{GL} + i\varepsilon_{CL}. \tag{6.71}$$

The real part of the complex elasticity of the complex production result for labor, which shows the contribution of labor to the change in the gross margin, is

$$\varepsilon_{GL} = b\frac{L^2}{K^2 + L^2} = \varepsilon\frac{L^2}{K^2 + L^2}. \tag{6.72}$$

Elasticity of the gross margin for labor is always positive and always lower than the total elasticity ε since a fraction of the expression (6.72) is always less than one. The positive nature of the gross margin elasticity coefficient for labor means that this function may be used when growth in labor resources leads to growth in the gross margin.

The imaginary component of the complex elasticity of the production result for labor (6.70), which shows the effect of labor on production costs, has the following form:

$$\varepsilon_{CL} = b\frac{LK}{K^2 + L^2}. \tag{6.73}$$

Since all the components of this part of the elasticity coefficient are positive, it is evident that growth in labor resources will inevitably lead to growth in production costs. This also corresponds exactly to actual production situations – the use of additional human resources or growth in payroll costs to stimulate the employed workers will increase total production costs.

Now, it is necessary to consider the contribution of the labor resource to changes in production on the whole, i.e., to changes in output volume. For that let us sum up the two parts of coefficient (6.70) and evaluate the contribution of the labor resource to the combined growth in both gross margin and production costs, with the assumption that the variables are properly scaled:

$$\varepsilon_{QL} = \varepsilon_{GL} + \varepsilon_{CL} = b\frac{L(L + K)}{K^2 + L^2}. \tag{6.74}$$

This means that a 1% increase in labor resources increases the production volume by the above-mentioned value.

If now we consider the complex elasticity of the production result for labor (6.70) in exponential form, the modulus of this complex value will be

$$R_{\varepsilon L} = bL, \tag{6.75}$$

and the polar angle will be

$$\varphi_{\varepsilon K} = arctg\frac{K}{L}. \tag{6.76}$$

Table 6.11 Elasticity coefficients of complex production function for Russian economy

Year	$b = \varepsilon$	ε_{GK}	ε_{CK}	ε_{GL}	ε_{CL}	ε_K	ε_L
1998	12.674	12.574	−1.124	0.018	1.124	11.449	1.225
1999	8.114	7.968	−1.081	0.068	1.081	6.886	1.228
2000	6.264	6.074	−1.074	0.205	1.074	4.999	1.264
2001	6.088	5.875	−1.118	0.365	1.118	4.757	1.331
2002	5.758	5.520	−1.146	0.674	1.146	4.375	1.384
2003	5.789	5.555	−1.141	1.009	1.141	4.414	1.375
2004	4.922	4.652	−1.120	1.910	1.120	3.532	1.390
2005	4.842	4.565	−1.123	2.822	1.123	3.442	1.400
2006	4.472	4.167	−1.129	4.893	1.129	3.038	1.435
2007	4.397	4.080	−1.137	8.101	1.137	2.943	1.454
2008	4.255	3.927	−1.135	12.821	1.135	2.792	1.463

To conclude our study of the elasticity of the exponential PFCV with real coefficients, we should mention one more important property of complex coefficients of elasticity of the production result for resources. If we sum up elasticity for capital (6.62) and elasticity for labor (6.70), we will obtain the total elasticity (6.57):

$$\varepsilon = \varepsilon_K + \varepsilon_L = b\left(\frac{K^2}{K^2+L^2} - i\frac{LK}{K^2+L^2}\right) + b\left(\frac{L^2}{K^2+L^2} + i\frac{LK}{K^2+L^2}\right) = b.$$

The result that we obtain on the basis of this chapter means that an exponential complex production function with real coefficients possesses much more extensive analytical capabilities for studying production processes than production functions with real variables. If a similar model of an exponential production function with real coefficients has only two elasticity coefficients, which are exponents of resource indices, the complex model allows for the following calculations:

Total elasticity (6.57);
Complex elasticity of the production result for capital (6.63), which consists of two real elasticity coefficients (6.64) and (6.66);
Elasticity of output Q for capital (6.69);
Complex elasticity of production result for labor (6.70) with two components (6.72) and (6.73);
Output elasticity Q for labor (6.74).

In total there are seven different elasticity coefficients showing the most diverse effects of the two resources on the complex production result and the gross output.

Let us show how to calculate the above-mentioned elasticity coefficients of a complex production function with real coefficients for real data. Here we will use previously mentioned data on the Russian economy from 1998 to 2008. The results of calculation of the coefficients are given in Table 6.11.

The obtained elasticity values show the following specific features of the development of the Russian economy for the period under consideration.

First of all, it should be noted that in 1998 Russia experienced a default that resulted in a devaluation of the ruble, a crucial reduction in the price of domestic goods, considerable growth in the demand for them, and an essential recovery in production and the economy in general. This is why the data for 1998 and 1999 are so different from those for subsequent years.

The 1990s were characterized by a sharp drop in production volume. Labor and capital resources were underutilized, basic production assets were mostly "frozen," and earnings of the citizens were extremely low. This is why the use of additional capital resources in the economy after the default of 1998 and the growth in material incentives resulted in a considerable increase in production volume, gross margin, and production costs. This is reflected in all the elasticity values.

It should be noted that investments in basic capital aimed at restoring production but not at modernization. For Russian entrepreneurs investments in innovations were pointless given the absence of competition among the national producers and a high price competitiveness of Russian products compared to exported products.

These specific features of the economic growth following the default of 1998 are reflected in the dynamics of the elasticity coefficients. It is evident from Table 6.11 that the effect of capital on the reduction in production costs is quite low: the elasticity coefficient εCK, reflecting this, demonstrates no pronounced dynamics since the efficiency of capital in GDP growth reflected by output elasticity for capital εK decreased. Investments in basic capital of the Russian economy, which in 1998 and 1999 led to excellent results in achieving high gross margins, as shown by the elasticity values εGK for these years, start playing less important role in the Russian economy.

Since at that time the elasticity coefficient of the gross margin for labor εGL increased considerably, the basic contribution of labor resources to the increased efficiency of the state economy becomes clear. It should also be noted that GDP elasticity for labor εL grew but not as fast as that of the gross margin for labor εGL. This means that labor productivity for the Russian economy in general during this period increased, but the rates of this increase were not significant.

We do not intend here to conduct a thorough analysis of the economic situation in the Russian economy. The calculation of elasticity coefficients using the Russian economy as an example aimed at demonstrating the possibility of such calculations and interpretation of the values of elasticity coefficients.

6.6 Power Production Function of Complex Variables with Complex Coefficients

It was shown in previous sections that the power PFCV with real coefficients possesses remarkable properties of compliance with actual production processes. Its seven elasticity coefficients considerably expand economists' analytical tools. But the diversity of production observed in the economy cannot be described in its entirety by only this power function. The reality is much richer and more diverse, and this diversity is partly described by the power model of a complex variable with real coefficients. It follows from the conclusions of the previous section that this model describes processes characterized by an increasing return on resources, a decreasing return on resources, and a constant return on resources. However, there are situations of crisis production or cyclical production, etc. This is why model (6.18), being in many ways universal, is not always able to provide the best results of production modeling.

Model (6.18) with real coefficients is one of the simplest among power PFCVs. Function (6.17) with complex variables is the most general among the possible power PFCVs:

$$G + iC = (a_0 + ia_1)(K + iL)^{(b_0+ib_1)}. \tag{6.77}$$

This general function can serve as the basis for the most varied types of function that can be obtained by varying the four coefficients. One of these types is the power production function with real coefficients (if imaginary parts of complex coefficients are equal to zero); another one is a linear function of complex variables (when the exponent is equal to a real number – unity).

The modeled production processes are diverse, which is why one can suppose that in various cases one of the types of power functions can be the best. Hence, the question arises as to how one selects the best option from among the above-mentioned variety of production functions. For that, we can recommend the following procedure. We use LSM to find parameters of a general power function with a complex proportionality coefficient and complex exponent (6.77). Based on the value of the coefficients a_0, a_1, b_0, and b_1 found by means of LSM, the researcher can choose the production function for modeling. For example, if $b_1 \to 0$ and $a_1 \to 0$, then, based on the principle of simplicity, one should use the power PFCV with real coefficients like

$$G + iC = a_0(K + iL)^{b_0}.$$

If the power coefficient values are close to zero, then one should use a linear function with complex variables:

$$G + iC = (a_0 + ia_1)(K + iL).$$

According to the above-mentioned statistical data, one can build, for example, a power production function with complex coefficients for the Diatom plant. Calculations give it the following form:

$$G + iC = (0.534 + i0.363)(K + iL)^{0.459 - i0.355}.$$

We see that all the components of complex coefficients (both real and imaginary parts) have huge values, which is why it would not be proper to use more simplified models for the purposes of approximation and economic analysis. It should be noted that this model is quite good at approximating real data (the average approximation error is 20%) taking into account the fact that investments K in fixed capital of the Diatom factory for the period in question vary gradually, their consequences being prolonged, which cannot help but reduce the accuracy of a description of the production dynamics by any model.

Everything is somewhat different with the power production function with complex coefficients for industrial production in Russia. Using LSM the following model for the available statistical data was obtained:

$$G + iC = (1.732 - i0.213)(K + iL)^{0.896+i0.351}.$$

Here, the imaginary component of the proportionality coefficient is small and closer to zero compared with its real part.

It is true that the imaginary part of the complex proportionality coefficient equal to (-0.213) in its absolute value is eight times smaller than its real part (1.732). This model is good at approximating real data – the average production result approximation error was (4.6%). However, for simplified calculations, due to small values of a_1, to describe industrial production in Russia, we can use the following model:

$$G + iC = a(K + iL)^{b_0+ib_1}.$$

Thus, calculation with the statistical data made in accordance with the proposed approaches allows for selecting power production functions with various coefficients – complex or real – in order to use a model of adequate complexity in each case.

Since complex-valued power functions with complex coefficients can turn to be the best for modeling some production processes, we should study their properties more carefully. For that, let us single out the real and the imaginary parts from equality (6.77):

$$G = f(K, L), \quad C = g(K, L).$$

In formula (6.77) the complex variable $(K + iL)^{(b_0+ib_1)}$ can be represented in the following way:

$$e^{(b_0+ib_1)\ln(K+iL)}.$$

Now formula (6.77) will have the form

$$G + iC = (a_0 + ia_1)\, e^{b_0 \ln(K+iL) + ib_1 \ln(K+iL)}. \tag{6.78}$$

The sum of the products in the exponent can be transformed taking account the properties of logarithms of complex numbers using their principal values:

$$G + iC = (a_0 + ia_1)\ e^{b_0 \ln\left(\sqrt{K^2+L^2}\right) + ib_0 \mathrm{arctg}\left(\frac{L}{K}\right) + ib_1 \ln\left(\sqrt{K^2+L^2}\right) - b_1 \mathrm{arctg}\left(\frac{L}{K}\right)}. \tag{6.79}$$

Now let us group the real and imaginary parts of the complex variable to obtain a formula that would be convenient for further study:

$$G + iC = (a_0 + ia_1)\ e^{b_0 \ln\left(\sqrt{K^2+L^2}\right) - b_1\ \mathrm{arctg}\left(\frac{L}{K}\right)} e^{i\left(b_0\ \mathrm{arctg}\left(\frac{L}{K}\right) + b_1 \ln\left(\sqrt{K^2+L^2}\right)\right)}. \tag{6.80}$$

On the right-hand side the complex coefficient $(a_0 + ia_1)$ is multiplied by a variable that can be designated as R. That is,

$$G + iC = (a_0 + ia_1)Re^{i\varphi}, \tag{6.81}$$

where

$$R = e^{b_0 \ln\left(\sqrt{K2+L2}\right) - b_1\ \mathrm{arctg}\left(\frac{L}{K}\right)}, \tag{6.82}$$

$$\varphi = b_0\ \mathrm{arctg}\left(\frac{L}{K}\right) + b_1 \ln\left(\sqrt{K^2 + L^2}\right). \tag{6.83}$$

Now if we present model (6.81) in trigonometric form, open the brackets, and group the real and imaginary parts, we obtain

$$G + iC = R((a_0 \cos\varphi - a_1 \sin\varphi) + i(a_0 \sin\varphi + a_1 \cos\varphi)), \tag{6.84}$$

which means that two equalities are satisfied:

$$G = R(a_0 \cos\varphi - a_1 \sin\varphi), \tag{6.85}$$

$$C = R(a_0 \sin\varphi + a_1 \cos\varphi). \tag{6.86}$$

Thus, to calculate profit G and production costs C, the researcher can use formulae (6.85) and (6.86).

According to the economic meaning of the problem under consideration, production costs cannot be negative but profit can (operating at a loss).

This means that (6.86) imposes restrictions on the limits of variation of the complex proportionality coefficient factors:

$$a_0 \sin\varphi + a_1 \cos\varphi > 0 \rightarrow tg\varphi > -\frac{a_1}{a_0}, \tag{6.87}$$

where φ is calculated from (6.83).

Finding function coefficients (6.77) is a simple task. In Chap. 4, we mentioned ways of solving them using LSM.

Let us give the following statement without proof (due to the considerable bulkiness of the conclusion): the coefficient of elasticity of a power complex-valued production function with complex coefficients is equal to the exponent of this function:

$$\varepsilon = b_0 + ib_1. \tag{6.88}$$

It follows from (6.39) that

$$\Delta y \quad = \quad \varepsilon_{yx}\, y\, \frac{\Delta x}{x} \tag{6.89}$$

or

$$\Delta(G + iC) = (b_0 + ib_1)(G + iC)\frac{\Delta(K + iL)}{K + iL}, \tag{6.90}$$

where we can find an answer to the question of what process of change in the gross margin and production costs does a power production function with complex coefficients model when production resources vary by 1%:

$$\begin{cases} \Delta G = (b_0G - b_1C)\dfrac{\Delta KK + \Delta LL}{K^2 + L^2} + (b_0C + b_1G)\dfrac{\Delta KL - \Delta LK}{K^2 + L^2}, \\ \Delta C = (b_0C + b_1G)\dfrac{\Delta KK + \Delta LL}{K^2 + L^2} + (b_0G - b_1C)\dfrac{(\Delta LK - \Delta KL)}{K^2 + L^2}. \end{cases} \tag{6.91}$$

It is seen from the obtained equalities that both the gross margin and production costs may decrease and increase with an increase in the resources given various combinations of the exponent coefficient values. In contrast to the power function with real coefficients, the elasticity of a function with a complex exponent gives little information on the direction of the complex result variation with a 1% increase in the production resources. Everything is determined by both the combination of values of the real and imaginary parts of the complex exponent and the values of the resources and the results.

However, since this variied combination makes it possible to model diverse production processes, the complex-valued power function with complex coefficients possesses rather high identification properties. It describes various production processes – from efficient to unprofitable, from processes with an increasing return on resources to those with a decreasing return.

If in practice calculations of production volumes are required but not a company's profits and costs, then one can derive a formula for the company's output volume for this function, taking into account that with correct scaling $Q = C + G$. Adding (6.85) and (6.86) we get

$$Q = G + C = R(a_0 \cos\varphi - a_1 \sin\varphi) + R(a_0 \sin\varphi + a_1 \cos\varphi)$$

or

$$Q = R((a_0 - a_1)\sin\varphi + (a_0 + a_1)\cos\varphi), \tag{6.92}$$

which is the same.

This function will have other elasticity coefficients than those calculated in the previous section for the power complex-valued model with real coefficients. Since they do not have as clear a meaning as those considered in the previous section, their calculation here is irrelevant.

The foregoing discussion shows that the complex-valued power production function with complex coefficients can be a powerful instrument for production process analysis and modeling.

6.7 Logarithmic Production Function of Complex Variables

In the theory of production functions, from among the whole set of nonlinear models, economists prefer power models, as they are convenient to use and have a simple economic interpretation. The same may be said of PFCVs: power functions are convenient to use and have a simple interpretation of their parameters. Nevertheless, in the real economy the most varied situations are possible when these models fail to approximate real production. These cases should use production functions of another form. One such alternative model is that of a logarithmic production function of complex variables. In this section we will focus on its properties and features for practical use.

The general form of a production logarithmic function has the following form:

$$G_t + iC_t = (a_0 + ia_1) + (b_0 + ib_1)\ln(K_t + iL_t). \tag{6.93}$$

The free term of this equality has a simple meaning – it corrects the initial model conditions with respect to the real values of the variables. This is its only contribution to modeling the production situation. This is why, without loss of generality, we consider the model without a free term:

$$G_t + iC_t = (b_0 + ib_1)\ln(K_t + iL_t). \tag{6.94}$$

The properties of this model are fully revealed if we present the logarithm of the complex variable in arithmetic form and multiply the value obtained by the complex proportionality coefficient:

$$G_t + iC_t = (b_0 + ib_1)(\ln\sqrt{K_t^2 + L_t^2} + iarctg\frac{L_t}{K_t}).$$

As a result, we obtain the following statement:

$$G_t + iC_t = (b_0 \ln \sqrt{K_t^2 + L_t^2} - b_1 arctg \frac{L_t}{K_t}) + i(b_1 \ln \sqrt{K_t^2 + L_t^2} + b_0 arctg \frac{L_t}{K_t}), \tag{6.95}$$

from which we derive two equalities characterizing the real and imaginary parts of the model:

$$\begin{cases} G_t = b_0 \ln \sqrt{K_t^2 + L_t^2} - b_1 arctg \dfrac{L_t}{K_t}, \\ C_t = b_1 \ln \sqrt{K_t^2 + L_t^2} + b_0 arctg \dfrac{L_t}{K_t}. \end{cases} \tag{6.96}$$

With positive coefficients and growing labor costs, production costs also grow and the gross margin may also grow depending on the coefficient values and variable scale, but much slower than the costs. The function may demonstrate another behavior under conditions where $b_0 < < b_1$ – then the gross margin decreases with an increase in labor resources.

If capital resources grow and the model coefficients are positive, then the gross margin and production costs will also increase.

However, a situation may arise where the model coefficients become negative. For example, with negative b_1 the gross margin will grow if there is a growth in labor resources and the effect of capital is uncertain – it may lead either to an increase or a reduction in the gross margin. Production costs may also behave ambiguously – with growing labor resources they may both increase and decrease. However, an increase in the capital resources will definitely result in a decrease in production costs.

This means that a logarithmic function of complex variables can be suitable for describing several different production situations.

Since model (6.94) has only two coefficients, it may mean, in reference to complex-valued functions, that there is a possibility of estimating their values with one statistical observation. Indeed, it definitely follows from (6.94) that

$$b_0 + ib_1 = \frac{G_t + iC_t}{\ln(K_t + iL_t)} = \frac{G_t + iC_t}{\ln \sqrt{K_t^2 + L_t^2} + iarctg \frac{L_t}{K_t}}. \tag{6.97}$$

Multiplying the numerator and denominator of the right-hand side of the equality by the value conjugate to the denominator we obtain

$$b_0 + ib_1 = \frac{G_t \ln \sqrt{K_t^2 + L_t^2} + C_t arctg \frac{L_t}{K_t} + i(C_t \ln \sqrt{K_t^2 + L_t^2} - G_t arctg \frac{L_t}{K_t})}{\ln^2 \sqrt{K_t^2 + L_t^2} + arctg^2 \frac{L_t}{K_t}}. \tag{6.98}$$

Since the left- and right-hand sides of the complex equality mean the simultaneous equality of the real and imaginary components, let us obtain the formulae for calculating each of the coefficients at observation t:

$$b_0 = \frac{G_t \ln \sqrt{K_t^2 + L_t^2} + C_t arctg \frac{L_t}{K_t}}{\ln^2 \sqrt{K_t^2 + L_t^2} + arctg^2 \frac{L_t}{K_t}},$$
$$b_1 = \frac{C_t \ln \sqrt{K_t^2 + L_t^2} - G_t arctg \frac{L_t}{K_t}}{\ln^2 \sqrt{K_t^2 + L_t^2} + arctg^2 \frac{L_t}{K_t}}. \qquad (6.99)$$

Using the previous data for the Inza Diatom factory, let us calculate these time-varying coefficients for the logarithmic production function of complex variables of this factory. First, to nullify the effect of scale, let us center the production performance and logarithms of the complex resource with reference to their averages, which follows from (6.94). The results are given in Table 6.12.

Analysis of the coefficient dynamics shows that both these coefficients tend to grow slightly with time. The dynamics of coefficient b_0 can be described by a linear trend, and its equation can be easily found by LSM:

$$b_{0t} = 0.1051t - 0.8496.$$

The other coefficient, b_1, has a less pronounced growth trend. Nevertheless, it can also be represented in the form of a linear trend. LSM allows us to find estimations of this trend:

$$b_{1t} = 0.0044t + 0.6521.$$

The proportionality coefficient of the latter trend is close to zero, which confirms that the dynamic of variation of the imaginary constituent b_1 of the complex regression is quite low, which is why we can consider this coefficient almost constant.

We see that the logarithmic complex-valued model (7.5.2) is not applicable for modeling the production of this factory, which is confirmed by systematic variation of one of the model coefficients, but it can be used, with the trends obtained, for purposes of forecasting and multivariate modeling. Taking into account the linear variation in time of one of the coefficients, the logarithmic complex-valued model of the Diatom factory production will have the form

$$G_t + iC_t = (0.1051t - 0.8496 + i0.6521)\ln(K_t + iL_t).$$

Now, setting different alternatives for the economic development of the factory resources, one can obtain alternative complex production results.

A similar method was used to calculate the coefficients of this model with respect to the statistical manufacturing data in Russia. To avoid calculating the

Table 6.12 Coefficients of logarithmic complex variables production function of Inza Diatom factory

t	b_0	b_1
1	−0.09084	−0.22095
2	−0.3334	−1.05475
3	−0.99276	3.642535
4	0.125047	0.568678
5	0.14192	3.308633
6	−3.03984	0.386848
7	−0.19265	−0.67284
8	1.018661	−0.58261
9	−0.32793	−1.0547
10	0.084627	0.177843
11	1.197226	2.964839

Table 6.13 Coefficients of logarithmic production function of complex variables for Russian industry

t	b_0	b_1
1998	2.150242	3.502649
1999	1.284403	3.327341
2000	−0.34767	9.019879
2001	0.783986	−0.1516
2002	−1.11559	1.390457
2003	0.08035	3.233731
2004	4.527272	5.703955

free term, we center the gross margin and production costs with reference to their averages. In the same way we center the principal logarithmic values. The coefficient variation dynamics is given in Table 6.13, which shows clearly that variation of this dynamics is crucial, without any visual trend of coefficients, which shows that a logarithmic model could not be applied to this case.

If a complex-value function with complex coefficients is applicable to virtually any production process with a particular degree of accuracy, the logarithmic function, as we can see from the above-mentioned example, is more "whimsical" and not as universal.

An example of a successful application of a logarithmic production function of complex variables to a company in St. Petersburg is SPb ZPS. The example was found and calculations made by D. Duhanina. Since the company works under conditions of unstable market conditions and its production cycle is quite significant, production activity performance in months is unstable. These data are given in Table 6.14. The company is a small business, which is reflected in a slight variation of the resources.

Duhanina expressed these data in a nondimensional value and centered them around their averages, after which she calculated the dynamics of each coefficient of the complex-valued logarithmic production function.

Table 6.14 Monthly data on production activity of SPb ZPS for 2005–2007 and model coefficients

	Revenue	Profit	Costs	Fixed assets	Model coefficients (6.94)	
Time, t	Q	G	C	K	b_0	b_1
1	2,231,714	−30,930	2,262,644	3,775,895	0.0541	0.0301
2	2,015,300	−147,251	2,162,551	3,746,729	0.0711	0.0329
3	1,769,635	−29,932	1,799,567	3,756,818	0.0417	0.0190
4	8,539,410	−919,909	9,459,319	3,756,818	0.2092	0.0683
5	3,670,696	−441,546	4,112,242	3,727,652	0.0920	0.0170
6	2,884,686	−101,970	2,986,656	3,698,485	0.0381	−0.0090
7	3,004,009	−105,777	3,109,786	3,669,318	0.0378	−0.0099
8	4,024,720	29,734	3,994,986	3,640,152	0.0165	−0.0048
9	2,098,340	146,384	1,951,956	3,610,985	−0.0010	0.0000
10	5,340,695	580,477	4,760,218	3,581,818	−0.0429	0.0025
11	1,709,550	351,809	1,357,741	3,552,652	−0.0324	−0.0136
12	2,923,416	131,948	2,791,468	3,523,485	0.0006	−0.0003
13	3,247,150	−60,978	3,308,128	3,494,318	0.0404	−0.0237
14	4,750,283	56,260	4,694,023	3,465,152	0.0197	−0.0130
15	1,838,114	223,327	1,614,787	3,508,333	−0.0164	0.0090
16	8,742,325	783,340	7,958,985	3,479,167	−0.1553	0.1147
17	7,783,112	887,750	6,895,362	3,450,000	−0.2038	0.1420
18	1,748,450	437,723	1,310,727	3,420,833	−0.1285	0.1000
19	4,445,690	94,670	4,351,020,	3,391,667	0.0234	−0.0194
20	2,935,020	366,858	2,568,162	3,362,500	−0.1771	0.1582
21	1,865,884	455,873	1,410,011	3,333,333	0.3221	−0.3457
22	534,629	−671,575	1,206,204	3,304,167	−0.4863	0.5474
23	4,858,625	1,121,220	3,737,405	3,275,000	0.4300	−0.5057

It is evident from the table that the complex proportionality coefficient varies from observation to observation, but this variation does not represent a trend, in contrast to the case with the Diatom factory. These variations are very slight and characterize deviations with reference to some average values. The last three observations are exceptions, as the coefficients calculated for them differ significantly from the whole series.

This means that on a given set of production results, a complex-valued logarithmic model of the production function will satisfactorily describe the production process. The deviations of the last three observations are due either to random factors or to changes in the development trends. Since thorough research of the situation at this small business is not the object of our study, let us focus on a formal problem – the possibility of modeling production by means of this model.

To build a model on the available statistical data, one can use the values from Table 6.14. Duhanina used LSM to estimate the coefficients of the complex-valued logarithmic model of the production function. This model has the form

$$G_t + iC_t = (0.0909 - i0.1715) + (0.0436 - i0.0389)(K_t + iL_t). \qquad (6.100)$$

To draw a conclusion regarding the applicability of the model in question to production modeling, let us compare the properties of this model with those of other real-variable models. Duhanina used LSM to find, based on the same data, coefficients of the Cobb-Douglas production function, which has the form

$$Q_t = 1.2603K_t^{0.3326}L_t^{0.6674}. \tag{6.101}$$

Now we can compare the models' suitability for describing production processes.

The logarithmic production function makes it possible to calculate both the gross margin and production costs. The approximation error was $G = 58.07\%$ for the gross margin. It is evident from Table 6.14 that the profit is subject to a high variance; this is why this approximation error should not mislead the reader. The model is much more accurate when it describes production-cost dynamics. The average approximation error for this parameter in the original series was $C = 8.2\%$.

Since the sum of the gross margin and production costs gives an output value, we use the logarithmic production function of complex variables to approximate this parameter too. The average approximation error for Q was 33.14%.

The use of the Cobb-Douglas function for the purposes of modeling gives an average approximation error of 62.49%, which is almost twice as high as the approximation error of the complex-value model.

Again, we will not focus here on the problem of selecting the best model for describing the given production process. It is not of interest here. It is important to note that from the given example it follows that for some production processes a logarithmic complex-valued function can be successfully applied.

We will not calculate the elasticity of logarithmic complex-valued production functions because they do not have a very simple form or as clear an economic interpretation as in the case with the power production function with real coefficients. Therefore, the elasticity will not contribute to our understanding of the model properties.

6.8 Exponential Production Function of Complex Variables

To complete our study of elementary complex-valued functions that may be used as production function models, let us turn our attention to the properties and features of applying an exponential production function. This function can have the most varied bases but the function properties will not change. It is the degree of complexity of model usage that changes. It is evident that the exponential function, where e is the base, will be the least problematic since the exponential form of complex variables is practically always used to understand model properties. This is why we can use a model of an exponential complex-valued function as an example of exponential models in general, and it has the following form:

$$G_t + iC_t = (a_0 + ia_1)e^{(b_0+ib_1)(K_t+iL_t)}. \tag{6.102}$$

The simplest variant of this model is one with real coefficients. It represents a rather simple function, which is of little interest:

$$G_t + iC_t = a_0 e^{b_0(K_t+iL_t)}. \tag{6.103}$$

Indeed, grouping the variables the comprise the model and the polar angle of the right-hand side of the equality we obtain

$$G_t + iC_t = a_0 e^{b_0 K_t} e^{ib_0 L_t}. \tag{6.104}$$

If we turn to the trigonometric form we obtain the following equalities for the real and imaginary parts:

$$\begin{cases} G_t = a_0 e^{b_0 K_t} \cos(b_0 L_t), \\ C_t = a_0 e^{b_0 K_t} \sin(b_0 L_t). \end{cases} \tag{6.105}$$

This means that for positive b_0, an increase in capital from the zero value will be followed by an increase in the production scale and, therefore, in the gross margin; this growth will retain the same proportions between these variables. In other words, the growth in production volumes maintains an unchanged level of profitability. For real production this is possible when the capital-labor ratio is low and its growth significantly affects productivity. There is also some control over pricing.

Growth in the labor resource from the zero point will be followed by an increase in the polar angle, the production scale remaining unchanged with unchanged capital. This means growth in the production costs and reduction in the gross margin. One should not forget that periodic functions can be both positive and negative. Therefore, for modeling production situations it is necessary either to center the original variables, in which case they will take both positive and negative values, or to impose restrictions on the b_0 coefficient, based on the economic meaning of the problem. This is all that can be said about this function.

If now instead of the real proportionality coefficient in the exponent we consider a situation where the proportionality coefficient is imaginary, i.e., $b_0 = 1$, $b_1 \neq 0$, we will see certain rather "symmetrical" changes in the function's properties. The production function in this case will have the form

$$G_t + iC_t = a_0 e^{ib_1(K_t+iL_t)}. \tag{6.106}$$

The trigonometric form of this model will be as follows:

$$\begin{cases} G_t = a_0 e^{-b_1 L_t} \cos(b_1 K_t), \\ C_t = a_0 e^{-b_1 L_t} \sin(b_1 K_t). \end{cases} \tag{6.107}$$

Growth in the labor resources definitely leads to a decrease in the production scale (if $b_1 > 0$) and a decrease in both gross margin and production cost values. Growth in capital from the zero point will lead to a decrease in the gross margin and an increase in production costs. On the whole, this model can describe crisis-situation production, when to attain the desired production performance it is necessary to reduce the number of employees and abandon noncore production, eliminating capital in this area of production.

It is understood that the use of complex coefficients makes it possible to synthesize these diverse properties into a single complex production relationship. To understand the influence of production resources on production performance in the complex coefficient model (6.102) the modulus and the polar angle in the right-hand side of the equality should be singled out. For that, we open the brackets in the exponent and express the complex proportionality coefficient in exponential form.

Then taking account the designations previously introduced we obtain

$$G_t + iC_t = ae^{b_0 K_t - b_1 L_t} e^{i(b_0 L_t + b_1 K_t + \alpha)}. \tag{6.108}$$

Now it is easy to obtain two equalities of real variables that model the real and imaginary parts of the model, i.e., describe the effect of production resources on the gross margin and production costs:

$$\begin{cases} G_t = ae^{b_0 K_t - b_1 L_t} \cos(b_0 L_t + b_1 K_t + \alpha), \\ C_t = ae^{b_0 K_t - b_1 L_t} \sin(b_0 L_t + b_1 K_t + \alpha). \end{cases} \tag{6.109}$$

With positive coefficients the model will possess the following properties. Capital growth entails growth in the gross margin and production costs. However, this growth is not similar. Thus, the gross margin's exponential growth induced by an increase in K_t in the exponent is somehow nullified as the cosine of the polar angle decreases with an increase in capital, and their product gives complex nonlinear dynamics.

Costs increase more intensively with the capital increase since with capital growth, an exponential increase is associated with an increase in the sine. Their product gives a particular multiplicative effect.

However, this general characteristic is corrected by the argument of the proportionality coefficient $\alpha = arctg \frac{a_1}{a_0}$. It characterizes a phase shift of the cosine and sinusoid. This shift may lead to inverse dependences.

The same complex character in this model characterizes the dependence of the production performance on labor resources. In the first approximation, the increase in labor resources is followed by a very marked decrease in the gross margin, as the decrease of the exponent in the first equality (6.109) is associated with a decrease in the cosine. Their product reinforces the trend.

The cost behavior is not as unambiguous – growth in the labor resource is followed by a decrease in the exponential component of the second equality (6.109); however, its harmonic – sinusoid – component grows.

Again, this complex character of the dependence is to a large extent corrected by the argument of the proportionality coefficient – its various values provide a phase shift in the harmonic factors, and these factors can behave in a different way compared to the original idea.

This is why an exponential complex-valued function with complex coefficients is able to describe various production types.

According to the manufacturing data in Russia that have already been used in this chapter as the basis for verifying the properties of PFCVs, A.M. Chuvazhov built an exponential model of the production function with coefficients estimated by LSM. The model has the form

$$G_t + iC_t = (1.656 + i0.534)e^{(0.265+i0.015)(K_t+iL_t)}. \tag{6.110}$$

It describes the original data quite accurately – profit approximation errors are 7.06% and those of production costs are 2.64%. Therefore, the exponential complex-valued model of the production function may rightfully be included in the range of production function models since there could definitely occur cases where this model turns out to be the best possible one.

6.9 Summary

If we compare complex-argument models (Chap. 5) and complex-variable models, we clearly see that the latter are much more interesting for economic researchers and more universal. In this chapter functions of complex variables have been used as production function models, and in a number of cases they demonstrate a clear advantage over models of real variables. Each of the complex-variable functions has its original properties and can be used as a production function model in various situations. The researcher investigating a particular production process should select the best one.

It is very important that models of complex variables, being compact, allow for modeling several production parameters simultaneously – production volume, gross margin, and production costs. This is an obvious advantage of the proposed models.

The complex-valued power production function with real coefficients possesses the most interesting properties from among the models of production functions, as it shows very well real production processes. In addition, such important characteristics as the elasticity of the gross margin, production costs and production volumes for each of the resources and for a complex resource can be easily calculated using this function. This means that this function has considerable analytic potential.

In conclusion, another important property of PFCVs should also be noted. Complex-variable models used as production models make it possible to solve some problems that are difficult to pose in the area of real variable models. The following problem is standard: determining the output volume for various resource combinations. Complex-variable models expand this problem since one can not only calculate the output but find the values for the gross margin and production costs for various resource combinations.

There are also more complicated problems in studying production processes and production planning: for example, understanding how efficient production is, determining how one can attain a particular production volume, reduce production costs, and achieve the largest profit.

In real-variable models, in solving this problem one should vary the resources and calculate the results finding the best resource combination for the set criterion.

PFCVs make it possible to solve these problems simply using the inverse function. By definition, building an inverse function means deriving a relationship like $x = F^{-1}(y)$, for which $x = F(x)$.

The inverse function of a real variable is possible only in the case of a monofactor dependence. And all production functions, even the simplest ones like Cobb-Douglas, are multifactorial, and it is not possible to calculate an inverse function.

In relation to complex-variable or complex-argument models, it is rather a simple and solvable problem, as we consider the dependence of one complex variable on another, i.e., monofactor dependence. Therefore, one can apply an inverse function to these dependences.

For the complex variable production function

$$G + iC = F(K + iL) \tag{6.111}$$

let us derive the function

$$K + iL = F^{-1}(G + iC). \tag{6.112}$$

Substituting the required values of the gross margin and costs into the obtained inverse function it is easy to determine what capital and labor resources are necessary to obtain the required values of the production performance.

Let us derive the following types of inverse functions for the above-mentioned models:

1. Power,
2. Logarithmic,
3. Exponential.

The complex-valued power production function (6.17) has the form

$$G_t + iC_t = (a_0 + ia_1)(K_t + iL_t)^{b_0 + ib_1}.$$

The inverse function is found in the following way:

$$K + iL = \left(\frac{G + iC}{a_0 + ia_1} \right)^{\frac{1}{b_0 + ib_1}}. \tag{6.113}$$

The logarithmic complex-valued production function (6.93) has the following form:

$$G_t + iC_t = (a_0 + ia_1) + (b_0 + ib_1) \ln(K_t + iL_t).$$

The inverse function (6.112) will have a bit more complex form:

$$\ln(K_t + iL_t) = \frac{G_t + iC_t - (a_0 + ia_1)}{b_0 + ib}, \tag{6.114}$$

though if necessary, it may be written as follows:

$$K_t + iL_t = e^{\frac{G_t + iC_t - (a_0 + ia_1)}{b_0 + ib}} \tag{6.115}$$

This form makes it possible to calculate the required values of capital and labor at the same time.

Exponential complex-valued function (6.102) considered in the last section of this chapter was given in the following form:

$$G_t + iC_t = (a_0 + ia_1)e^{(b_0 + ib_1)(K_t + iL_t)}.$$

It can be used to obtain the inverse function

$$K_t + iL_t = \frac{\ln \frac{G_t + iC_t}{a_0 + ia_1}}{b_0 + ib_1}. \tag{6.116}$$

Calculations with any of these functions are easy. First, any econometric method is used to find coefficients of the original production function of complex variables, after which they are substituted into the respective inverse function (6.114), (6.115), or (6.116). Now, setting the desired profit, costs, or production volume values one can estimate the required capital and labor resource values.

One can also solve a problem relating to the maintence of the same profitability level, as well as other problems.

We should point out one assumption that we used a priori without specifying it. The proposed production functions modeled production costs and profit, and the economic interpretation of the obtained results complies with these indicators. However, the company or industry gross margin is determined in a market economy not just by the internal company forces or the number of resources used, but by the market itself. And the latter was not taken into account in the models under consideration, and it is not incorporated into real-variable models. It is assumed with this approach that everything that is to be produced in a company will find a consumer and will be sold with the modeled profitability.

This circumstance should be taken into account for more careful scientific research.

However, despite that, real examples of participants in economic activity assured us that complex-valued production functions could be an efficient tool for the modeling and analysis of production processes and, therefore, could be included in the arsenal of the production function theory.

Reference

1. Svetunkov SG, Svetunkov IS (2008) Production functions of complex variables. LKI, Moscow

Chapter 7
Multifactor Complex-Valued Models of Economy

The economy as an object of research is a multifactor system. This is why to provide adequate modeling, multifactor models should be applied. This chapter considers multifactor, complex-valued models with respect to production functions. The production performance is represented in the form of a complex variable where the gross margin refers to the real part and the production costs to the imaginary part. The production resource is represented by two complex variables. The first variable characterizes capital expenses and is represented in complex form. Its real part includes costs of fixed production assets, and its imaginary part thecosts of fixed nonproduction assets. The second complex variable characterizes the labor costs and is also represented as a complex variable. Its real part comprises industrial production capital, and its imaginary part – nonproduction workers.

These multifactor production functions are considered in linear form, Cobb-Douglas power function form, and power form. With Russia as an example, it is shown how to use these models to model the development of the economy taking into account of its "shadow" part.

7.1 General Provisions of Complex-Valued Model Classification

There is no doubt that models of production functions of complex variables discussed in the previous chapter extend the instrumental base of the production function theory and in certain cases provide more adequate modeling results than the existing real-variable models. However, one should not overestimate the results of application of certain elements of the complex-variable function theory in the economy since complex-valued economics is not an alternative to existing models of real variables; it only extends the arsenal of economic and mathematical methods and models, supplying economists with new tools. Along with the obvious advantages, production functions of a complex argument and complex-variable functions have some disadvantages restricting their sphere of application. Some of them were shown in their respective chapters. However, there is another

S. Svetunkov, *Complex-Valued Modeling in Economics and Finance*,

DOI 10.1007/978-1-4614-5876-0_7,

circumstance that can be considered a drawback of the above-mentioned complex-valued functions.

Indeed, in general form, production functions of complex variables (complex-argument functions can be considered as their simplified analog) have the following form:

$$G_t + iC_t = F(K_t + iL_t). \tag{7.1}$$

An important advantage of such a model compared to real-variable models is the modeling of dependences of two economic variables on two other economic variables. Economics differs from other natural and exact subjects and sciences in that a set of variables is determined by the effect of other variables. In this case model (7.1) in the economy is a step ahead compared to real-variable models. However, if we try to extend, for example, the number of production resources in function (7.1) we will face a very complicated problem – how do we do it? A complex variable, by definition, consists of a pair of real ones. What will we do if we need to take into account the effect on production of one more resources, for example, the area of land for agriculture? If we consider it as a simple real variable, we do not take into account its influence on production costs, and if we take it as the imaginary part, we miss the influence of the new resource on the gross margin. On the other hand, it is quite difficult to add one more variable to model (7.1) without taking specific steps. This is the first essential shortcoming restricting the sphere of application of models like (7.1).

The second disadvantage is due to the necessity of very accurate work with the dimension and scale of each of the variables. And if everything is clear about the production results, since these variables are measured in the same units and have the same scale, it is not the case for the production resources as the complex-valued functions under consideration are nonuniform.

Indeed, if we take the power production function of real variables

$$Q_t = aK_t^{\alpha}L_t^{\beta} \tag{7.2}$$

and change, for example, the capital resource scale by multiplying the resource by λ, we will have

$$Q_t = a(\lambda K_t)^{\alpha}L_t^{\beta} = a\lambda^{\alpha}K_t^{\alpha}L_t^{\beta}. \tag{7.3}$$

That is, as a result of the change in the magnitude of the variable, only the aspect ratio will change; there is no need to adjust the other variables and coefficients. Complex-valued functions are another matter. The complex resource variable represented as $(K_t + iL_t)$ significantly changes the model in the scale of either of the two resources. For example, changing the scale of the variable capital share for the same factor λ leads to a substantial change in the most complex variable, as is easily seen:

$$\lambda K_t + iL_t = \sqrt{(\lambda K_t)^2 + L_t^2}e^{iarctg\frac{L_t}{\lambda K_t}},$$

which means that the complex variable of a production function is not uniform; therefore, any complex-valued function whose scale was changed will change all its coefficients and the accuracy of the description of the original variables.

Let us take the power production function with real coefficients as an example:

$$G_t + iC_t = a(K_t + iL_t)^b. \tag{7.4}$$

The gross margin is described by this function as follows:

$$G_t = a(\sqrt{K_t^2 + L_t^2})^b \cos(barctg\frac{L_t}{K_t}). \tag{7.5}$$

If we change the capital resource scale λ times we will have the following expression for the gross margin:

$$G_t = a(\sqrt{\lambda^2 K_t^2 + L_t^2})^b \cos(barctg\frac{L_t}{\lambda K_t}). \tag{7.6}$$

This means that an adjustment to variable K_t will cause changes in both the proportionality coefficient a and exponent b, and the accuracy of the description of the gross margin, which in practice will be influenced by random errors. And since in the previous chapter we discussed the analytical properties of this very model and an interpretation of exponent b, it is quite easy to understand how essential it is to have a cautious stance with respect to the dimension and scale of variables of production functions with complex variables.

Thus, two essential problems (besides numerous other, less important, ones) require the development of complex-valued production functions:

1. The need for new economic variables, which is impossible for model of the (7.1) type,
2. Nullifying the influence of change in the scale of the original variables on modeling results.

Before we show how to create economic models of complex variables that could be free from the above-mentioned defects, let us turn our attention to the very essence of economic indicators, which is best illustrated by the production process.

All the indicators used for modeling any economic process represent a result of some aggregation (and abstraction, which is taken for granted).

Indeed, the gross margin G_t consists of many components that can be divided into two parts – one part of the profit is assigned to the state budget in the form of income tax and the other stays in the company's coffers.

Production costs C_t also consist of many terms, but they can also be represented in the form of two components – fixed costs (depreciation, employee compensation, etc.) and variable costs (raw materials, semifinished products, energy for operations, etc.).

Capital K_t in any form can also be represented as two terms – core and noncore (for example, core and noncore assets).

In the same way, labor resources L_t consist of two large groups – industrial and production on the one hand and other personnel on the other hand.

The list could go on, but the idea is that practically each economic indicator, not only of production, can be represented as the sum of two terms based on some classification. As always, justification could be found to subdivide these two classes into active and passive parts depending on their different effects on production.

Indeed, labor resources that were used in the production function theory in general and in previous chapters in particular reflect, very roughly, the influence of this indicator on production performance, as an increase in the number of production and industrial workers has a different effect on production performance compared to other workers. Their summation nullifies this influence and leads to a deterioration of the model properties – both analytical and forecasting.

In the same way, the size of the active part of fixed production assets (machines, equipment, production lines, etc.) has a different effect on the production variant than their passive parts (buildings, construction, infrastructure, etc.).

Since the classification of practically every economic variable implies a division into active and passive groups, the natural conclusion is to represent them in a complex-valued economy in the form of a complex variable where the real part is associated with the active term and the imaginary part with the passive one. This is the rule we will stick to henceforth.

Then we can write the capital resource in the following way:

$$K_0 + iK_1, \tag{7.7}$$

where K_0 means fixed production assets and K_1 major nonproduction assets.

The labor resource is to be represented as follows:

$$L_0 + iL_1, \tag{7.8}$$

where L_0 is industrial and production personnel and L_1 nonproduction personnel.

Output can be represented, as previously, in the form of a complex variable including the gross margin and production costs, or as some classified variable, for example, gross output (including sold and unsold products):

$$Q_0 + iQ_1. \tag{7.9}$$

This idea of production modeling easily solves the first problem – that of adding a new variable. If it is necessary to add to a model a new production resource S, the latter can be represented as a complex variable consisting of an active and a passive part:

$$S_0 + iS_1. \tag{7.10}$$

For example, to model agricultural production, the land area can be represented as the area for crops. livestock, etc.

Taking into account all of the foregoing considerations, the general production model will have the following form:

$$Q_0 + iQ_1 = F[(K_0 + iK_1), (L_0 + iL_1), (S_0 + iS_1)]. \quad (7.11)$$

Extrapolation of this approach to all economic models, not only production ones, will result in the following general form:

$$y_0 + iy_1 = F(x_{0j} + ix_{1j}). \quad (7.12)$$

Here j is the number of the economic variable included in the model.

Model (7.12) is a multifactor model. Since this model opens up a new class of complex-valued models based on the classification of economic variables, we will call them *classification models*.

Classification models are free from the above-mentioned defect of nonuniformity of (7.1). Classification variables by definition have the same dimension since they represent two parts of one whole. This is why if you need to scale some complex variable, you should multiply both the real and the imaginary part of this variable by the scaling coefficient λ:

$$\lambda K_{0t} + i\lambda K_{1t} = \lambda\sqrt{K_{ot}{}^2 + K_{1t}^2}e^{iarctg\frac{K_{1t}}{K_{0t}}} = \lambda(K_{0t} + iK_{1t}).$$

We see that this is a uniform variable of the first degree. In this case uniformity or nonuniformity is determined not by the features of complex variables but by the peculiarities of the models used.

In this chapter we will discuss economic classification models with respect to production process modeling problems.

7.2 Linear Classification Production Function

Let us again consider a production function describing the behavior of a complex production result including the gross margin G_t and the production costs C_t subject to the use of capital and labor. In this case both capital and labor can be represented as complex classification variables (7.7) and (7.8):

$$K_0 + iK_1, \quad L_0 + iL_1.$$

A linear function is the simplest variant of the production classification function:

$$G_t + iC_t = (a_0 + ia_1) + (b_0 + ib_1)(K_{0t} + iK_{1t}) + (c_0 + ic_1)(L_{0t} + iL_{1t}). \quad (7.13)$$

Here K_0 means fixed production assets, K_1 fixed nonproduction assets, L_0 industrial and production personnel, and L_1 nonproduction personnel.

If in this function the imaginary parts of complex proportionality coefficients are equal to zero, then the function can be transformed into a system of elementary multifactor (two-factor) equations:

$$\begin{cases} G_t = a_0 + b_0 K_{0t} + c_0 L_{0t}, \\ C_t = a_1 + b_0 K_{1t} + c_0 L_{1t}. \end{cases} \tag{7.14}$$

This model will not be made more complex if the real parts of the complex proportionality coefficients are equal to zero:

$$\begin{cases} G_t = a_0 - b_1 K_{1t} - c_1 L_{1t}, \\ C_t = a_1 + b_1 K_{0t} + c_1 L_{0t}. \end{cases} \tag{7.15}$$

This is why, it is only model (7.13) that is reasonable to use in economic and mathematical modeling, and we can transform it into a system of two real equalities:

$$\begin{cases} G_t = a_0 + b_0 K_{0t} - b_1 K_{1t} + c_0 L_{0t} - c_1 L_{1t}, \\ C_t = a_1 + b_0 K_{1t} + b_1 K_{0t} + c_0 L_{1t} + c_1 L_{0t}. \end{cases} \tag{7.16}$$

Now we can give an interpretation to the processes that can be described by means of a complex-valued model like (7.13). It follows from the first equality of (7.16) that an increase in the fixed production assets K_1 and the number of nonproduction personnel L_1 leads to a reduction in the gross margin, and from the second one that growth in these resources is followed by growth in the production costs. This is the case in most real production processes – growth in noncore assets deteriorates the production parameters, as does growth in the administrative apparatus.

Growth in the fixed production assets K_0 and industrial and production personnel L_0 results in a linear growth in the gross margin and production costs. The nature of this growth is determined by the proportionality coefficients of the variables.

Since model (7.13) can be represented in the form of real variables (7.16), one can ask if it is reasonable to use a complex-valued function as it may be easier to use two real functions (7.16). The answer comes from a mere comparison of (7.13) and (7.16). Model (7.13) is compact, and it is easy to find the model coefficients by LSM. One need simply solve a system of six equations with six unknowns:

$$\begin{cases} \sum_t G_t = na_0 + b_0 \sum_t K_{0t} - b_1 \sum_t K_{1t} + c_0 \sum_t L_{0t} - c_1 \sum_t L_{1t}, \\ \sum_t C_t = na_1 + b_1 \sum_t K_{0t} + b_0 \sum_t K_{1t} + c_0 \sum_t L_{1t} + c_1 \sum_t L_{0t}, \\ \sum_t K_{0t}G_t - \sum_t K_{1t}C_t = a_0 \sum_t K_{0t} - a_1 \sum_t K_{1t} + b_0 \sum_t (K_{0t}^2 - K_{1t}^2) \\ \quad -2b_1 \sum_t K_{0t}K_{1t} + c_0(\sum_t K_{0t}L_{0t} - \sum_t K_{1t}L_{1t}) - c_1(\sum_t K_{0t}L_{1t} + \sum_t L_{0t}K_{1t}), \\ \sum_t K_{0t}C_t + \sum_t K_{1t}G_t = a_0 \sum_t K_{1t} + a_1 \sum_t K_{0t} + b_1 \sum_t (K_{0t}^2 - K_{1t}^2) \\ \quad +2b_0 \sum_t K_{0t}K_{1t} + c_0\left(\sum_t K_{0t}L_{1t} - \sum_t K_{1t}L_{0t}\right) + c_1\left(\sum_t K_{0t}L_{0t} + \sum_t L_{1t}K_{1t}\right), \\ \sum_t L_{0t}G_t - \sum_t L_{1t}C_t = a_0 \sum_t L_{0t} - a_1 \sum_t L_{1t} + b_0(\sum_t K_{0t}L_{0t} - \sum_t K_{1t}L_{1t}) \\ \quad -b_1(\sum_t K_{1t}L_{0t} + \sum_t K_{0t}L_{1t}) + c_0(\sum_t L_{0t}^2 - \sum_t L_{1t}^2) - 2c_1 \sum_t L_{0t}L_{1t}, \\ \sum_t L_{0t}C_t + \sum_t L_{1t}G_t = a_1 \sum_t L_{0t} + a_0 \sum_t L_{1t} + b_0(\sum_t K_{1t}L_{0t} - \sum_t K_{0t}L_{1t}) \\ \quad + b_1(\sum_t K_{0t}L_{0t} + \sum_t K_{1t}L_{1t}) + c_1(\sum_t L_{0t}^2 + \sum_t L_{1t}^2) + 2c_0 \sum_t L_{0t}L_{1t}. \end{cases}$$

Here, T is the number of observations, $t = 1,2,3,\ldots,T$.

In the case of model (7.16), which represents a system of two real linear equations, the situation is much more complicated – using LSM, it is necessary to estimate the coefficients of the first equation of the system and then those of the second one.

Application of LSM to the first equality of system (7.16) makes it necessary to solve a system of five such equations:

$$\begin{cases} \sum_t G_t = a_0 T + b_0 \sum_t K_{0t} - b_1 \sum_t K_{1t} + c_0 \sum_t L_{0t} - c_1 \sum_t L_{1t}, \\ \sum_t G_t K_{0t} = a_0 \sum_t K_{0t} + b_0 \sum_t K_{0t}^2 - b_1 \sum_t K_{0t}K_{1t} + c_0 \sum_t K_{0t}L_{0t} - c_1 \sum_t K_{0t}L_{1t}, \\ \sum_t G_t K_{1t} = a_0 \sum_t K_{1t} + b_0 \sum_t K_{0t}K_{1t} - b_1 \sum_t K_{1t}^2 + c_0 \sum_t K_{1t}L_{0t} - c_1 \sum_t K_{1t}L_{1t}, \\ \sum_t G_t L_{0t} = a_0 \sum_t L_{0t} + b_0 \sum_t K_{0t}L_{0t} - b_1 \sum_t K_{1t}L_{0t} + c_0 \sum_t L_{0t}^2 - c_1 \sum_t L_{1t}L_{0t}, \\ \sum_t G_t L_{1t} = a_0 \sum_t L_{1t} + b_0 \sum_t K_{0t}L_{1t} - b_1 \sum_t K_{1t}L_{1t} + c_0 \sum_t L_{0t}L_{1t} - c_1 \sum_t L_{1t}^2. \end{cases} \tag{7.17}$$

Here one should be sure that the obtained coefficients will correspond to those obtained in the solution of the other system of five LSM equations determined by the second equation of system (7.16):

$$\begin{cases} \sum_t C_t = a_1 T + b_0 \sum_t K_{1t} + b_1 \sum_t K_{0t} + c_0 \sum_t L_{1t} + c_1 \sum_t L_{0t}, \\ \sum_t C_t K_{1t} = a_1 \sum_t K_{1t} + b_0 \sum_t K_{1t}^2 + b_1 \sum_t K_{0t} K_{1t} + c_0 \sum_t K_{1t} L_{1t} + c_1 \sum_t K_{1t} L_{0t}, \\ \sum_t C_t K_{0t} = a_1 \sum_t K_{0t} + b_0 \sum_t K_{0t} K_{1t} + b_1 \sum_t K_{0t}^2 + c_0 \sum_t K_{0t} L_{1t} + c_1 \sum_t K_{0t} L_{0t}, \\ \sum_t C_t L_{1t} = a_1 \sum_t L_{1t} + b_0 \sum_t K_{1t} L_{1t} + b_1 \sum_t K_{0t} L_{1t} + c_0 \sum_t L_{1t}^2 + c_1 \sum_t L_{1t} L_{0t}, \\ \sum_t C_t L_{0t} = a_1 \sum_t L_{0t} + b_0 \sum_t K_{1t} L_{0t} + b_1 \sum_t K_{0t} L_{0t} + c_0 \sum_t L_{0t} L_{1t} + c_1 \sum_t L_{0t}^2. \end{cases} \tag{7.18}$$

However, we cannot be sure about this. Moreover, the coefficients resulting from the solution of system (7.17) will vary from those found by solving equation system (7.18), at least because all the original data represent the result of the implementation of a random process and therefore include random errors, which will lead to various coefficient values.

Also, from system (7.17) we see that its solution does not depend at all on changes in production costs, and from (7.18) we see that its solution does not depend on the nature of the change in the gross margin. In other words, probability that solving (7.17) and (7.18) will yield equal coefficients is practically equal to zero. This means that it is impossible to obtain system (7.16) by this method!

Even if we first solve system (7.17) and find coefficients a_0, b_0, b_1,, and c_1 by substituting the obtained values into system (7.18), we see that each equality of this system from the first to the fifth will give different values of coefficient a_1. To find the balance, specific coordination procedures are needed.

Solving system (7.17) we obtain model coefficient values:

$$G_t = a_0 + b_0 K_{0t} + b_1 K_{1t} + c_0 L_{0t} + c_1 L_{1t}. \tag{7.19}$$

Solving system (7.18) we obtain coefficients of another model:

$$C_t = a_1 + d_0 K_{1t} + d_1 K_{0t} + f_0 L_{1t} + f_1 L_{0t}. \tag{7.20}$$

Certainly, they can be synthesized in a model like (7.13):

$$G_t + iC_t = (a_0 + ia_1) + [b_0 K_{0t} + b_1 K_{1t} + i(d_0 K_{1t} + d_1 K_{0t})] + [c_0 L_{0t} + c_1 L_{1t} + i(f_0 L_{1t} + f_1 L_{0t})], \tag{7.21}$$

but it will be quite a different model! One can also transform active parts of the resources into the real part and passive part into the imaginary one. However, here the coefficients of the synthesized model will vary from those of the complex-valued model. In addition, the bulkiness of the procedure makes it pointless!

Since in this model we can use the procedure of preliminary centering of the original parameters with respect to their averages, calculation of the complex-valued model coefficients will be much simplified. The model involving centered parameters will have the following form:

$$G_t + iC_t = (b_0 + ib_1)(K_{0t} + iK_{1t}) + (c_0 + ic_1)(L_{0t} + iL_{1t}) \tag{7.22}$$

But a system of normal LSM equations for centered variables will contain four equations:

$$\begin{cases} \sum_t K_{0t}G_t - \sum_t K_{1t}C_t = b_0 \sum_t (K_{0t}^2 - K_{1t}^2) - 2b_1 \sum_t K_{0t}K_{1t} \\ \quad +c_0(\sum_t K_{0t}L_{0t} - \sum_t K_{1t}L_{1t}) - c_1(\sum_t K_{0t}L_{1t} + \sum_t L_{0t}K_{1t}), \\ \sum_t K_{0t}C_t + \sum_t K_{1t}G_t = b_1 \sum_t (K_{0t}^2 - K_{1t}^2) + 2b_0 \sum_t K_{0t}K_{1t} \\ \quad +c_0(\sum_t K_{0t}L_{1t} - \sum_t K_{1t}L_{0t}) + c_1(\sum_t K_{0t}L_{0t} + \sum_t L_{1t}K_{1t}), \\ \sum_t L_{0t}G_t - \sum_t L_{1t}C_t = b_0(\sum_t K_{0t}L_{0t} - \sum_t K_{1t}L_{1t}) - b_1(\sum_t K_{1t}L_{0t} + \sum_t K_{0t}L_{1t}) \\ \quad +c_0(\sum_t L_{0t}^2 - \sum_t L_{1t}^2) - 2c_1 \sum_t L_{0t}L_{1t}, \\ \sum_t L_{0t}C_t + \sum_t L_{1t}G_t = b_0(\sum_t K_{1t}L_{0t} - \sum_t K_{0t}L_{1t}) + b_1(\sum_t K_{0t}L_{0t} + \sum_t K_{1t}L_{1t}) \\ \quad +c_1(\sum_t L_{0t}^2 + \sum_t L_{1t}^2) + 2c_0 \sum_t L_{0t}L_{1t}. \end{cases} \tag{7.23}$$

If one works with software that provides mathematical operations with complex variables, the task will be even simpler.

If we record model (7.22) with complex variables:

$$\dot{z}_t = \dot{b}\dot{K}_t + \dot{c}\dot{L}_t \tag{7.24}$$

we will easily obtain a system of LSM estimations of the model coefficients:

$$\begin{cases} \sum \dot{z}_t\dot{K}_t = \dot{b}\sum \dot{K}_t^2 + \dot{c}\sum \dot{L}_t\dot{K}_t, \\ \sum \dot{z}_t\dot{L}_t = \dot{b}\sum \dot{L}_t\dot{K}_t + \dot{c}\sum \dot{L}_t^2. \end{cases} \tag{7.25}$$

The solution will allow us to obtain the required complex coefficients of the multifactor model.

The conclusion that follows is clear – model (7.13) extends the instrumental base of economic and mathematical modeling since the building of its analog in the sphere of real numbers requires complex computational procedures that do not ensure adequate modeling.

Linear complex-valued multifactor models possess all the drawbacks of linear production function models with real variables. Therefore, understanding that model (7.13) has the right to exist and practical use let us leave it for modeling simple production situations.

7.3 Classification Production Function of Cobb-Douglas Type

Linear production classification functions differ from similar real-variable functions in their meaning and convenience in usage. However, the properties of complex-valued functions are most prominent for nonlinear functions. Their analogs in the sphere of real variables are extremely complicated, as was pointed out in previous chapters. For that reason it is logical to consider nonlinear, complex-valued classification functions. The simplest variant of such a function that comes to mind is built according to the Cobb-Douglas production function.

Let us call models like

$$G_t + iC_t = (a_0 + ia_1)(K_0 + iK_1)^{\alpha}(L_0 + iL_1)^{1-\alpha} \tag{7.26}$$

Cobb-Douglas classification production functions.

Here, as in Cobb-Douglas functions, the exponent lies within the range

$$0 \leq \alpha \leq 1. \tag{7.27}$$

We know that functions are linearly uniform if the following condition is met:

$$f(\lambda x_1, \lambda x_2, \ldots, \lambda x_i) = \lambda f(x_1, x_2, \ldots x_i). \tag{7.28}$$

Let us check if this condition is met for model (7.26):

$$\begin{aligned}
&(a_0 + ia_1)(\lambda K_0 + i\lambda K_v)^{\alpha}(\lambda L_0 + i\lambda L_v)^{1-\alpha} \\
&\quad = \lambda^{\alpha}\lambda^{1-\alpha}(a_0 + ia_1) \times (K_0 + iK_v)^{\alpha}(L_0 + iL_v)^{1-\alpha} \\
&\quad = \lambda(a_0 + ia_1)(K_0 + iK_v)^{\alpha}(L_0 + iL_v)^{1-\alpha}.
\end{aligned} \tag{7.29}$$

This means that the function is linearly uniform.

If we represent the right-hand side of model (7.26) in exponential form we obtain

$$(a_0 + ia_1)(K_0 + iK_1)^{\alpha}(L_0 + iL_1)^{1-\alpha} = R_a e^{i\theta_a} R_K^{\alpha} e^{i\alpha\theta_K} R_L^{1-\alpha} e^{i(1-\alpha)\theta_L}. \quad (7.30)$$

Here we use the following designations:

$$R_a = \sqrt{a_0^2 + a_1^2},\ \theta_a = arctg\frac{a_1}{a_0}, \quad (7.31)$$

$$R_K = \sqrt{K_0^2 + K_1^2},\ \theta_K = arctg\frac{K_1}{K_0}, \quad (7.32)$$

$$R_L = \sqrt{L_0^2 + L_1^2},\ \theta_L = arctg\frac{L_1}{L_0}. \quad (7.33)$$

Then model (7.26) can be represented as a system of two real equations: (1) for the gross margin:

$$G_t = R_a R_K^{\alpha} R_L^{1-\alpha} \cos(\theta_a + \alpha\theta_K + (1-\alpha)\theta_L), \quad (7.34)$$

(2) for production costs:

$$C_t = R_a R_K^{\alpha} R_L^{1-\alpha} \sin(\theta_a + \alpha\theta_K + (1-\alpha)\theta_L) \quad (7.35)$$

Now the general characteristics of the model are quite clear. If all the other resources remain unchanged, then:

1. If investments in fixed capital K_0 grow, the absolute value of the capital resources increases (7.32) and their polar angle decreases. The cosine of the decreasing angle goes up. When the modulus of the capital resource grows, the gross margin (7.34) grows as well. For production costs in this situation there are two opposing trends – the factor, the modulus of capital resources (7.35), increases but the sine of the decreasing angle goes down. That is why it follows from (7.35) that depending on the values of the exponent α and the complex proportionality coefficient θ there could occur both growth (if α is close to zero) and decrease (if α is close to one) in production costs. Intervals are possible where costs remain unchanged.
2. If investments in nonfixed capital grow, the modulus of the capital resources will grow as well as the polar angle. This means that the gross margin is calculated as the product of two opposing trends – the increasing modulus of capital resources and decreasing cosine of the sum of angles. However, depending on the original data and exponent value, this can in some cases cause an increase in the resulting indicator, in other cases a decrease in its values, and in other cases stability of the indicator. The dynamics of the production costs in this situation is determined by

multiplying the increasing modulus by the increasing sine of the sum of polar angles. This means that production costs will uniquely increase.

Similar conclusions can be drawn with respect to the behavior of another complex variable – labor resources. With an increase in the number of industrial and production workers the gross margin grows and costs may increase, remain unchanged, or decrease. With an increase in nonproduction personnel costs will definitely rise – the profit dynamics being determined by the model coefficients growth, stability – or decrease.

Thus, we see that model (7.26) covers all the possible alternatives of production relationships, except perhaps production stagnation.

A thorough study of function (7.26)'s properties was carried out by E.V. Sirotina. Let us show some of the results of that study.

To find unknown parameters a_0, a_1, and α of model (7.26), let us use LSM estimation of nonlinear models of complex variables with complex coefficients suggested in Chap. 4.

For models of the Cobb-Douglas type, as with Cobb-Douglas production functions, it is a little easier than for power functions with unlimited exponent values. For this equation (7.26) should be transformed into linear form by taking the logarithm the left- and right-hand sides of the natural base:

$$\ln(G_t + iC_t) = \ln(a_0 + ia_1) + \alpha \ln(K_0 + iK_1) + (1 - \alpha)\ln(L_0 + iL_1). \quad (7.36)$$

Using simple transformations we can present this expression in the following form:

$$\ln\frac{G_t + iC_t}{L_0 + iL_1} = \ln(a_0 + ia_1) + \alpha \ln\frac{K_{0t} + iK_{1t}}{L_{0t} + iL_{1t}}. \quad (7.37)$$

We see that the model looks like a simple logarithmic equation of two complex variables. To simplify further our conclusions, let us introduce the following designations of these variables. The resulting variable under the logarithm sign, which E.V. Sirotina proposed calling "complex productivity," can be transformed into the following form, which is convenient for research:

$$g_t = \frac{G_t + iC_t}{L_0 + iL_1} = \frac{G_tL_{0t} + C_tL_{1t}}{L_0^2 + L_1^2} + i\frac{C_tL_{0t} - G_tL_{1t}}{L_0^2 + L_1^2}. \quad (7.38)$$

The complex resource, also under the logarithm sign on the right-hand side, will be designated in the following way:

$$k_t = \frac{K_{0t} + iK_{1t}}{L_{0t} + iL_{1t}} = \frac{K_{0t}L_{0t} + K_{1t}L_{1t}}{L_{0t}^2 + L_{1t}^2} + i\frac{K_{1t}L_{0t} - K_{0t}L_{1t}}{L_{0t}^2 + L_{1t}^2}. \quad (7.39)$$

Since this variable represents the capital-to- labor ratio, it can just be called that.

These two new economic parameters can provide the economist with an additional characteristic of the production processes taking place, but will not explore this idea further.

As was stated in the Chap. 1, in this study we use principal logarithmic values, which is why equality (7.37) can be presented in the following way:

$$\ln R_g + i\theta_g = A_0 + iA_1 + \alpha(\ln R_k + i\theta_k). \tag{7.40}$$

Here:

$$R_g = \sqrt{\left(\frac{G_tL_{0t} + C_tL_{1t}}{L_0^2 + L_1^2}\right)^2 + \left(\frac{C_tL_{0t} - G_tL_{1t}}{L_0^2 + L_1^2}\right)^2}, \theta_g = iarctg\frac{C_tL_{0t} - G_tL_{1t}}{G_tL_{0t} + C_tL_{1t}}, \tag{7.41}$$

$$R_k = \sqrt{\left(\frac{K_{0t}L_{0t} + K_{1t}L_{1t}}{L_{0t}^2 + L_{1t}^2}\right)^2 + \left(\frac{K_{1t}L_{0t} - K_{0t}L_{1t}}{L_{0t}^2 + L_{1t}^2}\right)^2}, \theta_k = iarctg\frac{K_{1t}L_{0t} - K_{0t}L_{1t}}{K_{0t}L_{0t} + K_{1t}L_{1t}}, \tag{7.42}$$

$$A_0 = \ln\sqrt{a_0^2 + a_1^2}, A_1 = arctg\frac{a_1}{a_2}. \tag{7.43}$$

Now coefficients of this model can be found by LSM. With reference to this model the problem becomes trivial, since it is necessary to solve a system of equations with three unknowns:

$$\begin{cases} \sum_t \ln R_g = TA_0 + \alpha\sum_t \ln R_k, \\ \sum_t \theta_g = TA_1 + \alpha\sum_t \theta_k, \\ \sum_t \ln R_g \ln R_k + \sum_t \theta_g\theta_k = A_0\sum_t \ln R_k + A_1\sum_t \theta_k + \alpha\left(\sum_t \ln^2 R_k + \sum_t \theta_k^2\right). \end{cases} \tag{7.44}$$

E.V. Sirotina built a Cobb-Douglas-type production classification function for statistical data for the period from 1999 to 2006 on the company Lenoblgaz plc. Since these data represent a commercial secret, we will not present them here. However, in accordance with the statistical data, after reduction of the variables to a single scale and calculation of all the intermediate variables, a system of LSM equations (7.44) was obtained, which, with respect to this company, will have the following form:

$$\begin{cases} 14.21 = 8A_0 + 19.07\alpha, \\ 7.37 = 8A_1 + 1.59\alpha, \\ 32.65 = 14.21A_0 - 1.59A_1 + 46.68\alpha, \end{cases}$$

where it is easy to find the required coefficient values: $A_0 = 0.084$, $A_1 = 1.062$, $\alpha = 0.71$. Using the known values of coefficients A_0 and A_1 one can find the original values of coefficients a_0 and a_1:

$$a_0 + ia_1 = e^{0.084+i1.062} = 0.53 + i0.95$$

Then the model of the Cobb-Douglas-type classification production function for Lenoblgazfor the period under consideration has the following form:

$$G_t + iC_t = (0.53 + i0.95)(K_0 + iK_1)^{0.71}(L_0 + iL_1)^{0.29}. \tag{7.45}$$

It is evident that there were no difficulties in building such a function with two complex variables. Moreover, the Cobb-Douglas-type classification production function turned out to be easy to apply in practice.

7.4 Elasticity and Other Characteristics of a Classification Production Complex-Valued Function

According to its properties, the Cobb-Douglas-type classification production function is more suitable for economic reality than the Cobb-Douglas function as it describes most actually existing production relationships between resources and production performance. It is obvious that it also describes the relationships that are unattainable in the sphere of real-variable models.

However, to consciously use this new mathematical tool for economic modeling it is necessary to examine its properties more carefully. For this we will use the standard set of tools used by specialists for studying properties of "neoclassical" production functions and the Cobb-Douglas function as a kind of "neoclassical" production function. The following conditions will hold for the power production function model with real variables with exponents lying within a range from zero to one:

1. In the absence of one of the resources, production is impossible in general, which is expressed mathematically as follows:

$$F(0, L) = F(K, 0) = 0. \tag{7.46}$$

2. With unlimited increase in one of the resources the output increases infinitely, i.e.,

$$F(K,\infty) = F(\infty, L) = \infty\sqrt{b^2 - 4ac}. \tag{7.47}$$

3. With growth in resources output grows, which means that the first derivatives are positive:

$$\frac{dQ}{dK} > 0, \ \frac{dQ}{dL} > 0. \tag{7.48}$$

4. With an increase in resources the rate of output increase slows down, which means that the second derivative is negative:

$$\frac{\partial^2 Q}{\partial K^2} < 0, \ \frac{\partial^2 Q}{\partial L^2} < 0. \tag{7.49}$$

These properties follow from the mathematical form of the power "neoclassical" production function, which is considered by economists as important arguments for practical applicability of the model.

Since in this section we are studying the properties of one of the simplest classification production functions – Cobb-Douglas-type of functions with complex variables (7.26) – let us look for similar properties of this function.

The Cobb-Douglas-type classification production function has the following form:

$$G_t + iC_t = (a_0 + ia_1)(K_0 + iK_1)^{\alpha}(L_0 + iL_1)^{1-\alpha}.$$

Here the exponent lies within the range $0 \leq \alpha \leq 1$, which makes it possible to consider the function as a certain analog of the well-known Cobb-Douglas function.

It is evident that if at least one of the complex resources is equal to zero, the output modeled by this function will also be equal to zero, i.e., condition (7.46) is satisfied.

Theoretically it can happen that one component of a complex resource is equal to zero and the other is not. In this case (7.46) does not hold, for example, $L_0 = 0$, $L_1 > 0$. However, these cases do not occur in actual economic practice. It is impossible to imagine a situation where industrial and production workers are absent at a company and nonproduction personnel works, or, conversely, where a company has industrial and production personnel but does not have managers, accountants, or a general director. The same can be said about the fixed assets of an enterprise – if there is no path (K_0) to the factory floor (K_0), the workers will not get to the floor.

This is why the real economy reflects a situation where both the real and imaginary components of a complex result are equal or not equal to zero simultaneously. Other variants refer to idealization where the object models are associated with properties that are not inherent in those models. While building models, we use abstraction but not idealization.

The second property i obvious: with an unlimited increase in one of the complex resources – capital or labor –output increases infinitely. Growth in the modulus of the complex resource in the model is shown by growth in the modulus of the production result.

This means that the first and second properties inherent in "neoclassical" models are also inherent in the model under consideration, which makes the analogy between them more sound.

Now let us find the signs of the first and second derivatives of function (7.26) to verify the feasibility of conditions (7.48) and (7.49). E.V. Sirotina made certain calculations, which we can use. First, to simplify calculation of the derivatives, let us represent the model of production function (7.26) in exponential form:

$$G + iC = R_a e^{i\theta_a} (R_K e^{i\theta_K})^{\alpha} (R_L e^{i\theta_L})^{1-\alpha}, \tag{7.50}$$

where

$$R_a = \sqrt{a_0^2 + a_1^2}, \quad \theta_a = arctg\frac{a_1}{a_0}, \quad R_K = \sqrt{K_0^2 + K_1^2}, \quad \theta_K = arctg\frac{K_1}{K_0},$$
$$R_L = \sqrt{L_0^2 + L_1^2}, \quad \theta_L = arctg\frac{L_1}{L_0}.$$

In this form (7.50), the model's modulus and polar angle are easily grouped:

$$G + iC = (R_a R_K^{\alpha} R_L^{1-\alpha}) e^{i(\theta_a + i\alpha\theta_K + i(1-\alpha)\theta_L)}. \tag{7.51}$$

This is why the model of a Cobb-Douglas-type classification production function can be represented in trigonometric form:

$$G + iC = R_a R_K^{\alpha} R_L^{1-\alpha} [\cos(\theta_a + \alpha\theta_K + (1-\alpha)\theta_L) + i\sin(\theta_a + \alpha\theta_K + (1-\alpha)\theta_L)]. \tag{7.52}$$

This makes it possible to calculate the first and second partial derivatives of the function with respect to resources.

According to the d'Alembert-Euler (Riemann-Cauchy) condition, to find a derivative of a complex-valued function, it is sufficient to take the derivatives with respect to their real or imaginary part. The real part of model (7.52) has the following form:

$$G + iC = R\cos\theta = U. \tag{7.53}$$

This is why the derivative of function (7.26) with respect to the complex capital resource, for example, can be found in the following way:

$$\frac{\partial(G+iC)}{\partial K}=\frac{\partial U}{\partial K_0}-i\frac{\partial U}{\partial K_1}=\frac{\partial(R\cos\theta)}{\partial K_0}-i\frac{\partial(R\cos\theta)}{\partial K_1}. \tag{7.54}$$

Let us calculate the first component of derivative (7.54), namely, $\frac{\partial(R\cos\theta)}{\partial K_0}$, as a derivative of a complex function:

$$\frac{\partial(R\cos\theta)}{\partial K_0}=\frac{\partial R}{\partial K_0}\cos\theta+\frac{\partial\cos\theta}{\partial K_0}R. \tag{7.55}$$

The first term is

$$\frac{\partial R}{\partial K_0}\cos\theta=\frac{\partial R_a R_K^{\alpha}R_L^{1-\alpha}}{\partial K_0}\cos\theta=\alpha R_a R_L^{1-\alpha}R_K^{\alpha-1}\frac{K_0}{\sqrt{K_0^2+K_1^2}}\cos\theta. \tag{7.56}$$

Since $\sqrt{K_0^2+K_1^2}=R_K$, we finally get

$$\frac{\partial R}{\partial K_0}\cos\theta=\frac{\partial R_a R_K^{\alpha}R_L^{1-\alpha}}{\partial K_0}\cos\theta=\alpha R_a R_L^{1-\alpha}R_K^{\alpha-2}K_0\cos\theta. \tag{7.57}$$

The second term (7.56) is a derivative of the cosine of the argument with respect to K_O. It can also be found:

$$\begin{aligned}\frac{\partial\cos\theta}{\partial K_0}R&=\frac{\partial\cos(\theta_a+\alpha\theta_K+(1-\alpha)\theta_L)}{\partial K_0}R=-R\sin\theta\frac{\partial(\alpha\theta_K)}{\partial K_0}\\&=-R\sin\theta\alpha\frac{\partial(arctg\frac{K_0}{K_1})}{\partial K_0},\end{aligned}$$

where we finally obtain

$$\begin{aligned}\frac{\partial\cos\theta}{\partial K_0}R&=-\alpha R\sin\theta\frac{-K_1}{K_0^2(1+\frac{K_1^2}{K_0^2})}=\alpha R_a R_K^{\alpha}R_L^{1-\alpha}\sin\theta\frac{K_1}{K_0^2+K_1^2}\\&=\alpha R_a R_K^{\alpha-2}R_L^{1-\alpha}K_1\sin\theta.\end{aligned} \tag{7.58}$$

Substituting (7.57) and (7.58) to (7.55) we obtain a partial derivative of the classification production function with respect to the fixed capital:

$$\begin{aligned}\frac{\partial(R\cos\theta)}{\partial K_0}&=\alpha R_a R_L^{1-\alpha}R_K^{\alpha-2}K_0\cos\theta+\alpha R_a R_K^{\alpha-2}R_L^{1-\alpha}K_1\sin\theta\\&=\alpha R_a R_L^{1-\alpha}R_K^{\alpha-2}(K_0\cos\theta+K_1\sin\theta).\end{aligned} \tag{7.59}$$

In the same way we can find a partial derivative of the Cobb-Douglas-type classification production function: $\frac{\partial(R\cos\theta)}{\partial K_1}$. Since this derivative is a derivative of a complex function, it can be calculated as follows:

$$\frac{\partial(R\cos\theta)}{\partial K_1} = \frac{\partial R}{\partial K_1}\cos\theta + \frac{\partial\cos\theta}{\partial K_1}R. \tag{7.60}$$

The first term (7.60) as the modulus derivative will be recorded as follows:

$$\frac{\partial R}{\partial K_1}\cos\theta = \alpha R_a R_L^{1-\alpha} R_K^{\alpha-1}\frac{K_1}{\sqrt{K_0^2+K_1^2}}\cos\theta = \alpha R_a R_L^{1-\alpha} R_K^{\alpha-2} K_1\cos\theta. \tag{7.61}$$

The second term (7.60) represents the derivative of the cosine of the argument with respect to subsidiary capital K_I. This derivative also can be derived as follows:

$$\frac{\partial\cos\theta}{\partial K_1}R = \frac{\partial\cos(\theta_a+\alpha\theta_K+(1-\alpha)\theta_L)}{\partial K_1}R = -R\sin\theta\alpha\frac{\partial(arctg\frac{K_0}{K_1})}{\partial K_1}.$$

After calculating the derivative of the arctangent we finally have

$$\begin{aligned}\frac{\partial\cos\theta}{\partial K_1}R &= -\alpha R\sin\theta\frac{1}{K_0\left(1+\frac{K_1^2}{K_0^2}\right)} = -\alpha R_a R_K^{\alpha} R_L^{1-\alpha}\sin\theta\frac{K_0}{K_0^2+K_1^2}\\ &= -\alpha R_a R_K^{\alpha-2} R_L^{1-\alpha} K_0\sin\theta.\end{aligned} \tag{7.62}$$

Substituting (7.61) and (7.62) into (7.60) we obtain a partial derivative of the classification production function with respect to subsidiary capital:

$$\begin{aligned}\frac{\partial(R\cos\theta)}{\partial K_1} &= \alpha R_a R_L^{1-\alpha} R_K^{\alpha-2} K_1\cos\theta - \alpha R_a R_K^{\alpha-2} R_L^{1-\alpha} K_0\sin\theta\\ &= \alpha R_a R_L^{1-\alpha} R_K^{\alpha-2}(K_1\cos\theta - K_0\sin\theta).\end{aligned} \tag{7.63}$$

Since we have all the terms we need to obtain a derivative of the classification function with respect to complex capital, we can do it:

$$\begin{aligned}\frac{\partial(G+iC)}{\partial K} &= \frac{\partial U}{\partial K_0} - i\frac{\partial U}{\partial K_1} = \frac{\partial R\cos\theta}{\partial K_0} - i\frac{\partial R\cos\theta}{\partial K_1}\\ &= \alpha R_a R_L^{1-\alpha} R_K^{\alpha-2}(K_0\cos\theta + K_1\sin\theta) - i\alpha R_a R_L^{1-\alpha} R_K^{\alpha-2}(K_1\cos\theta - K_0\sin\theta)\\ &= \alpha R_a R_L^{1-\alpha} R_K^{\alpha-2}[(K_0 - iK_1)\cos\theta + (K_1 + iK_0)\sin\theta].\end{aligned} \tag{7.64}$$

In the second term of the last factor of the resulting expression we can take the imaginary unit out and get

$$\begin{aligned}\frac{\partial(G+iC)}{\partial K} &= \alpha R_a R_L^{1-\alpha} R_K^{\alpha-2}[(K_0 - iK_1)\cos\theta + i(K_0 - iK_1)\sin\theta] \\ &= \alpha R_a R_L^{1-\alpha} R_K^{\alpha-2}(K_0 - iK_1)(\cos\theta + i\sin\theta).\end{aligned} \tag{7.65}$$

Since $\cos\theta + i\sin\theta = e^{i\theta}$, (7.47) can be written in more convenient form:

$$\frac{\partial(G+iC)}{\partial K} = (a_0 + ia_1)(K_0 + iK_1)^{\alpha}(L_0 + iL_1)^{1-\alpha}[\alpha R_K^{-2}(K_0 - iK_1)]. \tag{7.66}$$

Let us simplify the expression in square brackets:

$$\alpha R_K^{-2}(K_0 - iK_1) = \alpha\frac{K_0 - iK_1}{K_0^2 + K_1^2} = \frac{\alpha}{K_0 + iK_1} = \alpha(K_0 + iK_1)^{-1}. \tag{7.67}$$

Substituting this value into (7.66) we obtain the final formula for the first derivative of the classification production function with respect to complex capital:

$$\frac{\partial(G+iC)}{\partial K} = \alpha(a_0 + ia_1)(K_0 + iK_1)^{\alpha-1}(L_0 + iL_1)^{1-\alpha}. \tag{7.68}$$

After carrying out similar difficult calculations we obtain the resulting formula of the first derivative of the classification production function of complex variables with respect to complex labor:

$$\frac{\partial(G+iC)}{\partial L} = (1-\alpha)(a_0 + ia_1)(K_0 + iK_1)^{\alpha}(L_0 + iL_1)^{-\alpha}. \tag{7.69}$$

Both (7.68) and (7.69) represent a complex number consisting of real and imaginary parts. And, as is known, complex numbers can be neither positive nor negative. This requirement (of positivity, for example) can be made either with respect to the modulus of the complex variable or to its polar angle, to its real or imaginary part. This means that with reference to this complex-valued production function, condition (7.48) on its first derivative will not hold.

Therefore, let us consider the essence of the first derivatives of the function and then draw a conclusion.

Since by definition $0 \le \alpha \le 1$ and all the production resources lie in the first quadrant of the complex plane, the modeled result will lie in the first quadrant of the complex plane of the production result, subject to the nonnegativity of each of the coefficients of the complex proportionality coefficient:

$$a_0 \ge 0, a_1 \ge 0. \tag{7.70}$$

One should remember that positivity of the first derivatives of the "neoclassical" production function testifies to the fact that positive growth in resources leads to positive growth in production performance.

In the case under consideration, if any of the conditions (7.70) in the Cobb-Douglas-type classification production function of complex variables is violated, the modeled result will not fall into the first quadrant. Therefore, the use of some resource worsens the production result and vice versa – a reduction in the labor or capital resource has a favorable effect on production performance. Such cases exist in the economy, which is why one can say that the model of a Cobb-Douglas-type classification production function of complex-variable models more diverse alternatives than real-variable models. But if researchers are going to follow a strict rule – an increase in resources should lead to an increase in production performance, then the Cobb-Douglas-type classification model, besides restrictions on the exponent $0 \leq \alpha \leq 1$, should be supplemented by condition (7.70). But, for Cobb-Douglas functions the condition of positivity of the proportionality coefficient is extended in and of itself.

Let us calculate the second derivative with respect to complex resources. Omitting those calculations that are similar to the preceding ones, we present the final formulas of the second derivative of the Cobb-Douglas-type of the classification production function of complex variables with respect to its complex resources.

The second derivative of this function with respect to the complex capital will have the following form:

$$\frac{\partial^2(G+iC)}{\partial K^2} = \alpha(\alpha-1)(a_0+ia_1)(K_0+iK_1)^{\alpha-2}(L_0+iL_1)^{1-\alpha}. \tag{7.71}$$

Since the exponent of this function is positive $0 \leq \alpha \leq 1$, as the coefficients of the complex proportionality coefficient are positive (7.70), the second term (7.71) will be negative, and with all the other factors positive, the modeled result will fall into the third quadrant of the complex plane, i.e., both the real and imaginary parts of the second production complex-valued function are negative.

The second derivative of the function under consideration with respect to complex labor resources will be equal to

$$\frac{\partial^2(G+iC)}{\partial L^2} = -\alpha(1-\alpha)(a_0+ia_1)(K_0+iK_1)^{\alpha}(L_0+iL_1)^{-\alpha-1}. \tag{7.72}$$

For the same reasons, all the factors of (7.72) are positive, and since there is a minus sign in their product, both the real and imaginary parts of the second derivative with respect to complex labor will be negative.

Thus, when we set conditions

$$0 \leq \alpha \leq 1, \quad a_0 \geq 0, a_1 \geq 0, \tag{7.73}$$

for the Cobb-Douglas-type classification production function of complex variables, it acquires properties similar to those of "neoclassical" production functions of real variables.

Then it is necessary to calculate the elasticity of the Cobb-Douglas-type classification production function of complex variables. This is easy to do since the calculation of the first derivatives of this function with respect to complex capital and complex labor was performed previously.

Since the formula of the elasticity in the sphere of real numbers has the following form:

$$\varepsilon_x = \frac{\partial y}{\partial x}\frac{x}{y},$$

it can also be used for models of complex variables, as was done in previous chapters, if we know the first derivatives of the complex-valued function.

Then the elasticity of the Cobb-Douglas-type classification production function of complex variables with respect to complex capital will have the following form:

$$\varepsilon_K = \frac{\partial(G+iC)}{\partial K}\frac{K}{G+iC} = \alpha(a_0+ia_1)(K_0+iK_1)^{\alpha-1}(L_0+iL_1)^{1-\alpha}\frac{K}{G+iC} \tag{7.74}$$

or, with obvious reductions,

$$\varepsilon_K = \alpha. \tag{7.75}$$

Similarly, we can find the elasticity with respect to complex labor:

$$\varepsilon_L = \frac{\partial(G+iC)}{\partial L}\frac{L}{G+iC} = (1-\alpha)(a_0+ia_1)(K_0+iK_1)^{\alpha-1}(L_0+iL_1)^{-\alpha}\frac{L}{G+iC}, \tag{7.76}$$

where finally we have

$$\varepsilon_L = (1-\alpha). \tag{7.77}$$

The conclusion that follows from the foregoing results is clear – in the Cobb-Douglas-type classification production function of complex variables the exponents play the same role as the exponents in the Cobb-Douglas production function: they are parameters of the elasticity of the production result with respect to the appropriate complex resource.

Since the proposed Cobb-Douglas-type classification production function represents each research as a complex variable, one can say that the function itself, besides elasticity of the complex resource as a whole, also has elasticity of each of the resource components, real and imaginary.

Let us derive a formula of elasticity of the Cobb-Douglas-type classification production function with respect to fixed capital. For that, let us first calculate the first partial derivative of the production complex-valued function with respect to the fixed capital using the same designations as previously:

$$\frac{\partial(G+iC)}{\partial K_0}=\frac{\partial(R\cos\theta)}{\partial K_0}+i\frac{\partial(R\sin\theta)}{\partial K_0}. \tag{7.78}$$

Previously, in (7.59), for the first term we had

$$\frac{\partial(R\cos\theta)}{\partial K_0}=\alpha R_a R_L^{1-\alpha}R_K^{\alpha-2}(K_0\cos\theta+K_1\sin\theta)$$

and for the second term (7.78)

$$\frac{\partial(R\sin\theta)}{\partial K_0}=\alpha R_a R_L^{1-\alpha}R_K^{\alpha-2}(K_0\sin\theta-K_1\cos\theta). \tag{7.79}$$

Substituting the results obtained into (7.78) we can find the first partial derivative of the Cobb-Douglas-type classification production function with respect to fixed capital:

$$\frac{\partial(G+iC)}{\partial K_0}=\alpha R_a R_L^{1-\alpha}R_K^{\alpha-2}[(K_0\cos\theta+K_1\sin\theta)+i(K_0\sin\theta-K_1\cos\theta)]. \tag{7.80}$$

Simplifying the expression we have

$$\frac{\partial(G+iC)}{\partial K_0}=\alpha(a_0+ia_1)(K_0+iK_1)^{\alpha}(L_0+iL_1)^{1-\alpha}\frac{K_0-iK_1}{K_0^2+K_1^2}, \tag{7.81}$$

where we can find the elasticity of the Cobb-Douglas-type classification production function of complex variables with respect to fixed capital:

$$\varepsilon_{K_0}=\frac{\partial(G+iC)}{\partial K_0}\frac{K_0}{G+iC}=\alpha\frac{K_0^2-iK_0K_1}{K_0^2+K_1^2}=\alpha\left(\frac{K_0^2}{K_0^2+K_1^2}-i\frac{K_0K_1}{K_0^2+K_1^2}\right). \tag{7.82}$$

This is a complex coefficient. Let us give it an economic interpretation.

If the fixed capital increases by one, the real part of the complex production result, i.e., the gross margin, increases by less than α since the real part of the complex coefficient is less than unity. With an increase in the fixed capital to infinity, an increase in the fixed capital by unity leads to an increase in the gross margin by α.

The imaginary component of the complex coefficient of elasticity of the production result for the fixed capital is negative. This means that the imaginary component of the production result, i.e., gross costs, decreases by a value lower than α. When the fixed capital tends to infinity, the costs stop changing.

In a similar way we can calculate the first derivative of the Cobb-Douglas-type classification production function for subsidiary capital and on the basis of these calculations determine the formula for calculating the elasticity of this resource. Omitting intermediary calculations, we can present the final formula for the elasticity:

$$\varepsilon_{K_1} = \frac{\partial(G+iC)}{\partial K_1}\frac{K_1}{G+iC} = \alpha\frac{K_1^2+iK_0K_1}{K_0^2+K_1^2} = \alpha\left(\frac{K_1^2}{K_0^2+K_1^2}+i\frac{K_0K_1}{K_0^2+K_1^2}\right). \quad (7.83)$$

This can also receive an economic interpretation.

An increase in the subsidiary capital leads to an increase in the gross margin as the real component and the gross cost as the imaginary component of the production result. However, with an increase in the subsidiary capital to infinity, the real component of the complex coefficient of elasticity tends to a, and the imaginary one tends to zero. This means that with high values of the subsidiary capital its growth leads to growth in the gross margin by α without influencing the production costs.

The sum of the complex coefficient of the elasticity of fixed capital (7.82) and that of the subsidiary capital (7.83) will give the value of the total elasticity of complex capital:

$$\varepsilon_{K_0} + \varepsilon_{K_1} = \alpha\left(\frac{K_0^2}{K_0^2+K_1^2} - i\frac{K_0K_1}{K_0^2+K_1^2} + \frac{K_1^2}{K_0^2+K_1^2} + i\frac{K_0K_1}{K_0^2+K_1^2}\right) = \alpha.$$

In a similar way we derive the formulas for elasticity of the real part of complex labor:

$$\varepsilon_{Lo} = \frac{(1-\alpha)\cdot L_o{}^2}{L_o{}^2+L_1{}^2} - i\frac{(1-\alpha)\cdot L_o\cdot L_1}{L_o{}^2+L_1{}^2} \quad (7.84)$$

and of the imaginary part:

$$\varepsilon_{L1} = \frac{(1-\alpha)\cdot L_1{}^2}{L_o{}^2+L_1{}^2} + i\frac{(1-\alpha)\cdot L_o\cdot L_1}{L_o{}^2+L_1{}^2}. \quad (7.85)$$

Then these coefficients show how an increase in industrial and production workers or nonproduction workers affect the gross margin and costs. The contribution of each component is determined by both the exponents and particular values of the hired labor.

The sum of these partial coefficients of elasticity will give us the total elasticity of complex labor:

$$\varepsilon_L = \varepsilon_{Lo} + \varepsilon_{L1} = \frac{(1-\alpha)\cdot L_o{}^2 + (1-\alpha)\cdot L_1{}^2}{L_o{}^2+L_1{}^2} = 1-\alpha.$$

Analysis of partial coefficients of elasticity shows that they vary with variation in the volumes of resources use, i.e., return on resources varies with the resource scale, their sum remaining unchanged:

$$\varepsilon_{K_0} + \varepsilon_{K_1} = \alpha, \quad \varepsilon_{Lo} + \varepsilon_{L1} = 1 - \alpha.$$

We will not give an economic interpretation to this interesting property since careful study of the economic meaning of model parameters is not the focus of this work.

7.5 Classification Power Production Function

The artificial restriction imposed on the limits of variation of exponent α in the Cobb-Douglas-type classification production function also limits the application of this model. This restriction (7.27) must simplify considerably the problem of finding model coefficients, but such simplifications almost always worsens the model approximation properties. This is why actual production and other economic processes will be more adequately described by models that do not impose any a priori restrictions on their coefficients, unless it follows from the core of the modeled economic situation. Removal of limitations on the domain of the exponents of the power classification function extends its capacities.

Thus, E.V. Sirotina estimated the coefficients of one of the alternatives of the power classification production function using as an example the company Lengaz – a function with real exponents – and obtained the following model:

$$G_t + iC_t = (2.34 + i3.65)(K_0 + iK_1)^{0.44}(L_0 + iL_1)^{0.50}.$$

This model describes the company's production process more precisely than the system of power models of real variables describing the gross margin and production costs separately.

We will not concentrate on this type of model since its properties are similar to those determined for the Cobb-Douglas-type classification production function considered in the previous section, but we will pay attention to a model of another type – classification power production functions with complex exponents:

$$G_t + iC_t = (a_0 + ia_1)(K_0 + iK_1)^{(b_0+ib_1)}(L_0 + iL_1)^{(c_0+ic_1)}. \tag{7.86}$$

Let us verify the uniformity of this function in accordance with rule (7.28):

$$\begin{aligned}&(a_0 + ia_1)(\lambda K_0 + i\lambda K_v)^{(b_0+ib_1)}(\lambda L_0 + i\lambda L_v)^{(c_0+ic_1)} = \\ &\lambda^{(b_0+ib_1)}\lambda^{(c_0+ic_1)}(a_0 + ia_1)(K_0 + iK_v)^{(b_0+ib_1)}(L_0 + iL_v)^{(c_0+ic_1)} = \\ &\lambda^{(b_0+ib_1)+(c_0+ic_1)}(a_0 + ia_1)(K_0 + iK_v)^{(b_0+ib_1)}(L_0 + iL_v)^{(c_0+ic_1)},\end{aligned} \tag{7.87}$$

i.e., this function is nonuniform with reference to the complex exponent

$$(b_0 + ib_1) + (c_0 + ic_1). \tag{7.88}$$

Setting various exponent values, one can obtain its most diverse forms. For example, a Cobb-Douglas-type classification production function will be obtained if the sum (7.88) is equal to unity and it will become uniform to the first power.

It is logical to consider production functions that are linearly pseudo uniform, namely, uniform to power i. This can happen if (7.88) gives in sum an imaginary unity:

$$(b_0 + ib_1) + (c_0 + ic_1) = i.$$

Let us designate the exponent of this function as β. Then the pseudo-uniform Cobb-Douglas-type classification production function will have another form:

$$G_t + iC_t = (a_0 + ia_1)(K_0 + iK_1)^{i\beta}(L_0 + iL_1)^{i(1-\beta)} \tag{7.89}$$

with the restrictions

$$0 \leq \beta \leq 1. \tag{7.90}$$

To understand the core of the behavior of the pseudo-uniform Cobb-Douglas-type classification production function, let us represent it in exponential form:

$$G_t + iC_t = R_a e^{i\theta_a} R_K^{i\beta} e^{i\theta_K{}^{i\beta}} R_L^{i(1-\beta)} e^{i\theta_L{}^{i(1-\beta)}}. \tag{7.91}$$

Here we use previously introduced designations:

$$R_a = \sqrt{a_0^2 + a_1^2},\ \theta_a = arctg\frac{a_1}{a_0}, \tag{7.92}$$

$$R_K = \sqrt{K_0^2 + K_1^2},\ \theta_K = arctg\frac{K_1}{K_0}, \tag{7.93}$$

$$R_L = \sqrt{L_0^2 + L_1^2},\ \theta_L = arctg\frac{L_1}{L_0}. \tag{7.94}$$

If we transform the right-hand side of model (7.91) so that we can calculate the modulus of the right-hand side of the complex variable and its polar angle, we will have

$$G_t + iC_t = R_a e^{-\beta\theta_K} e^{\theta_L(\beta-1)} e^{\ln R_K^{i\beta}} e^{\ln R_L^{i(1-\beta)}} e^{i\theta_a}. \tag{7.95}$$

Since

$$e^{\ln R_K^{i\beta}} = e^{i\beta \ln \sqrt{K_0^2+K_1^2}}, \quad \text{and} \quad e^{\ln R_L^{i(1-\beta)}} = e^{i(1-\beta)\ln \sqrt{L_0^2+L_1^2}},$$

the gross margin, according to the pseudo-uniform Cobb-Douglas-type classification production function, will be calculated as follows:

$$G_t = R_a e^{(\beta-1)arctg\frac{L_1}{L_0} - \beta arctg\frac{K_1}{K_0}} \cos\left(\theta_a + \beta \ln \sqrt{K_0^2 + K_1^2} + (1-\beta)\ln \sqrt{L_0^2 + L_1^2}\right). \tag{7.96}$$

Accordingly, the production costs will be calculated in compliance with the formula

$$C_t = R_a e^{(\beta-1)arctg\frac{L_1}{L_0} - \beta arctg\frac{K_1}{K_0}} \sin\left(\theta_a + \beta \ln \sqrt{K_0^2 + K_1^2} + (1-\beta)\ln \sqrt{L_0^2 + L_1^2}\right). \tag{7.97}$$

Let us now study what aspect of production is being modeled by this pseudo-uniform Cobb-Douglas-type classification production function. With the assumption of unchanged complex labor resources we will get the following character of the effect of the complex capital resource on production performance:

1. Growth in investments in fixed capital K_0 leads to a decrease in the polar angle of these resources and an increase in the modulus of the capital resources. The gross margin in (7.96) represents the multiplication of two varying factors – growing exponent $e^{(\beta-1)arctg\frac{L_1}{L_0} - \beta arctg\frac{K_1}{K_0}}$ and decreasing cosine $\cos(\theta_a + \beta \ln \sqrt{K_0^2 + K_1^2} + (1-\beta)\ln \sqrt{L_0^2 + L_1^2})$. Depending on the ratio between the fixed and noncore capital, as well as on the values of the complex proportionality coefficient and exponent β, it is possible to model an increase in profits, the stability of its values, and a decrease in the volumes. However, for the production costs in this situation two trends are multiplied – the exponent grows with the growing sine of the angle. This is why it follows from (7.96) that costs will definitely grow.
2. If investments in nonfixed capital grow, the polar angle grows; therefore, the exponent where this angle is used with the "minus" sign decreases. The argument of the cosine component increases with the increase in the modulus of capital resources, which means a decrease in this very factor. Thus, two decreasing trends are multiplied. This means that investments that made in nonfixed capital lead to a decrease in the gross margin. As for production costs, it is shown that if the exponential component decreases, the sine of the increasing angle increases. Therefore, multiplication of these two factors can result in different trends for different coefficients.

The same conclusions can be drawn on the character of influence of the other complex variable – labor – on production performance. With an increase in the number of industrial and production workers, costs will grow, but changes in the gross margin can vary widely but an increase in nonproduction workers will lead to a decrease in the gross margin and a varied behavior of costs.

Since the relationships modeled by the pseudo-uniform Cobb-Douglas-type classification production function are in agreement with many actual production processes, one can say that this function can be used in economic modeling. However, since in this model a priori assumptions are made about its uniformity and restrictions on the exponent signs (7.90), application of this model will be limited. This is why we will not pay too much attention to its properties, as we did with respect to the pseudo-uniform Cobb-Douglas-type classification production function in the previous section.

A model where any restrictions on the uniformity of model (7.86) and the coefficient signs are removed will be considered universal. The application area of this model is extended though it gets much more complicated and the procedure for calculating complex coefficients becomes more difficult.

Taking logarithms of the right- and left-hand sides of equality (7.86) we obtain the following model:

$$\ln(G_t + iC_t) = \ln(a_0 + ia_1) + (b_0 + ib_1)\ln(K_0 + iK_1) + (c_0 + ic_1)\ln(L_0 + iL_1). \tag{7.98}$$

We do not foresee any theoretical problems in connection with the estimation of coefficients of this model on the basis of statistical data – everything is quite simple. On the other hand, to do this, let us solve a system of six equations with six unknowns (if we work with real variables) or a system of three complex-valued LSM equations. Here, it is necessary to observe the polar angles of the variables since most modeling programs calculate angles by cutting off their values in a range of 0–2π, which can cause errors in the estimation of model coefficients.

In any case, if economists end up working with the power classification production function of complex variables, they will achieve more adequate modeling of production processes and of the influence of the active and passive parts of the production resources than they will if they use real-variable models.

It is obvious that besides power models other classification types can be used – exponential, logarithmic, or a combination of the two. It is crucial that the number of resources used is limited only by the computational capacities of the technology and problem context.

7.6 The Shadow Economy and Its Modeling by Means of Complex-Valued Functions

Using the approach of singling out active and passive parts, one can classify economic processes into two groups – legal and shadow economies. It is evident that legal economy indicators should be assigned to the real parts of economic variables and shadow economy parameters to the imaginary components. On this basis we can build diverse models describing the shadow economy. Let us consider one of them.

Let us introduce the following variables:

K_0 – cost of fixed assets shown in statistical reports;
K_1 – cost of fixed assets used in illegal production;
L_0 – number of workers involved in production;
L_1 – number of employees involved in the shadow economy;
Q_0 – gross domestic product as shown in official reports;
Q_1 – gross domestic product of the shadow economy.

The simplest power model taking into account the shadow business has the following form:

$$Q_0 + iQ_1 = a(K_0 + iK_1)^{\alpha}(L_0 + iL_1)^{\beta}, \tag{7.99}$$

where α and β are function exponents without any restrictions on values and signs.

From the point of view of best approximation of socioeconomic phenomena and taking into account shadow processes, the power production function of complex variables with complex coefficients will be more precise:

$$Q_0 + iQ_1 = (a_0 + ia_1)(K_0 + iK_1)^{b_0+ib_1}(L_0 + iL_1)^{c_0+ic_1}. \tag{7.100}$$

Its coefficients can be found by LSM, as was already mentioned. However, from the point of view of the economic core of the modeled process, the real coefficient model (7.99) turns out to be more feasible.

Indeed, if the values for the fixed assets and workers involved in the shadow economy are equal to zero, the output in the shadow economy will also be equal to zero. Model (7.99) explains this fact: if $K_1 = 0$ and $L_1 = 0$, then it is evident that $Q_1 = 0$. And in a model with complex-valued coefficients (7.100), if the shadow resources are equal to zero, the real variable, put to a complex power, will result in a complex variable and calculate the volume of shadow products associated with the imaginary part and not equal to zero. This means that this model is not as precise in showing real processes. This is the first reason.

The second reason for preferring model (7.99) over model (7.100) stems from the nature of the original data. Indeed, statistical data on the legal economy, that is, their order and trends, though full of mistakes, are not very distorted by these mistakes. However, information on the shadow economy and its components cannot be taken into account in a similar way by definition. Here we must use expert evaluations whose accuracy is, of course, not great. Under these conditions, when the original data are not exact, it is pointless to complicate the model in order to increase its accuracy. This is another reason why we would prefer the simple model to model (7.100).

Let us make a preliminary analysis of the properties of the proposed model of economic dynamics taking into account the shadow economy. In exponential form, model (7.99) is transformed into

$$Q_0 + iQ_1 = a(K_0^2 + K_1^2)^{\frac{\alpha}{2}}(L_0^2 + L_1^2)^{\frac{\beta}{2}} e^{i(\alpha arctg\frac{K_1}{K_0} + \beta\alpha arctg\frac{L_1}{L_0})}. \quad (7.101)$$

Here we can determine how the model takes into account the influence of the factors on the legal and shadow economies. The volume of officially shown production is

$$Q_0 = a(K_0^2 + K_1^2)^{\frac{\alpha}{2}}(L_0^2 + L_1^2)^{\frac{\beta}{2}} \cos\left(\alpha arctg\frac{K_1}{K_0} + \beta\alpha arctg\frac{L_1}{L_0}\right). \quad (7.102)$$

Similarly, one can determine how the proposed model describes illegal output:

$$Q_1 = a(K_0^2 + K_1^2)^{\frac{\alpha}{2}}(L_0^2 + L_1^2)^{\frac{\beta}{2}} \sin\left(\alpha arctg\frac{K_1}{K_0} + \beta\alpha arctg\frac{L_1}{L_0}\right). \quad (7.103)$$

We see that the model describes the influence of both legal and illegal resources on both the official and shadow economies. Does this occur in the real economy?

Indeed, L_1 who are employed in the shadow economy produce an illegal product that is legalized through official sales and shown as a part of real GDP Q_0, and those employed in legal production often produce, without suspecting as much, products hidden from taxation and included in the turnover in the shadow economy. In addition, the remuneration of people involved in illegal business cause them to direct most their earnings to satisfy their demands using legally produced goods.

Similarly, on fixed assets K_0 officially shown in balance sheets, workers produce products for shadow turnover, though in various "backstreet" shops and with the equipment assigned to the fixed assets of the shadow business K_1 they produce goods distributed in official turnover. Complex-valued power functions show these complex relationships.

Let us see what trends are described by the proposed model. Suppose that the situation in the region under study leads to an increase in the number of entrepreneurs involved in the shadow economy and the resources involved in legal production remain unchanged. This means that both capital K_1 used in illegal production and the number of employees in the shadow economy L_1 increase. According to (7.102), this leads to an increase in both factors $(K_0^2 + K_1^2)^{\frac{\alpha}{2}}$ and $(L_0^2 + L_1^2)^{\frac{\beta}{2}}$ and a decrease in the trigonometric factor – the cosine of the increasing angle. Then, how does strengthening of the shadow economy influence the legal economy? The answer to this question is determined by α and β exponent values. If they are less than one, then growth in the power components will occur to a lesser degree than the cosine growth modeling either stability or a slight decrease in legal production. The closer these parameters are to zero, the more negative is the influence of the shadow economy on the legal one.

If the exponents are greater than one, the power factors increase faster than the cosine decreases. Therefore, in this case we model a positive influence of the shadow economy on the results of the legal economy.

How do the resources involved in the shadow business influence its scope? The answer can be derived from the second component of model (7.103). The power factors grow with the growth in resources K_1 and L_1. The trigonometric factor also grows – the sine curve increases its values with the growth in the angle. Therefore, we model the implicit growth in shadow production whether the exponent is close to zero or greater than one.

Let us now assume again that the production situation makes pointless the development of the shadow production. Then entrepreneurs will fix it at a particular level and invest the resources in legal industries. This means that capital K_0 involved in legal production and the number of those employed in the economy L_0 increase. How will model (7.99) react to such resource variation?

The dynamics of legal production determined by the product of power factors and trigonometric factor (7.102) will be as follows. The power factors increase, and this increase is determined by the exponents. With the growth in K_0 and L_0 the cosine argument will decrease. The cosine of the decreasing angle will increase. Therefore, the scope of legal production will increase in any case.

And what happens with the shadow economy? The power components also increase, but the trigonometric factor – the sine curve – decreases. Therefore, depending on the exponent values and the ratio of the resources of the legal to those of the illegal economy, the scope of illegal production could increase, decrease, or remain unchanged. If the exponents are less than one, then growth in the power factors occurs to a smaller extent than the sine curve decrease.

As a result, the model correctly shows the possible alternatives of influence of the resources of the legal and shadow economies on each other.

According to the available data it is not difficult to build model (7.99) – we already did it in previous sections. However, the main difficulty we face when building model (7.99) is the absence of reliable information on the variables assigned to the shadow economy in Russia and other countries. To overcome this problem, I.S. Savkov suggested relying on the available expert information on Russia's shadow economy. Since we know how many people the experts believe are involved in Russia's shadow economy (the average estimate is 41 %) and how big the shadow economy is compared to the official GDP (the average expert estimate is 45 %), we can use these data for model building. However, the cost of the fixed assets (capital) involved in the shadow economy is unknown. We have not seen expert estimates of this value. To determine it we propose the following approach. According to the available statistical data on Russia we calculate the parameters of the power production function of real variables, which has the following form:

$$Q_t = 0.5241K_t^{0.569}L_t^{0.579}. \tag{7.104}$$

Since the exponents reflect the contribution of each resource to production performance, the shadow economy's technology differs very little from that of official production, especially since most of the shadow products are manufactured on the same production lines, we can use these coefficients for model (7.99):

$$Q_{0t} + iQ_{1t} = 0.5241(K_{0t} + iK_{1t})^{0.569}(L_{0t} + iL_{1t})^{0.579}, \tag{7.105}$$

where we can obtain an estimation of the fixed assets used in Russia's shadow economy. For that let us take the logarithms of the right- and left-hand sides of this equality (let us use the principal logarithm values):

$$\ln\sqrt{Q_{0t}^2 + Q_{1t}^2} + iarctg\frac{Q_{1t}}{Q_{0t}} = \ln 0.5241 + 0.569\left[\ln\sqrt{K_{0t}^2 + K_{1t}^2} + iarctg\frac{K_{1t}}{K_{0t}}\right]$$
$$+ 0.579\left[\ln\sqrt{L_{0t}^2 + iL_{1t}^2} + iarctg\frac{L_{1t}}{L_{0t}}\right].$$

Now we can calculate the unknown value of the capital in Russia's shadow economy.

Incidentally, the specific properties of complex variables allow for simplifying this problem. Since two complex numbers are equal to each other only when their real and imaginary parts are equal, then with reference to the exponential form, this means that the moduli of the complex numbers and their polar angles should be equal to each other. This is why from (7.105) one can obtain formulae that are more convenient for calculation:

$$\sqrt{Q_{0t}^2 + Q_{1t}^2}e^{iarctg\frac{Q_{1t}}{Q_{0t}}} = 0.5241(K_{0t}^2 + K_{1t}^2)^{\frac{0.569}{2}}(L_{0t}^2 + L_{1t}^2)^{\frac{0.579}{2}} \tag{7.106}$$

or

$$arctg\frac{Q_{1t}}{Q_{0t}} = 0.569 arctg\frac{K_{1t}}{K_{0t}} + 0.579 arctg\frac{L_{1t}}{L_{0t}}. \tag{7.107}$$

Solving any of the foregoing equations with respect to the fixed assets of the shadow economy K_{0t}, we see that lately they have made up 37.5% of the official value of the fixed assets of Russia. Thus, the model shows that about one-third of the fixed assets of the Russian economy exists in the shadow economy.

Indeed, the calculation results for the model obtained are mainly conditional since they are based on a number of simplifications and assumptions; and should relatively reliable variable values assigned to the shadow economy emerge, the model coefficients (7.105) should be recalculated and will certainly change. However, what is really important is that the model for classifying a complex-valued function can be used for economic modeling taking into account its shadow component. Models of real variables represent in this case a less fine tool for research and analysis.

7.7 Formation of Complex, Multifactor Models of Complex Variables

Classification models of complex variables expand considerably the capacities of a complex-valued economy. So far we have been looking at models of one type – linear and power models. However, in the real economy factors that affects performance need not follow only one particular law. This is why in real-variable models we see complex multifactor models such as, for example,

$$y_t = ax_{1t}^{\alpha}x_{2t}^{\beta} + bx_{3t}.$$

It is difficult to apply LSM directly to this model since we would obtain a system of four nonlinear equations with four unknown coefficients. This system can be solved by numeric methods, but it is easier, using the same numeric methods available from software packages, to find, by direct iteration, satisfactory coefficients corresponding to an LSM criterion. For that, for example, there is the "Search for Solution" function in Microsoft Excel.

However, it is not possible to use this function or a built-in algorithm in the software to search for coefficient values of a similar complex-valued model:

$$y_{rt} + iy_{it} = (a_0 + ia_1)(x_{1rt} + ix_{1it})^{\alpha}(x_{2rt} + ix_{2it})^{\beta} + (b_0 + ib_1)(x_{3rt} + ix_{3it}).$$

There is a need for building complex, nonlinear, multifactor, complex-valued models. For example, the volume of produced crops and livestock in agriculture is determined by the number of workers in these industries, land area allocated for land and pastures, number of machines for working the land, stock breeding, and other factors. To correctly model agricultural production it is necessary to build multifactor, complex-valued models, which will definitely not be linear.

In the fullness of time, a method of synthesizing monofactor models into multifactor ones was proposed to provide solutions to similar problems in the field of real variables.

Let us assume that a researcher has used some method, for example LSM, to build several monofactor dependences of parameter y on factors $k = 1,2,3,\ldots n$,

$$y = f_k(x_k) + \varepsilon_k. \tag{7.108}$$

Each of these models describes the behavior of the parameter with variance:

$$\sigma_k^2 = \frac{1}{n-1}\sum_{k=1}^{n}(y_k - f_k(x_k))^2.$$

Since in all n models (7.108) the left-hand sides of the equalities are equal to each other and equal to y, let us sum up their left- and right-hand sides n times:

$$ny = \sum_{k=1}^{n} f_k(x_k) + \sum_{k=1}^{n} \varepsilon_k, \tag{7.109}$$

where

$$y = \frac{1}{n}\sum_{k=1}^{n} f_k(x_k) + \frac{1}{n}\sum_{k=1}^{n} \varepsilon_k. \tag{7.110}$$

We have a multifactor model as a synthesis of monofactor ones. From the record obtained it follows that each monofactor model is incorporated into a multifactor one with the same weight equal to *1/n*. Since the variances of each model differ from each other, to minimize the total approximation and variance error following synthesis into a common multifactor one, we introduce the weight of each model v_i. Here it is obvious that the higher the variance of monofactor models, the less weight it should have when incorporated into a common multifactor one.

Taking into account the foregoing discussion, the multifactor model as a synthesis of monofactor models will have the following form:

$$y = \sum_{k=1}^{n} v_k f_k(x_k) + \sum_{k=1}^{n} v_k \varepsilon_k, \tag{7.111}$$

where

$$v_l = \frac{1}{n-1} \frac{\sum_{k=1}^{n} \sigma_k - \sigma_l}{\sum_{i=1}^{n} \sigma_k}. \tag{7.112}$$

We see that the sum of weights given by formula (7.112) is equal to one. There could be other methods of establishing the weights of each monofactor model in a multifactor one.

Let us use this method to build multifactor, complex-valued models. Let us assume the construction of several monofactor, complex-valued dependences of some complex parameter $(y_r + iy_i)$ on complex factors $(x_{rk} + \mathrm{i}x_{ik})$, $k = 1,2,3,\ldots l, \ldots n$, each of the models describing the behavior of a complex parameter with the average approximation error $(\varepsilon_{rk} + i\varepsilon_{ik})$:

$$y_r + iy_i = f_i(x_{rk} + ix_{ik}) + (\varepsilon_{rk} + i\varepsilon_{ik}). \tag{7.113}$$

Let us synthesize these monofactor models into a multifactor one with the corresponding weights:

$$y_r + iy_i = \sum_{k=1}^{n} v_k f_k(x_{rk} + ix_{ik}) + \sum_{k=1}^{n} v_k(\varepsilon_{rk} + i\varepsilon_{ik}), \tag{7.114}$$

where

$$v_l = \frac{1}{n-1} \frac{\sum_{k=1}^{n} (\sigma_{rk} + i\sigma_{ik}) - (\sigma_{rl} + i\sigma_{il})}{\sum_{k=1}^{n} (\sigma_{rk} + i\sigma_{ik})} \tag{7.115}$$

is the complex weight of each factor in the common multifactor, complex-valued model.

Where it is assumed an additive multifactor model will be built, the proposed method significantly simplifies the process forming multifactor, complex-valued models and reduces its labor intensiveness.

It is a bit more complicated when researchers propose building a multifactor multiplicative model. In this case each monofactor model should be considered through the multiplicative approximation error:

$$\mu_k = \frac{y}{f_k(x_k)}. \tag{7.116}$$

If the model uniquely describes a parameter, the multiplicative approximation error will always be equal to one. If the description provides some variance, the multiplicative approximation error will vary around one, and the worse the model describes the parameter, the stronger the variation of the error. Therefore, the error can estimate the model accuracy:

$$\varepsilon_k = 1 - \mu_k. \tag{7.117}$$

Here the variance can be calculated easily.

Again the following rule is valid: the higher the variance, the lower the weight the model should have. However, if in the additive case the sum of weights equals one, then in the multiplicative case their product should be equal to one. Then each weight can be found by the formula

$$v_l = \frac{\left(\prod_{k=1}^{n} \sigma_k\right)^{\frac{1}{n}}}{\sigma_l}. \tag{7.118}$$

Taking into account the foregoing discussion, the multifactor multiplicative model will have the following form:

$$y = \left(\prod_{k=1}^{n} v_k f_k(x_k)\right)^{\frac{1}{n}}. \tag{7.119}$$

The same method can be used to build a multifactor, complex-valued model:

$$y_r + iy_i = \left(\prod_{k=1}^{n} v_k f_k (x_{rk} + ix_{ik}) \right)^{\frac{1}{n}}, \tag{7.120}$$

where complex weights are calculated by the formula

$$v_l = \frac{\left(\prod_{k=1}^{n} (\sigma_{rk} + i\sigma_{ik}) \right)^{\frac{1}{n}}}{\sigma_{rl} + i\sigma_{il}}. \tag{7.121}$$

Thus, one can avoid serious computational difficulties and build a complex-valued model of a given accuracy. In the final chapter this method will be used to build complex, multifactor, complex-valued models.

Of course, one should carefully watch the limits of the parameter and coefficient variations since many complex-valued functions are multivalent and periodic, and as a result the pursuit of complexity of a model can produce a meaningless model.

7.8 Summary

The TFCV is limited to the study of monofactor, complex variables. The economy represents a complex object, and attempts to maximally take into account the actual complexity of the relationships involved in modeling will inevitably lead to the building of multifactor dependences. In this chapter we have shown that multifactor, complex-valued models do expand researchers' mathematical apparatus. Using this apparatus, it is possible to solve problems that could never even be posed in the domain of real variables. Classification models where each complex variable represents two parts of one whole economic indicator or factor make it possible to model various contributions of these components to the performance. The efficiency of these models was demonstrated using production functions of an actually operating economic entity and a complex model of the Russian economy taking into account its illegal part.

Bearing in mind the highly promising nature of multifactor, complex-valued modeling let us nevertheless point out that building these models involves rather labor-intensive problems for estimating model parameters since the number of unknown coefficients in these models is large.

In this chapter we mainly looked at power multifactor, complex-valued models, as linear and power multiplicative multifactor models prevail in economic analysis with real variables.

The last section showed how to build complex, nonlinear, multifactor, complex-variable models by synthesizing monofactor complex-valued ones.

The richness of the TFCV (and this chapter shows that it is possible to work with models of not one but several complex variables) opens up multidimensional perspectives for practical usage of complex-valued models in economic modeling.

Chapter 8
Modeling Economic Conditions of the Stock Market

The models and methods of a complex-valued economy can be used not only in the modeling of production in the form of production functions. One area of application of the TFCV is the stock market. This chapter shows how to build a complex index and compare it with the index of real variables. An important result was obtained from an analysis of the stock market in the phase plane of a complex index modulus and the polar angle of the complex index. In this case it is possible to reveal and describe a complex nonlinear relationship between stock sales volume and stock prices. We called this relationship the "K-pattern." Methods for determining K-patterns and for modeling them are given in this chapter. The main results of the research are demonstrated for the Russian stock market.

8.1 Stock Market Indexes

One of the directions specified in early studies aimed at using the TFCV in economic was for modeling economic conditions in stock markets [1]. Economics has been working on this problem for over 100 years. Considerable progress has been made, and economists everywhere are using some standard set of methods to move forward.

Two approaches are used to model economic conditions: the first one determines the parameters of economic conditions as a certain dependence on the condition-creating factors, and the second one implies aggregation of the parameters into some general value – index – which shows the status of the conditions.

The basis for using complex-valued models with the first approach are stated in previous chapters. Using complex-variable models it is possible to build more complex and probably more adequate models of dependence of the parameters of economic conditions on the condition-creating factors. We do not expect methodological difficulties here. As for the use of complex-valued structures with the second approach, i.e., for building indexes for economic conditions, there are no ready-made solutions.

S. Svetunkov, *Complex-Valued Modeling in Economics and Finance*,
DOI 10.1007/978-1-4614-5876-0_8,

In order to be able to judge the behavior of some entity on the whole, one should use not an aggregate of parameters but some generalized one that includes the diagnostic properties of the aggregate of the parameters. This is clear as it is difficult to compare a set of parameters of one entity with a set of parameters of another entity, or the same one but in the prior time period, and draw some general conclusion. More often, certain parameters of one set testify to an entity's advantages, while others demonstrate the advantages of the other one. A generalized parameter involves the main features of the majority of parameters showing the entity's average status.

Since the state of economic conditions of any market is reflected by a set of possible parameters, an index appears to be more preferable here as it aggregates varied information. In this chapter we will not consider the modeling of market conditions in general, focusing our attention on one type of market – the stock or securities market.

Every major national security market usually has its own stock index or several indexes. Stock traders are guided in their activities by their knowledge of those markets. A stock market index is a number that characterizes its state. As a rule, the value of this number bears no crucial information. It is not the mere value of this number that is important, but its correlation with its previous values. The index can characterize the stock market in general, the market of groups of securities (state securities, bonds, stocks, etc.), the market of securities of certain industries (oil and gas, telecommunications, transport, banks, etc.), and others.

Comparing the dynamics of the behavior of these indexes we see how the state of an industry changes with respect to the market on the whole and therefore market conditions or the dynamics of changes taking place there.

The index theory has clear logical parallels with indifference curves and surfaces known in economic theory. In accordance with the conclusions of the latter, the sums of the cost of goods in a closed system given various prices with other conditions unchanged, will remain the same (constant consumption level):

$$\sum_{j} \mathrm{P}^{\mathrm{j}}\mathrm{Q}^{\mathrm{j}} = const.$$

If the situation changes in this closed system over time, the aggregate cost will also change. An index shows the variation in this aggregate cost over time. Therefore, the aggregate cost of all the purchases in a given securities market or ratio of this cost to the same value at the previous moment of time will be used as the generalizing value at each moment t:

$$\mathrm{I_t} = \frac{\sum_{\mathrm{j=1}}^{\mathrm{m}} \mathrm{P_t^j Q_t^j}}{\sum_{\mathrm{j=1}}^{\mathrm{m}} \mathrm{P_{t-1}^j Q_{t-1}^j}}, \tag{8.1}$$

where P_t^j is the price for the *j*th stock sold on the market; Q_t^j is the volume of the *j*th stock sold on the market; *j* is the stock number (or that of the enterprise selling the goods) on the market, $j = 1, 2, 3, \ldots m;$ and *t* is time.

The index helps to compare aggregate costs at a given moment *t* with the aggregate cost at the previous moment ($t-1$). If economic conditions in the market have improved compared to the previous moment, then business activity has increased and the number of deals has grown compared to the previous moment. Therefore, the aggregate cost of sales has grown as well, and index (8.1) is greater than one. If market conditions have deteriorated, then the activities of market participants have decreased, and the number of deals and sales volume has fallen and aggregate costs have decreased as well. Therefore, the numerator in (8.1) is lower than the denominator, the index being less than one. If the market conditions do not change, then the index is equal to one.

In this way, various values of index (8.1) allow for interpreting the market conditions at one moment compared to the previous one. Market condition indexes have the great advantage of generalizing massive amounts of data since their numerator and denominator represent the sums of products of prices for goods and their sales volumes. It is not difficult to add another term to this sum; therefore, an index can take into account and generalize information on changes in the costs of all goods sold and bought on this market. Therefore, the index gives a unique opportunity to use all the information at the researcher's disposal. However, the possibility of generalizing a great amount of data is in turn a condition of the index's existence as the principle of index calculation implies a generalization of large amounts of data. This objective necessity represents the disadvantages of an index – a general index is not able to signal system disproportions whose trends are gaining strength in the market nowadays.

Indeed, due to the fact that the numerator and denominator of (8.1) comprise the sums of products, there could be cases where a decrease in one parameter in the sum will be compensated by an increase of another one, for example, halving of the stock price ("goods" means stocks) will be compensated by doubling of the sales volume of these securities. A sharp decrease in the price of a firm's stock might be the sign of a crisis at the company, and if it is part of a chain of economic interconnections, it may result in further collapse of the entire chain. This may lead to various consequences for the market, even the collapse of the entire market.

There could be another case where a decrease in the price and sales volumes of a stock in the whole aggregate will be compensated by an increase in price and sales volumes of another stock. In general, the index itself does not change. This change may lead to a number of unfavorable circumstances – we know of several "dark days," which means sudden collapses at stock exchanges, though the indexes of economic circumstances provided no hint of collapse. That is why attempts are made to limit the number of parameters included in the index to the most important ones.

Lately, practicing economists tend not to use an index as a ratio of the amount of sales at the given moment to that at the previous moment. They use the sales volume of some selected set of stocks and analyze how it varies over time. Similar stock indexes are calculated by various information agencies, stock exchanges, rating agencies, etc. They are calculated as the average of stock prices of companies included in the selection. Index developers apply various approaches – for example, simple average, geometric mean, weighted average.

The general formula for calculating an index's average has the following form:

$$\mathrm{I} = \frac{\sum_{j=1}^{m} P_j}{\mathrm{m}}, \tag{8.2}$$

where I is the index, P_j is the price of stock j sold on the market, and m is the number of companies.

The index's geometric mean has a more complex calculation:

$$I = \sqrt[m]{I_1 \times I_2 \times \ldots I_m} \;, \tag{8.3}$$

where I is the composite index; I_1, I_2, $\mathrm{I}_3, \ldots\ldots \mathrm{I}_\mathrm{n}$ are individual indexes of companies, and m is the number of companies in the selection.

The weighted-average method is the best known among index calculations. It takes into account the company size and scale of operations in the stock market. As a rule, the market capitalization of a company, i.e., the total market value of a company's stocks, is taken as the weight.

According to the weighted-average method, index calculation has the following form:

$$I = \frac{\sum_{j=1}^{m} P_j^t Q_j^t}{\sum_{j=1}^{m} P_j^0 Q_j^0} \times I_0, \tag{8.4}$$

where P_j^0 and P_j^t are stock prices of company i in the basic and reporting periods;
Q_j^0 and Q_j^t are the number of stocks in circulation in the basic and reporting periods;
$i = 1, 2, \ldots, m$ is number of companies in the selection; and
I_0 is the index's basic value.

The variety of approaches to index calculation is due to the fact that none of them possesses sufficient diagnostic properties. This is why analysts improve the indexes using the most varied methods.

Let us see how the general principles and approaches to a complex-valued economy that are stated in this study can be used to develop tools for stock market analysis.

As a product, a stock has two components: consumer properties inherent in goods and price, which is a monetary assessment of consumer properties of this product by a particular consumer. According to our study, these two aspects characterize the product, and therefore these two parameters characterize the product in general, and they should not be considered separately from each other but in tandem, i.e., as a complex variable:

$$z_{jt} = q_{jt} + ip_{jt}. \tag{8.5}$$

Here

q is the sales volume,
p is the unit price, and
j is the stock number.

The original values of the price and volume must be represented as dimensionless quantities; otherwise a complex variable cannot be created.

The above-mentioned expression (8.5) makes it possible to fully describe the properties of a particular stock and mathematically correctly work both with each of the components and with the aggregate in general if the original variables are brought to one scale and dimension.

Complex variable (8.5) is characterized by its modulus and polar angle:

$$R_{jt} = \sqrt{q_{jt}^2 + p_{jt}^2}, \quad \theta_{jt} = arctg\frac{p_{jt}}{q_{jt}}. \tag{8.6}$$

Therefore, it can be represented in both arithmetic (8.5) and trigonometric and exponential forms. Later on we will need the exponential representation of this complex variable:

$$z_{jt} = R_{jt}e^{i\theta_{jt}}. \tag{8.7}$$

If a researcher possesses data on stock sales at moment t and the previous moment $t - 1$, then by comparison one can judge the change in the market for these stocks. In our case a comparison can be carried out either by subtraction from z_{jt} of its previous value or by division of z_{jt} by z_{jt-1}.

In the first case the real part of the obtained difference will characterize the change in the sales volume and the imaginary one the change in the unit price. The quotient of the two complex numbers will be characterized by the ratio of the modules and angles of the two complex variables. The modulus will be less or greater than one depending on whether the modulus of the variable has increased or decreased, and the polar angle will characterize variation in the price with respect to the volume change. Since information on each stock is in its complex model (8.5), then, besides activity with individual stocks one can create a parameter generalizing information for all stocks, in other words, an index. This index should contain information on all the relevant sales activity in the market, i.e., for this it is necessary to use m complex variables (8.7).

A simple sum of complex variables (8.5) would be meaningless since we would sum up the real and imaginary parts separately and the resulting complex number would not have any crucial significance. However, it is obvious that the properties of each individual stock will get lost, with no new properties appearing for the complex value as a result of this generalization. Multiplication of the complex sales variables of all m stocks in the given market will have the following meaning:

$$Z_t = \prod_{j=1}^{m} z_{jt} = \prod_{j=1}^{m} (R_{jt} e^{i\theta_{jt}}) = e^{i\sum_{j=1}^{m} \theta_{jt}} \prod_{j=1}^{m} R_{jt}. \tag{8.8}$$

Similarly, one can find the product of the complex sales variables in the same market at the previous moment $t - 1$:

$$Z_{t-1} = \prod_{j=1}^{m} z_{jt-1} = \prod_{j=1}^{m} (R_{jt-1} e^{i\theta_{jt-1}}) = e^{i\sum_{j=1}^{m} \theta_{jt-1}} \prod_{j=1}^{m} R_{jt-1}. \tag{8.9}$$

The ratio of (8.8) to (8.9) will also be a complex variable and will characterize the situation in the market, i.e., it will be some index:

$$I_{St} = \left(\frac{\prod_{j=1}^{m} z_{jt}}{\prod_{j=1}^{m} z_{jt-1}} \right)^{\frac{1}{m}} = e^{i\frac{1}{m}\left(\sum_{j=1}^{m}\theta_{jt} - \sum_{j=1}^{m}\theta_{jt-1}\right)} \left(\frac{\prod_{j=1}^{m} R_{jt}}{\prod_{j=1}^{m} R_{jt-1}} \right)^{\frac{1}{m}}. \tag{8.10}$$

According to complex-variable properties, this very index I_S is a complex variable with real and imaginary parts. It also can be presented in exponential form using modulus R_z and polar angle φ_z.

The polar angle of index I_S is found as the exponent in (8.10):

$$e^{i\frac{1}{m}\left(\sum_{j=1}^{m}\theta_{jt} - \sum_{j=1}^{m}\theta_{jt-1}\right)}, \tag{8.11}$$

i.e., it is equal to the difference

$$\varphi_z = \frac{1}{m} \left(\sum_{j=1}^{m} \theta_{jt} - \sum_{j=1}^{m} \theta_{jt-1} \right). \tag{8.12}$$

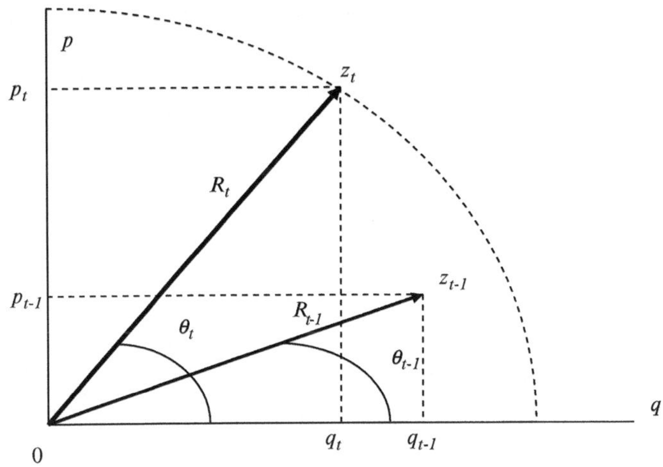

Fig. 8.1 Stock behavior shown in the complex plane

The modulus of index I_S is determined according to the formula

$$R_{zt} = \left(\frac{\prod_{j=1}^{m} R_{jt}}{\prod_{j=1}^{m} R_{jt-1}} \right)^{\frac{1}{m}} = \left(\prod_{j=1}^{m} \frac{R_{jt}}{R_{jt-1}} \right)^{\frac{1}{m}}. \tag{8.13}$$

To understand the meaning of the modulus and polar angle of the index, let us turn to its graphic interpretation in a complex plane. By definition, the coordinate axes will be the scaled unit price and stock sales volume.

Let us first consider the situation where $m = 1$, i.e., where we study the behavior of only one stock. Figure 8.1 shows the position of the complex variable at moment t and at the previous moment $t - 1$.

In the figure the complex variable z_t is shown in circumstances where the sales volume of this stock q_t fell somewhat compared to the previous moment q_{t-1}, but the unit price p_t considerably increased: $p_t > p_{t-1}$. This is shown in the model by the fact that the modulus of the complex variable has increased: $R_t > R_{t-1}$, and the polar angle, $\theta_t > \theta_{t-1}$, has increased as well.

It is evident that for the polar angle of any security lies within a range of 0 to $\pi/2$.

If now we relate the complex variable at moment t to its values at the previous moment, we obtain a new complex variable (private index):

$$I_t = \frac{z_t}{z_{t-1}} = \frac{R_t}{R_{t-1}} e^{i(\theta_t - \theta_{t-1})} = \frac{R_t}{R_{t-1}} \cos(\theta_t - \theta_{t-1}) + i \frac{R_t}{R_{t-1}} \sin(\theta_t - \theta_{t-1}). \tag{8.14}$$

Since the polar angle of this complex index is equal to

$$\theta_{It} = \theta_t - \theta_{t-1} = arctg\frac{p_t}{q_t} - arctg\frac{p_{t-1}}{q_{t-1}}, \tag{8.15}$$

and the modulus to

$$R_{It} = \frac{\sqrt{q_t^2 + p_t^2}}{\sqrt{q_{t-1}^2 + p_{t-1}^2}}, \tag{8.16}$$

one can make a preliminary conclusion on the variations in the complex index's polar angle (8.14) and its modulus.

First of all it should be noted that the complex index's polar angle lies within the limits

$$-\pi/2 < \theta_{\mathrm{It}} < \pi/2. \tag{8.17}$$

Polar angle (8.15) of partial index (8.14) is equal to zero if the proportion between the price and sales volume has not changed; this follows from (8.15). A proportionate increase in price with an equivalent concomitant growth in sales volume means increasing interest in stocks and buoyancy of the market. A proportionate decrease in price with an equivalent concomitant decrease in the sales volume means stockholders' tendency to hold onto stocks with falling prices. Therefore, the fact that the polar angle of the private index (8.14) is equal to zero suggests a certain stability of this stock in the market – an increase in its price leads to an increase in trading in these securities, while a decrease in the price leads to a decrease in trading.

With reference to the arithmetic form of the complex index, its imaginary part is equal to zero.

The polar angle (8.15) is greater than zero if the stock price has grown but the sales volume has remained unchanged, or if the price has not changed but the sales volume has fallen. The first suggests a growth in interest in these securities and a reluctance on the part of a number of shareholders to get rid of them, while the second suggests that this price does not satisfy shareholders and they refrain from selling. This is a situation of expectations where market participants consider these securities promising.

The positive character of the polar angle means that both cosine and sine are positive, as are the real and imaginary parts of index (8.14). The greater the imaginary part of this index, the more expectations the market has with respect to future changes in the stock's behavior, meaning price increases, which is why stock prices have grown compared to the previous observation. The positive character of the polar angle of the private index reflects an increase in the interest in the securities on the market.

If the polar angle of the private index is less than zero, this means that at fixed prices the sales volume of this stock is increasing or that at fixed sales volumes the unit price falls. The first means that shareholders are in a hurry to get rid of this stock, expecting a deterioration in its positions. The second means that the price is starting to decrease but the stock is not held onto because demand for it is not high. Thus, a negative polar angle of the private index reflects a decrease in interest in this stock.

The arithmetic form means that the imaginary part of the complex private index is negative, and the higher the modulus, the worse the positions of this stock in the market.

Now let us consider possible dynamics of modulus (8.16) of the private complex index. It can remain stable, increase, or decrease. Figure 8.1 shows a dotted circle with radius R_t. At any point of this circle the radius will be the same – for an increasing polar angle (increase in price and decrease in sales volume) and a decreasing polar angle (decrease in price and increase in the sales volume). Therefore, this index does not show the expectations of market participants. It shows only the scale (not volume!) of activity with this stock.

The sales volume of this stock represents the product of the price and volume or, in the case under consideration, the product of the real part of the private complex index and the real component of its imaginary part, i.e.,

$$p_t q_t = R_t \cos\theta_t \times R_t \sin\theta_t = \frac{1}{2} R_t^2 \sin(2\theta_t). \tag{8.18}$$

A decrease in the modulus of the private complex index with permanent polar angle reflects a decrease in sales volume, and its increase an increase in sales volume. Simultaneous change in both the polar angle and the modulus can lead to constant sales volume but a dramatic change in the attitude of market participants toward this stock (Fig. 8.2).

Sales volume can remain constant if, for example, there is an increase in the modulus of the complex variable but its polar angle decreases. Figure 8.2 show that this corresponds only to a situation where the price falls and the sales volume increases.

Can this describe a situation where stock owners dump a stock? For this reason, sales volume conveys little information about market conditions, though a change in the polar angle and the imaginary part of the complex variable are more informative.

Let us summarize the information with reference to the private complex index:

1. Proximity to zero of the imaginary part of an index reflects a stable market. An increase in the real part indicates growing interest in the securities, while a decrease means that interest has fallen off.
2. If the imaginary part of the complex private index is positive, this reflects the fact that the stock has generated some interest. If the imaginary part is positive and far from zero values, this means that there is a demand for the stock in the

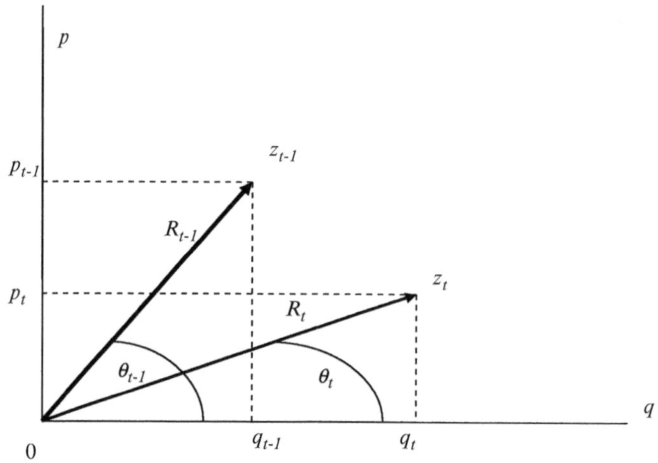

Fig. 8.2 Stability of sales volumes with changing market conditions for a given stock

Table 8.1 Conditional example for index calculation (dimensionless units)

	Stock 1		Stock 2		Stock 3		Stock 4	
	Price, *p*	Sales volume, *q*	Price, *p*	Sales volume, *q*	Price, *p*	Sales volume, *q*	Price, *p*	Sales volume, *q*
t	11	10	10	8	7	5	15	10
t–1	10	11	8	10	5	7	10	15

market that is close to exuberance. If the real part is high, then we observe a boom; if the real part is low, then sales volume has decreased in the expectation of further price increases of prices.

3. If the imaginary part of the index is negative, this means a decrease in interest in this stock, and the greater its modulus, the sharper the decrease in interest. A high value of the real part will reflect the fact that the sales volume has increased –market participants are dumping the stock; if not, market participants are holding this stock in the expectation of improved conditions.

Since the generalizing index (8.10) represents a product of private indexes, the above-mentioned conclusions also extend to its properties. Table 8.1 presents a conditional example demonstrating properties of the index of complex variables compared to the standard approach.

The stock prices and volume are selected to have the classic index (8.1) equal to 1, showing stability in the market. One can easily see that

$$I_t = \frac{11 * 10 + 10 * 8 + 7 * 5 + 15 * 10}{10 * 11 + 8 * 10 + 5 * 7 + 10 * 15} = 1. \tag{8.19}$$

Table 8.2 Calculated values of stock moduli for each time moment

	Stock 1	Stock 2	Stock 3	Stock 4
	R_1	R_2	R_3	R_4
t	14.87	12.80	8.60	18.03
t–1	14.87	12.80	8.60	18.03

Table 8.3 Calculated values of polar angle (in radians)

	Stock 1	Stock 2	Stock 3	Stock 4
	θ_1	θ_2	θ_3	θ_4
t	0.83	0.90	0.95	0.98
t–1	0.74	0.67	0.62	0.59

Now for each stock and at each moment of time, one can calculate the modulus of the complex variable (Table 8.2).

Then we can determine the polar angles of each variable at various moments of time (Table 8.3).

Let us now consider the values of private complex indexes.

The first one is

$$I_{1t} = \frac{14.87}{14.87} e^{i(0.83-0.74)} = 1\cos 0.09 + i1\sin 0.09 = 0.995952733 + i0.089878549,$$

the second one is

$$I_{2t} = \frac{12.80}{12.80} e^{i(0.90-0.67)} = \cos 0.23 + i\sin 0.23 = 0.973666395 + i0.227977524,$$

the third one

$$I_{3t} = \frac{8.60}{8.60} e^{i(0.95-0.62)} = \cos 0.33 + i\sin 0.33 = 0.946042344 + i0.324043028,$$

and the fourth and last one is

$$I_{4t} = \frac{18.03}{18.03} e^{i(0.98-0.59)} = \cos 0.39 + i\sin 0.39 = 0.92490906 + i0.380188415.$$

The imaginary parts of all the indexes are positive, which reflects a growing interest in the stocks. The imaginary part of the first index is close to zero, which is why interest in this stock can be considered quite normal. However, the fourth stock has a considerable imaginary part – 0.380188415. This reflects surge in the stock. Indeed, we see from Table 8.1 that the price for this stock grew by 50%.

Now let us calculate the summarizing complex index (8.10) characterizing the market in general. The index modulus is

$$R_t = \left(\frac{14.87 \times 12.80 \times 8.6 \times 18.03}{14.87 \times 12.80 \times 8.6 \times 18.03}\right)^{\frac{1}{4}} = \left(\frac{29.513}{29.513}\right)^{\frac{1}{4}} = 1. \tag{8.20}$$

Its polar angle is

$$\theta_t = \frac{1}{4}[(0.83 - 0.74) + (0.9 - 0.67) + (0.95 - 0.62) + (0.98 - 0.59)] = 0.26. \tag{8.21}$$

Then the arithmetic form of the complex index is

$$I_{St} = 0.966389978 + i0.257080552. \tag{8.22}$$

The imaginary part of this coefficient is positive and greater than zero, which reflects a favorable economic situation and buoyancy of the market.

The classic index for the conditional example under study is equal to one, which means a stable situation, though we can see a clear change. Again, we see that usage of complex variables makes it possible to model the economic entity in a different way and draw different conclusions compared to those obtained by models and indexes of real variables.

Very often, to show market conditions, economists take into account some "summary" index representing the sum of sales volumes of a certain set of stocks ("blue chips"):

$$I_t = \sum_j \mathrm{P}_t^j \mathrm{Q}_t^j = const. \tag{8.23}$$

To analyze the situation in the market, we do not divide the current index value by the previous one but just analyze the index dynamics. It goes without saying that the dynamic range of this index makes it possible to compare both neighboring and remote indexes with each other.

If by analogy to (8.23) we use index (8.8),

$$Z_t = \prod_{j=1}^{m} z_{jt} = \prod_{j=1}^{m} (R_{jt} e^{i\theta_{jt}}) = e^{i\sum_{j=1}^{m} \theta_{jt}} \prod_{j=1}^{m} R_{jt},$$

then we could face a problem – as a complex number this index increases its modulus with an increase in the number of data used and increases the polar angle, which is equal to the sum of polar angles, each of them being positive:

$$\theta = \sum_{j=1}^{m} \theta_{jt}.$$

This is why this index is a periodic value, since for various combinations of the stocks comprising the index one can obtain polar angles that differ from each other by $2\pi k$; therefore, the ratio of their real and imaginary parts will be similar, though the processes are different. Index (8.10), which includes the quotient of the complex indexes, is free from this shortcoming since the polar angles of the same stocks are subtracted from each other. The question is whether it is possible to obtain a certain analog of the consolidated index (8.23) in the sphere of complex variables.

The geometric mean can meet these requirements (8.8):

$$I_t = \sqrt[m]{e^{i\sum_{j=1}^{m}\theta_{jt}} \prod_{j=1}^{m} R_{jt}} = e^{i\frac{1}{m}\sum_{j=1}^{m}\theta_{jt}} \sqrt[m]{\prod_{j=1}^{m} R_{jt}}. \tag{8.24}$$

Calculation of this index will allow us to analyze separately the dynamics of its four characteristics:

1. Polar angle:

$$\varphi_t = \frac{1}{m}\sum_{j=1}^{m}\theta_{jt},$$

2. Modulus:

$$R_t = \sqrt[m]{\prod_{j=1}^{m} R_{jt}},$$

3. Real part:

$$I_{rt} = R_t \cos\varphi_t,$$

4. Imaginary part:

$$I_{it} = R_t \sin\varphi_t.$$

Looking at the complex index as a supplementary tool of stock market analysis, but not an alternative to real-variable tools, we have every reason to assert that such a development of the instrumental base of the economy is quite productive.

8.2 Phase Plane and *K*-patterns

The stock market is a special market where one will not find vendors and end users. Its goods do not meet immediate needs, vendors quickly become buyers, and vice versa. Limiting usefulness or labor value theories fail to explain their behavior. Models and methods of economic theory applied to this market do not work – the stock market operates by other laws, other interrelations, other behavior styles. This market more than others is governed by psychological and even sociological laws. Mass panic in the market leads to its collapse, and expectations represent the main driving force. Much effort is expended on forecasting price movements with little to show for it.

Stock market theory is fully completed. One should remember that the goal of any theory is to explain the phenomenon under study including the models involved.

Nowadays there are two approaches to studying the stock market. The first is known as fundamental analysis and represents verbal and graphic models to describe complex relationships of causes and factors affecting the market. The second one is called "technical analysis" and involves various mathematical models trying to describe and forecast numerical characteristics of the market without getting into the system of cause-and-effect relationships and without taking into account the influence of factors measured on a nominal or ordinal scale. Synthesis of these two directions makes up a kind of stock market theory.

Since in this market the unit price for one share of stock and the sales volume thereof are the main indicators studied by stock market theory, it is quite natural to determine and understand the relationships between these two economic parameters. However, so far their use in stock market theory has been limited to the calculation of various indexes like (8.1) or (8.2). Nothing more has been accomplished.

However, economists understand that between the unit price and the sales volume of a given product, even a stock in the stock market, there is an economic relationship that is intuitively felt but not thus far inexplicable. This is why nowadays stock market theory recommends analyzing visually the dynamics of price p_t and sales volume q_t, placing one graph under the other and correlating the time axis scale. This combined positioning of the dynamics of economic parameters gives economists some intuitive impression on the process under study. There is also a certain understanding of the mutual influence of these two parameters, though no model is capable of describing either the direction or the core significance of this influence.

Since both the stock price and the sales volume are interrelated and distributed in time, one could combine them in one graph representing what we know from the natural sciences as a phase plane, as each value of the price and sales volume has an index – time – providing their precise identification. This is why, if we place price on one axis and sales volume on the other, we will obtain a phase plane whereon we can plot a set of points, which often looks like a law and is therefore called a "phase

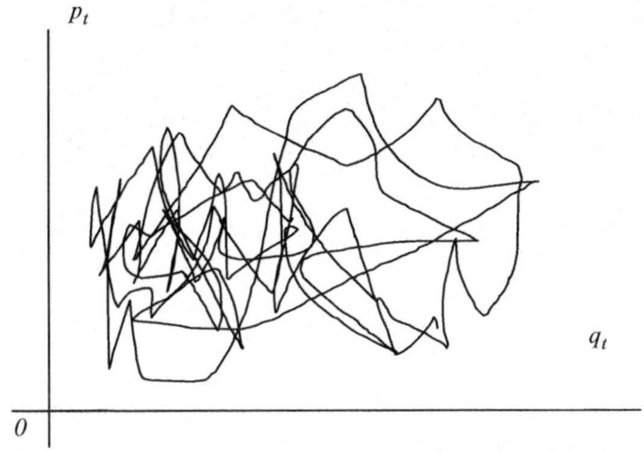

Fig. 8.3 Typical phase portraits of a stock on the stock market

portrait." The "Hysteresis loop," well-known in physics, is the most vivid example of the phase portrait.

Attempts to build phase portraits of securities listed on various stock markets fail. We obtain not a portrait but a chaotic accumulation of points and their connecting lines. Figure 8.3 gives an example of the phase plane of a certain stock.

Therefore, we should acknowledge that using real-variable instruments and well-known approaches it will not be possible to solve the problem of determining and describing relationships between the unit price of this stock and its circulation volume in the stock market. This means that science is not able to solve the main task of revealing the market's driving force as demonstrated by its laws. As a matter of fact, nobody argues that modern stock market theory has developed a large body of empirical materials and discovered its numerous dynamic laws. However, the empirical-deductive method used is not able to reach a level of theoretical generalizations. The hypothesis of existence of a relationship between price and volume in this market has not been confirmed so far.

Since this research is aimed at demonstrating only a small part of those rich possibilities that open up for economists when using the TFCV, let us show how to solve the set problem and test the hypothesis on the existence of interrelations between stock prices and sales volume in the stock market. For that let us use the previously introduced variable (8.5):

$$z_{jt} = q_{jt} + ip_{jt}.$$

This complex variable can be written both in exponential and trigonometric forms for which we calculate its modulus and polar angle:

$$R_{jt} = \sqrt{q_{jt}^2 + p_{jt}^2}, \quad \theta_{jt} = arctg\frac{p_{jt}}{q_{jt}}. \tag{8.25}$$

Both the modulus of the complex variable and its polar angle characterizing the stock (price and sales volume) undergo changes in time since the original complex

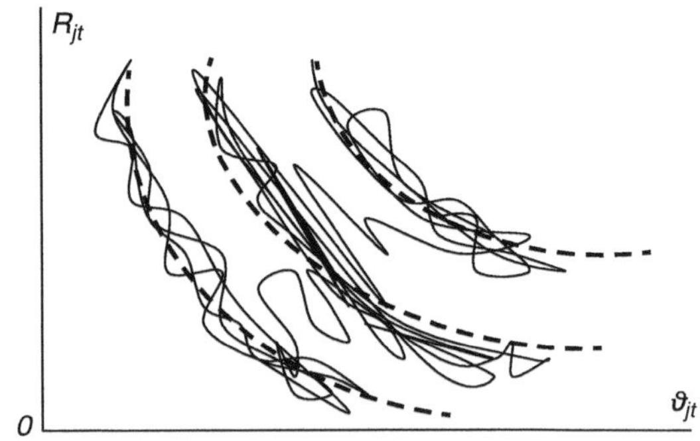

Fig. 8.4 Typical phase portrait showing behavior of polar coordinates of securities on stock market

variable is dynamic. This is why graphs of change in time of both the polar angle and the modulus can give additional characteristics for the stock. The economic meaning of an increase or decrease in these parameters was stated in the previous section (Figs. 8.1 and 8.2). This is why, on the one hand, in building these graphs economists obtain additional information on changes in the stock with time.

On the other hand, it definitely follows from (8.25) that these two variables are related to each other since the same pair – price and volume – are used to calculate the modulus and the polar angle. This means that the dynamic variables (8.25) can be placed on one graph – phase plane graph – the relationship between them, if any, characterizing the phase portrait of the process under study.

T.V. Koretskaya built several tens of these securities listed on MICEX from 2007 to 2009. In the phase planes of all the studied securities one can clearly see the phase portrait of a typical shape (Fig. 8.4).

In the phase portrait we see that at a certain period of time, the points of the securities' modulus and polar angle lie on a concave curve (in Fig. 8.4 we see these curves as a bold dotted line). When considered in their dynamics, the position of these points represents a swinging movement along the bold line up and down. Then, when the market dynamics changes, the phase portrait points "break off" these stable lines and move to another, lower level, where they stay for a rather long period of time on a similar concave line. Since T.V. Koretskaya considered only the period characterized by a global economic crisis, resulting in a collapse of stock markets and withdrawal from the market of the funds invested in securities by financial companies , the transition from one position in the phase plane to another has a downward trend. I am quite sure that with the first signs of improvement in the global economy and the available resources reappear in the stock market, the phase planes of securities will demonstrate a transition from one phase state to another with an upward trend.

We call the phase portrait lines *K*-patterns patterns [2]. Graphic analysis of the phase state of a security shown in the graph as a *K*-pattern made it possible to formulate a hypothesis on parallel *K*-patterns of the same securities. To test this hypothesis, Koretskaya proposed a simple but very productive approach – taking the logarithm of the securities' moduli and polar angles and further building the

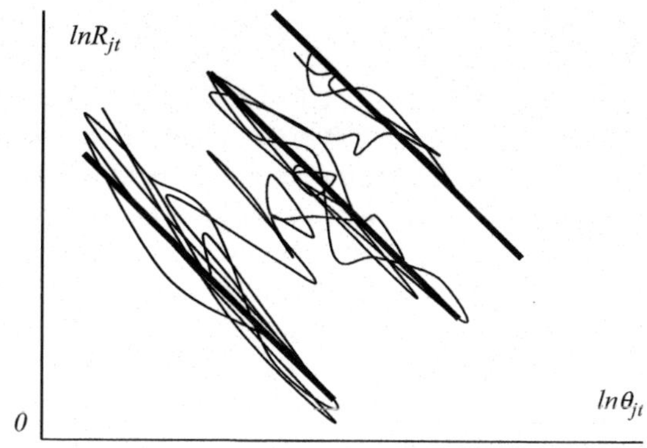

Fig. 8.5 Phase portrait of security in logarithmic phase plane

phase plane in logarithms. This transformation of the complex variables and the phase plane yielded phase portraits in the form of parallel straight lines (Fig. 8.5).

Of course, since graphic models are built according to sample parameter values, this happens under the influence of numerous arbitrary factors that allow some lines to deviate from the general location in the phase plane. However, the parallel nature, on average, of the straight lines in a logarithmic phase plane under visual analysis of tens of security graphs is evident. This was confirmed by statistical analysis when the slope angle of linear models fluctuated around some average value.

Having solved the problem of determining the *K*-pattern relationship of the unit price and the sales volume of a security in the stock market it is necessary to give an economic interpretation to the results obtained.

The relationship between price and volume is complex and nonlinear since the *K*-pattern represents a nonlinear relationship of the modulus of a complex variable (square root of the sum of squares of price and volume) and its polar angle (arctangent of price-to-volume ratio). The complex form of this relationship does not provide an unambiguous interpretation like, for example, that a price increase leads to a particular trend. These explicit conclusions do not suggest themselves; others do. *K*-patterns as an aggregation of points of a phase plane may have something in common that determines their particular position in the plane. But what is the nature of these causes that are reflected in the same position of phase portrait points and why are phase transitions from one state to another observed? Is the assumption true that during periods of stability in the stock market stocks are arranged in *K*-patterns, and in the times of instability they "break away from" these patterns?

To answer this question, let us turn our attention to the processes taking place in the stock market. Let us consider the MICEX stock market of 2008 as an example.

On January 21, 2008 auctions were not held in the USA. However, on that very day Asian stock exchanges collapsed along with the Russian stock market. In that crisis situation of instability, many market participants hurried to unload their securities and withdraw funds from the market, leading to a considerable decrease in the volume of funds circulating in the market. This situation in the MICEX market did not change for 3 and a half months up to May 4, 2008. If we build a phase portrait of any of the MICEX "blue chips," we see that this period of "stability" manifests itself in the phase plane of this security by a *K*-pattern. Of course, there are obvious deviations from the line induced by random factors; however, this dispersion is not critical. The period shown by the *K*-pattern is characterized by a similar attitude of market participants toward the securities and – what is important – by the relative stability in the market in general.

Beginning on May 5 the points of the phase plane started a gradual transition to a new state. They started going lower and lower than the *K*-pattern. This decreasing trend coincided with the events of August 18, 2008, which was marked by a new fall of Russian stocks against the background of the Georgia-South Ossetia conflict. On the day of the summit of NATO ministers of foreign affairs in Brussels on August 19, 2008, when the question of how to punish Russia for its prolonged occupation of Georgia, the Russian stock market collapsed, which is shown in the phase plane by a chaotic fluctuation of the points with a clear tendency to a new level.

On September 16, 2008, influenced by the bankruptcy of the American investment bank Leman Brothers, the Russian stock market underwent another collapse. This period saw an active withdrawal of funds from deposit accounts in Russia, this instability being demonstrated by a downward trend of the polar angle and modulus.

On Monday, September 29, the Russian stock market again collapsed. Nervousness reached a fever pitch, which showed on October 6, 2008, that adoption of [U.S. Treasury Secretary Hank] Paulson's plan did not relieve the tension from the American stock market and did not dispel doubts about its efficiency. As a result, on Monday, October 6, 2008, the market collapsed again.

After the collapse, a measure of stability returned to the market from October 29, when the U.S. Federal Reserve announced a 0.5 % decrease in the discount rate. This pushed the Asian and European markets up on October 30, and the Russian stock market also showed some upward movement. The volume of the funds circulating in the market was fixed at some stable level. From that day until the end of 2008, the points of the phase plane accurately arranged themselves into a new *K*-pattern.

Analysis of other phase portraits shows that the periods of time characterized by stable markets show all points in one *K*-pattern. As soon as there is a change in the stock market, the points of the phase portrait start moving toward another *K*-pattern, which takes shape in the subsequent period of stock market stability.

It is interesting to note that in the early 2000s Lukoil stock was arranged in a *K*-pattern, and beginning in 2005 stopped behaving in this way. A.M. Chuvazhov, who discovered this phenomenon, assumed that the absence of *K*-patterns in this period was due to certain political and economic causes, including the acquisition of stocks by American companies, as a result of which changes in the Russian stock market stopped influencing the behavior of the Lukoil's stock.

8.3 Mathematical Models of *K*-Patterns

The linear character of *K*-patterns in the logarithmic phase plane predetermines the type of mathematical model of the *K*-pattern. In logarithms it has the following form:

$$\ln R_{jt} = a_0 + a_1 \ln(arctg\theta_{jt}). \tag{8.26}$$

Moving from the logarithmic to the multiplicative form we have

$$R_{jt} = e^{a_0} arctg\theta_{jt}{}^{a_1}. \tag{8.27}$$

If now we substitute price and volume for the modulus and polar angle, we have the following *K*-pattern equation:

$$\sqrt{p_{jt}^2 + q_{jt}^2} = e^{a_0}\left(artg\frac{p_{jt}}{q_{jt}}\right)^{a_1}. \tag{8.28}$$

It is difficult to transform this model into an explicitly given dependence of price on volume or volume on price. However, since for the given polar angle we calculate the complex-variable modulus according to (8.27), with these two parameters it will be quite possible to calculate the sales volume as the real part of the complex variable:

$$q_{jt} = e^{a_0} arctg\theta_{jt}{}^{a_1} \cos\theta_{jt} \tag{8.29}$$

and the unit stock price as the imaginary part of the complex variable:

$$p_{jt} = e^{a_0} arctg\theta_{jt}{}^{a_1} \sin\theta_{jt}. \tag{8.30}$$

There is an obvious dependence between the unit stock price and sales volume, which follows from complex-variable properties:

$$p_{jt} = q_{jt} tg\theta_{jt}. \tag{8.31}$$

Now one can propose an algorithm of stock market modeling. If there are grounds to consider the polar angle equal to some set (for example, forecasted) value, then under stable market conditions, when K-patterns are observed, it is possible to calculate the sales volume according to (8.29) and the unit stock price according to (8.30).

For practical purposes model (8.26), or its analog (8.27), is quite suitable since this model's coefficients are easily found from statistical data using any method of statistical estimation, for example, LSM.

According to the data on changes in quotations of Aeroflot at MICEX for 2008, T.V. Koretskaya found several K-patterns using LSM equations. Let us consider two of them. The first K-pattern corresponds to a relatively stable period from January 21 to May 4, 2008, the second one from October 29, 2008 to the end of that year.

The model Eq. (8.27) for the first K-pattern has the following form:

$$\ln \hat{R}_t = 0.2366 - 0.8261 \ln \theta_t. \tag{8.32}$$

The model determination coefficient with factual data is 0.955.

The same equation for the second K-pattern is

$$\ln \hat{R}_t = -0.9945 - 0.9396 \ln \theta_t. \tag{8.33}$$

This model describes the original data just as well since the determination coefficient with factual data is 0.9613.

It is easy to see that the free term decreased and became negative, which indicates a downward shift in the K-pattern. The model proportionality coefficient also changed, which is due to the influence of random factors, but the scale and sign did not change. The regression equations of all the K-patterns are statistically significant.

How can we use K-pattern equations calculated by LSM in practice? Assume that we know that in December 2008 the market situation is stable and we know that one of the payers plans to sell a large stake $Q_{t+1} = 100{,}000$ of Aeroflot. We have a K-pattern equation at our disposal with coefficients calculated by LSM (8.33). Expressing this volume as a basic value we get, for example, $q_{t+1} = 1.28$.

To make the right decision we should forecast a possible price for a share of Aeroflot stock that will take shape as a result of the auctions given the volume of stock up for sale. The market does not change; therefore, the K-pattern equation does not change either, but all the K-pattern points are characterized by a changed polar angle and modulus. This is why the calculation algorithm (8.29)–(8.30) cannot be applied in this case.

To solve this problem, it is necessary to use a mathematical equation of a K-pattern. Substituting the volume of stocks put on sale into the equation of the K-pattern model we obtain

$$\sqrt{p_{t+1}^2 + 1.28^2} = e^{-0.9945}\left(artg\frac{p_{t+1}}{1.28}\right)^{-0.9396} = 2.703\left(artg\frac{p_{t+1}}{1.28}\right)^{-0.9396}$$

It is impossible to derive the value of the forecasted price and calculate it from the preceding equation, but it is not impossible to solve the given problem since the equation

$$\sqrt{p_{t+1}^2 + 1.28^2} - 2.703\left(artg\frac{p_{t+1}}{1.28}\right)^{-0.9396} = 0,$$

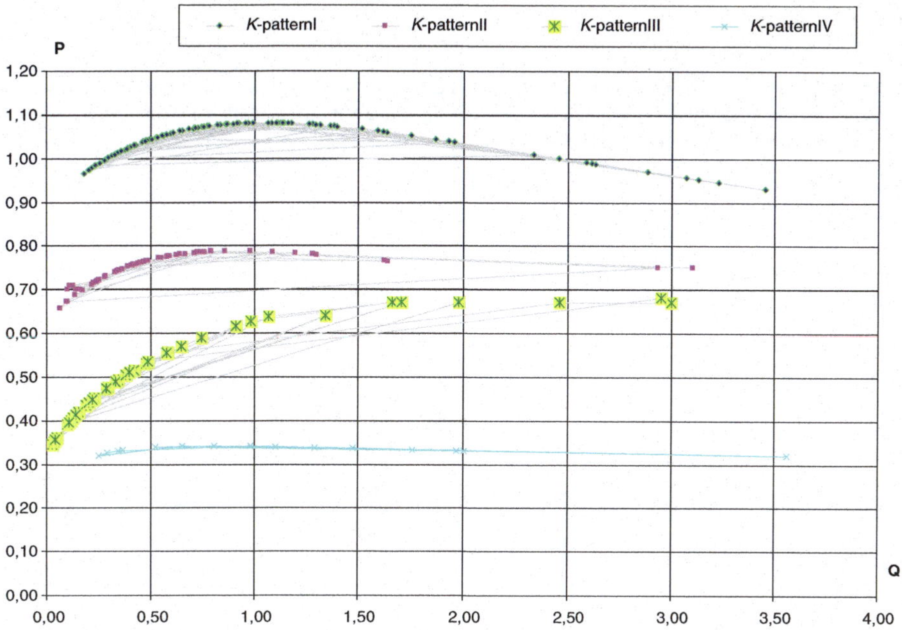

Fig. 8.6 *K*-patterns for the Aeroflot shares

with respect to the only unknown p_{t+1}, can be solved using one of the numerical methods.

The obtained stock price value can be used as the forecasted one if the stake is sold.

According to the results stated in this chapter, we could draw the conclusion that the complex index (8.10) used along with other indexes makes it possible for researchers to obtain additional information about stock market conditions. New information on the stock market can also be obtained calculating the complex index (8.24) and analyzing its main characteristics.

Identification of a *K*-pattern and justification of the form of its mathematical model (8.26) allows for developing theoretical ideas about the stock market since it represents a mathematical model of the relationship of a unit stock price and sales volume, which is quite complex and unidentifiable using real-variable models. The general idea of this relationship is given by Fig. 8.6

The transition process from one *K*-pattern to another is of particular interest. Since *K*-patterns reflect periods of stable markets, it is most interesting to forecast instability – the transition from one state to another. Perhaps combining *K*-patterns with market cycles would make it possible to obtain the desired results – but that is a task for the future. The most important thing is that has become possible to identify these *K*-patterns and show that they correspond to periods of stability in the markets.

References

1. Svetunkov SG (2006) Complex variables in the index theory. Theory of function of complex variable in economic and mathematical modeling: materials of All-Russian Scientific Seminar (19 December 2005). SUEF, St. Petersburg, pp 15–37
2. Svetunkov SG, Koretsksya TV (2009) Comparative research of classical index and complex variable index for stocks dynamics at MICEX. Bulletin of Orenburg State University 5(2009):78–81

Chapter 9
Modeling and Forecasting of Economic Dynamics by Complex-Valued Models

This chapter is devoted to one goal – modeling economic dynamics using models of complex variables. The chapter opens with the model of Ivan Svetunkov, which represents a complex-valued analog of the model of short-term forecasting of exponential smoothing (Brown's model). This model possesses remarkable properties and ushers in a new class of models of short-term economic forecasting. It shows how to use a complex-valued analog of Solow's model of economic dynamics, and its particular properties are demonstrated. The chapter and book conclude with a section devoted to diagnostics of regional socioeconomic development.

9.1 Ivan Svetunkov's Model for Short-Term Forecasting

Ivan Svetunkov's short-term forecasting model is one of the remarkable results of the application of the methods of complex-valued economic modeling. It successfully combines the advantages of Brown's short-term forecasting model and properties of complex-valued models. The core of the model is as follows [1].

In 80 % of the cases of practical application, the problem of forecasting socioeconomic dynamics in the short term is solved by means of Brown's method (also known as the exponential smoothing method). The main idea is that the forecasted value is defined as the weighted average of the previous range, which in compact form is written as follows:

$$\hat{Y}_{t+1} = \alpha Y_t + (1 - \alpha)\hat{Y}_t. \tag{9.1}$$

The properties that made the model rather popular among forecasters are evident for such a grouping:

$$\hat{Y}_{t+1} = \hat{Y}_t + \alpha\left(Y_t - \hat{Y}_t\right). \tag{9.2}$$

S. Svetunkov, *Complex-Valued Modeling in Economics and Finance*,

DOI 10.1007/978-1-4614-5876-0_9,

From this way of writing BrownBrown's model we see that the forecasted value is calculated using the previously forecasted one but corrected by the deviation of fact from forecast.

The accuracy of Brown's model for forecasting problems is determined by values of coefficient α, called the smoothing constant. Smoothing constant α values lie within the range

$$0 < \alpha < 2. \tag{9.3}$$

Since numerous values of Brown's model lying within

$$1 \leq \alpha < 2 \tag{9.4}$$

have been ignored by economists up till now, and as early as 1997 it was called the transcendental set of Brown's method [2], we will not dwell on it. We merely mention that situations with transcendental cases of Brown's method are quite frequent in modern economics, especially when the researcher is faced with the problem of forecasting irreversible processes.

Ivan Svetunkov developed the logic of Brown's method and proposed the use of features of complex-valued economics to forecasting of two parameters simultaneously: parameter Y_t values and its deviation from the actual one ε_t:

$$\varepsilon_t = Y_t - \hat{Y}_t.$$

To do this the forecasted parameter and deviation from it can be represented in the form of a single complex variable:

$$Y_t + i\varepsilon_t, \tag{9.5}$$

which is considered to be the subject of short-term forecasting by Brown's method.

Since both the parameter and the deviation have the same dimension and show various sides of the same process, this variable has a right to exist. Let us designate the calculated forecasting value of this complex variable as

$$\hat{Y}_t + i\hat{\varepsilon}_t.$$

Then, similarly to Brown's model (9.1), we can obtain the following forecasted model:

$$\hat{Y}_{t+1} + i\hat{\varepsilon}_{t+1} = (\alpha_0 + i\alpha_1)(Y_t + i\varepsilon_t) + ((1+i) - (\alpha_0 + i\alpha_1))\left(\hat{Y}_t + i\hat{\varepsilon}_t\right). \tag{9.6}$$

Unlike Brown's model, in the case under consideration we use the complex smoothing constant:

$$\alpha_0 + i\alpha_1. \tag{9.7}$$

However, the variation limits of this smoothing constant need not coincide with those determined for the original Brown's model (9.3). The boundaries will be examined later, but for now we will study the properties of Ivan Svetunkov's short-term forecasting model.

Taking into account properties of complex variables, namely, their real and imaginary parts, the complex-valued model (9.6) can be reduced to the following system of real equations:

$$\begin{cases} \hat{Y}_{t+1} = \left(\alpha_0 Y_t + (1-\alpha_0)\hat{Y}_t\right) - \left(\alpha_1 \varepsilon_t + (1-\alpha_1)\hat{\varepsilon}_t\right) = \left(\hat{Y}_t^0\right) - \left(\hat{\varepsilon}_t^1\right), \\ \hat{\varepsilon}_{t+1} = \left(\alpha_0 \varepsilon_t + (1-\alpha_0)\hat{\varepsilon}_t\right) + \left(\alpha_1 Y_t + (1-\alpha_1)\hat{Y}_t\right) = \left(\hat{\varepsilon}_t^0\right) + \left(\hat{Y}_t^1\right). \end{cases} \tag{9.8}$$

It is seen from (9.8) that the forecasted value $\hat{Y}_{t+1}$ is determined as some forecasted value $\hat{Y}_t^0$ found by Brown's method and corrected for some value $\hat{\varepsilon}_t^1$, also forecasted by Brown's method $\hat{\varepsilon}_t^1$. For its part, the forecasted value of the correction index $\hat{\varepsilon}_{t+1}$ is also determined by two constituents found by the same Brown's method, only by addition: the forecasted correction index $\hat{\varepsilon}_t^0$ and the forecasted value $\hat{Y}_t^1$. Here, the top indexes "0" and "1" show which value of the two is used to calculate these values (α_0 or α_1).

Obviously, in model (9.6) the forecasted values are created by the previous factual ones with some complex weights set by the algorithm, which is similar to exponential smoothing in Brown's method but somewhat more complicated. Let us represent in formula (9.6) the calculated value $\hat{Y}_t + i\hat{\varepsilon}_t$ via the previous actual Y_{t-1} $+i\varepsilon_{t-1}$:

$$\hat{Y}_{t+1} + i\hat{\varepsilon}_{t+1} = (\alpha_0 + i\alpha_1)(Y_t + i\varepsilon_t) + (\alpha_0 + i\alpha_1)[(1-\alpha_0) + i(1-\alpha_1)]((Y_{t-1} + i\varepsilon_{t-2}) + \ldots$$

Then it is evident that this case also involves the process of calculating the weighted average, but using complex numbers and variables. Complex variables of (9.5) are multiplied by weight coefficients, which in this case are complex. The series of these complex weights from the above-mentioned equality can be represented as follows:

$$(\alpha_0 + i\alpha_1),\ (\alpha_0 + i\alpha_1)[(1-\alpha_0) + i(1-\alpha_1)],\ (\alpha_0 + i\alpha_1)[(1-\alpha_0) + i(1-\alpha_1)]^2, \ldots \tag{9.9}$$

which is simply the geometric progression series of complex numbers.

With this range we can determine the variation boundaries of the complex smoothing variable (9.7). By definition this series should converge since otherwise older observations would carry more weight than the new ones, which could result in inaccuracy of the model forecast.

This means that for the complex series of weights (9.9), the following equality should hold:

$$\lim_{n\to\infty} (\alpha_0 + i\alpha_1)[(1-\alpha_0) + i(1-\alpha_1)]^n = 0. \tag{9.10}$$

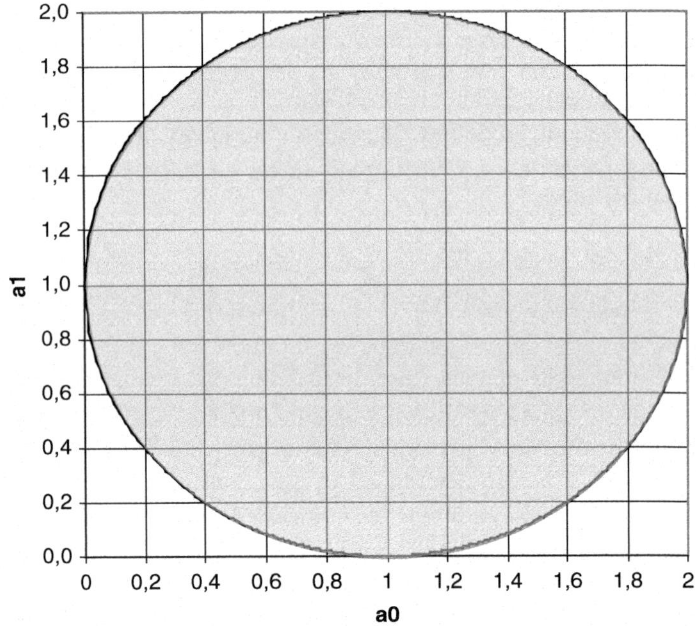

Fig. 9.1 Domain of complex smoothing constant

With constant values of the complex smoothing constant and obviously existing inequality

$$(\alpha_0 + i\alpha_1) \neq 0,$$

equality (9.10), as well as for the series of the geometric progression of real numbers, converges to a certain number when the following condition is met:

$$\sqrt{(1-\alpha_0)^2 + (1-\alpha_1)^2} < 1 \tag{9.11}$$

In the a_0 and a_1 plane inequality (9.11) represents a circle with its center in the point (1;1) and unit radius (Fig. 9.1). If the value of the complex smoothing constant is inside the circle, then series (9.9) converges to a complex number. If the value appears to be on the boundary or goes beyond the circle, then series (9.9) diverges.

From (9.11), by elementary transformations one can obtain the boundaries within which a_1 should lie for the series to converge:

$$1 - \sqrt{1 - (1-\alpha_0)^2} < \alpha_1 < 1 + \sqrt{1 - (1-\alpha_0)^2}.$$

For its part, this term is clearly true only when the radicand is positive (the situation where α_I is a complex number is not considered):

$$1-(1-\alpha_0)^2>0. \tag{9.12}$$

From the restriction (9.12) it is easy to derive the boundaries within which α_0 and α_1 should lie:

$$\begin{cases} 0<\alpha_0<2, \\ 0<\alpha_1<2, \\ 1-\sqrt{1-(1-\alpha_0)^2}<\alpha_1<1+\sqrt{1-(1-\alpha_0)^2}. \end{cases} \tag{9.13}$$

Thus, we have the boundaries within which the real and imaginary parts of the complex smoothing coefficient should lie for the series of complex weights (9.9) to converge to a certain number. In addition, Ivan Svetunkov calculated what this number is. It is calculated according to the formula

$$S=\frac{(\alpha_0^2-\alpha_1+\alpha_1^2)+i(\alpha_0)}{\alpha_0^2+(1-\alpha_1)^2}. \tag{9.14}$$

One can see that in (9.4) the real part can be represented by any real number (positive or negative) but the imaginary part can be represented only by a positive real number. This means that in model (9.6) the series of weights (9.9) has a more complex meaning than in Brown's model, where for the model to exist the series of weights should definitely converge to *1*.

The specific properties of model (9.6) also appear in the fact that the series of weights (9.9) converge to (9.14) with varying rate depending on the constant smoothing value. Research has shown that the closer the constant smoothing value is to the circle edge (9.11), the slower is the convergence. The modulus of the number is in this case a certain characteristic of the convergence rate:

$$v=\sqrt{(1-\alpha_0)^2+(1-\alpha_1)^2}. \tag{9.15}$$

According to (9.11) v lies within a range of 0 to 1, and the closer it is to 1, the more inert is the model, i.e., the forecast is influenced by obsolete data, and the closer it is to 0, the less important are the old data in the creation of the forecast, and the model becomes more adaptive.

We have data on the magnitude of electric power generation by windmills in one U.S. states. This time series consists of 20,000 observations taken every half hour. Since these data are intended exclusively for official use, it was agreed that the American side would not provide even selective values of this series but demonstrate only the results of the calculations.

If one uses Brown's method directly for this series, the optimal value of α is 0.28, which shows that the model adapts slowly to incoming information. The mean relative approximation error in this case was 7.54 %, the determination coefficient was 0.1485, and the compliance coefficient was 94.22 %. These parameters show that the model does not describe and forecast wind generation series very well.

If we apply to the same series Ivan Svetunkov's model (9.6), we can also find the optimal value of the complex smoothing, which is equal to

$$\alpha_0 + i\alpha_1 = 0.59 + 1.00i$$

The mean relative approximation error is 5.06 %, the determination coefficient is 0.6051, and the compliance coefficient is 95.68 %. According to these parameters, the results forecast by the correction method appear to be more accurate than was possible with Brown's method or its identifications.

It should be noted that Ivan Svetunkov's model differs from all existing short-term forecasting models by its ability to detect trends in series data that make it possible to describe them. The correction model "predicts" the series dynamics while Brown's model always "procrastinates."

Ivan Svetunkov made a comparative analysis of the proposed model with various modifications to Brown's model (Holt method, Holt modification method, etc.). This analysis showed that in a series of cases model (9.6) gives much more accurate results than these modifications. This reflects the fact that Ivan Svetunkov's model can be included in the arsenal of tools for forecasting socioeconomic activity along with the existing models.

In conclusion, it should be noted that additional research by E. Tsedyakova showed that Ivan Svetunkov's model gives better forecasts in data series with chaotic dynamics for which the optimal smoothing coefficient, according to Brown's, model is close to 0.

9.2 Complex-Valued Autoregression Models

If application of Brown's model of complex variables leads to such interesting results, it is useful to consider other forecasting models to which complex variables could be applied.

The complex-valued production functions considered in Chaps. 5 and 6 can well be used for these purposes as forecasting models – linear and nonlinear. These models' properties were studied earlier and there is no sense in repeating them here.

From these points of view autoregression models seem more interesting if one considers autoregression of a complex variable.

The first-order autoregression model of a complex variable can be written as follows:

$$y_{rt} + iy_{it} = (a_0 + ia_1)(y_{rt-1} + iy_{it-1}), \quad t = 1, 2, 3, \ldots \tag{9.16}$$

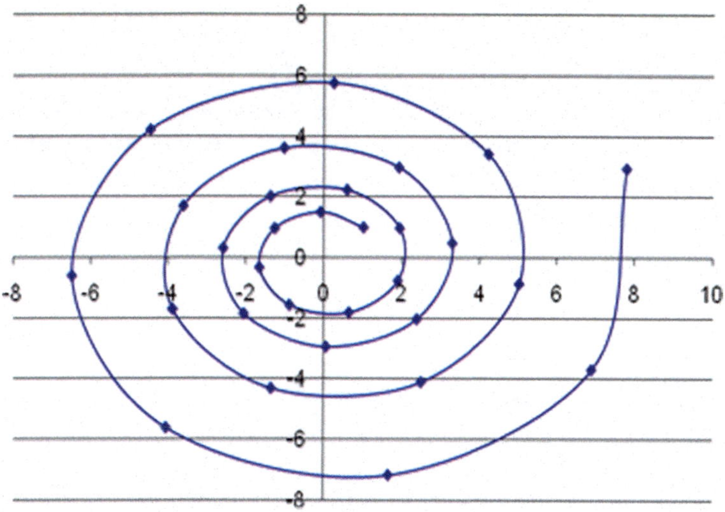

Fig. 9.2 Complex-valued autoregression model at power base with modulus higher than 1 (9.20)

It is important to recall here that the first-order autoregression model of a real variable is written as follows:

$$y_t = ay_{t-1} \tag{9.17}$$

and ultimately represents the following function:

$$y_t = a^t y_0. \tag{9.18}$$

The dynamic of variation in time of this exponential function is completely determined by the function base – coefficient a.

It is easy to show that in a similar way the complex-valued autoregression model (9.16) can be represented as a power function:

$$y_{rt} + iy_{it} = (a_0 + ia_1)^t (y_{r0} + iy_{i0}). \tag{9.19}$$

It is known that the complex-valued power function is periodic and diverges in the form of a spiral with increases in the exponent if the modulus of the base is greater than one and converges along the spiral to zero if the complex-valued modulus in the power base is less than one.

The graph in Fig 9.2 shows the variation in the autoregression function:

$$y_{rt+1} + iy_{it+1} = (0.7 + i0.8)(y_{ro} + iy_{i0}), \quad y_{ro} = 1, y_{i0} = 1. \tag{9.20}$$

The complex coefficient autoregression modulus is equal to $1.063 > 1$. Therefore, the graph shows how the parameter modeled by autoregression diverges from

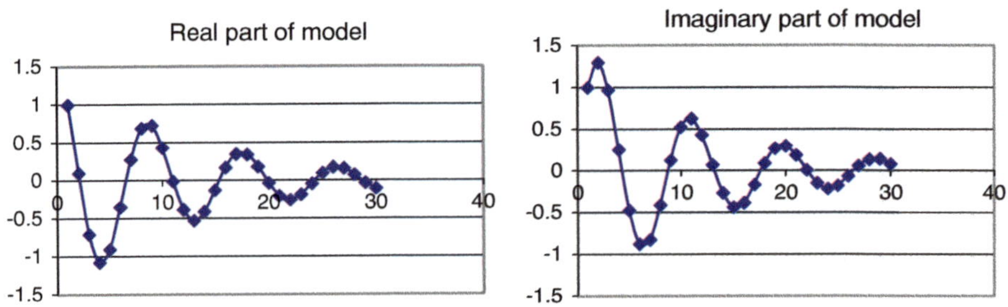

Fig. 9.3 Variation in time of real and imaginary parts of complex-valued autoregression model (9.20)

the initial point in the complex plane with coordinates (1;1) along a spiral. It is evident that one can set any initial point and the character of the modeled relationship will not change, except for the initial point's location.

Since in economics phase planes like the one shown in Fig. 9.2 are not practically applied, using mainly time variations of the parameters, Fig. 9.3 shows the variation in time of the real and imaginary parts of the autoregression model (9.20). Their dynamics have a fluctuating character that diverges with time.

If in the complex-valued autoregression model the modulus of the complex coefficient of autoregression is less than one, then the model will generate a process converging to the zero point along a spiral. Thus, for the model

$$y_{rt} + iy_{it} = (0.7 + i0.6)(y_{rt-1} + iy_{it-1}), \quad y_{ro} = 1, \ y_{i0} = 1, \tag{9.21}$$

the modulus of the complex-valued proportionality coefficient is equal to $0.922 < 1$.

The autoregression model generates a series converging to zero from the initial point with coordinates (1;1) in the complex plane (Fig. 9.4).

Variation in time of the real and imaginary parts of the autoregression model (9.21) will have an oscillatory and damped character, and over time, both parts tend to zero.

Second- and higher-order autoregression models are more complicated. Let us consider the second-order autoregression model of a real variable:

$$y_t = ay_{t-2}, \quad t = 2, 3, 4, \ldots \tag{9.22}$$

For $t = 2$ it is easy to see that $y_2 = ay_0$. For $t = 3$, $y_3 = ay_1$ is calculated. We see that these two calculated values do not depend on each other. Further, we see that for $t = 4$ $y_4 = ay_2 = a^2y_0$, for $t = 5$ $y_5 = ay_3 = a^2y_1$ are calculated.

In general, for the second-order autoregression model (9.22):

$$\begin{cases} y_t = a^{\frac{t}{2}}y_0, & \text{if } t \text{ is an even number,} \\ y_t = a^{\frac{t-1}{2}}y_1, & \text{if } t \text{ is odd.} \end{cases} \tag{9.23}$$

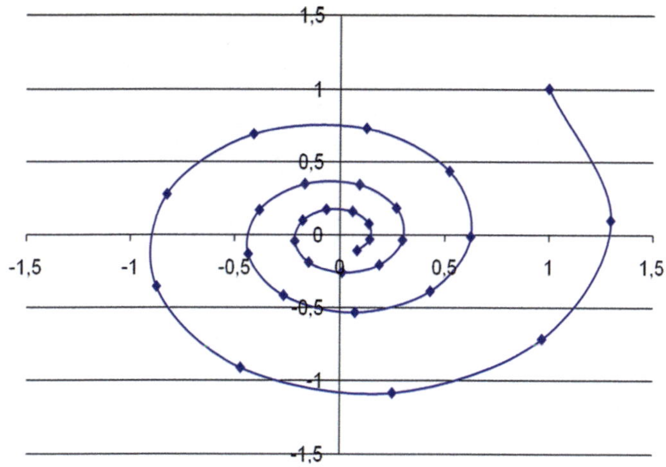

Fig. 9.4 Complex-valued autoregression model at power base with modulus less than one

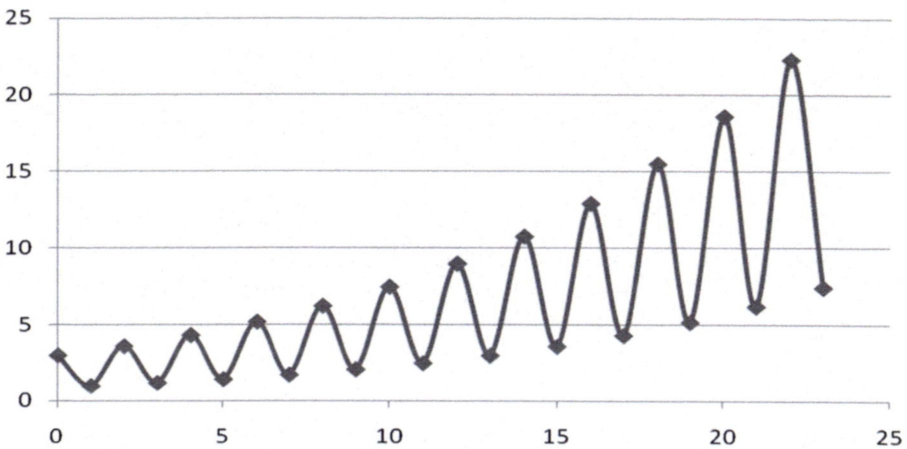

Fig. 9.5 Second-order autoregression model $y_t = 1.3y_{t-2}, \quad y_0 = 3, \quad y_1 = 1$

Then, if the modulus of the autoregression coefficient is higher than one, the model generates an oscillatory process with increasing amplitude, as shown in Fig. 9.5, and if the modulus is less than one, the model will generate a damped oscillatory process.

The character of the modeled process for second-order autoregression is determined by the autoregression coefficient and the first two values of the modeled series.

Similarly, the second-order autoregression model for the complex series

$$y_{rt} + iy_{it} = (a_0 + ia_1)(y_{rt-2} + iy_{it-2}), \quad t = 2, 3, \ldots \tag{9.24}$$

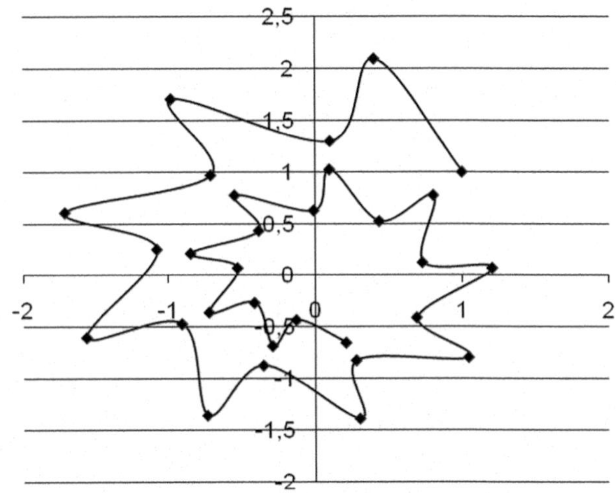

Fig. 9.6 Second-order complex-valued autoregression model (9.25)

will have a more complicated dynamic than that of the first-order model. The character of the dynamics is determined by both values of the complex auto regression coefficient and the first two values of the complex-valued series.

As an example, let us consider the dynamics of the second-order complex-valued autoregression model

$$y_{rt} + iy_{it} = (0.7 + i0.6)(y_{rt-2} + iy_{it-2}), \quad y_{ro} = 1, y_{i0} = 1, \quad y_{r1} = 0.4, y_{i1} = 2.1. \tag{9.25}$$

The phase portrait of this autoregression model is given in Fig. 9.6.

Since the modulus of the autoregression coefficient is less than one, the model converges to zero. If it is higher than one, the model will diverge in the complex plane.

Figure 9.7 demonstrates the variation in time of the real and imaginary parts of this second-order complex-valued autoregression model.

The figure demonstrates that both parts of the model show a damped process with a complex oscillatory structure. It is evident that their oscillation period is the same and they are shifted with respect to each other by the same lag.

Autoregression models of a more complex order will generate more complex variants of the dynamics.

This dynamic can be demonstrated by the autoregression model with lags distributed for one and two observations:

$$\begin{aligned} y_{rt} + iy_{it} &= (-0.3 - i0.4)(y_{rt-1} + iy_{it-1}) + (0.6 - i0.8)(y_{rt-2} + iy_{it-2}), \\ y_{ro} &= 1, y_{i0} = 1, \quad y_{r1} = 1.2, y_{i1} = 0.8 \end{aligned} \tag{9.26}$$

The dynamics of variation of the real and imaginary parts of these parameters are given in Fig. 9.8.

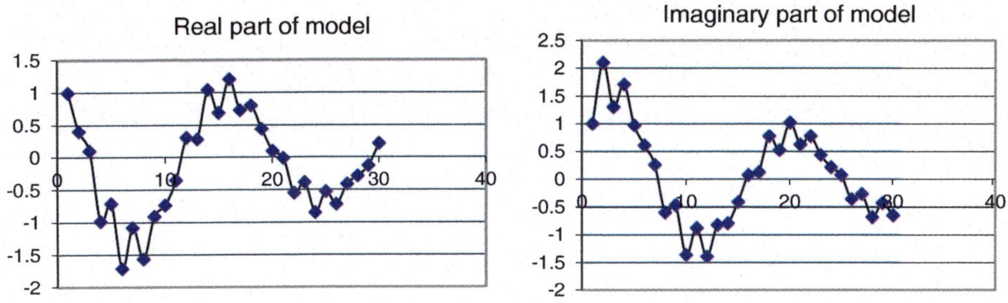

Fig. 9.7 Real and imaginary parts of second-order complex-valued autoregression model (9.25)

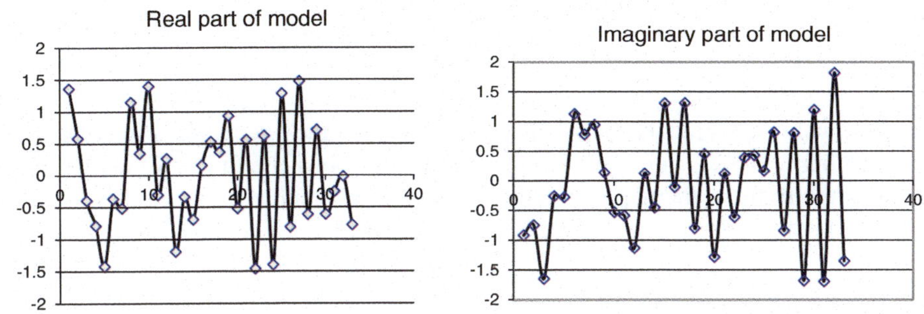

Fig. 9.8 Real and imaginary parts of complex-valued autoregression model with distributed lags (9.26)

The give graph of complex-valued autoregression models show that they can be applied to the modeling of many processes with a seasonal component or, for example, in stock markets. This means that complex-valued autoregression models can take their rightful place in the range of socioeconomic dynamic forecasting models. Coefficients of these models for the available statistical data can be found using LSM, especially as described in Chap. 4.

9.3 Solow's Model of Economic Dynamics and Its Complex-Valued Analog

Solow's model underlies numerous modern economic dynamics models. It is not necessary to delve deep into the specifics of the economic dynamics modeling apparatus. We will only show how complex-valued economics can enrich the apparatus of economic and mathematical modeling using Solow's fundamental model as an example.

Solow's model has the following form:

In the model, the final product is determined the by Cobbs-Duglas function:

$$Y_t = aK_t^{\alpha}L_t^{1-\alpha}. \tag{9.27}$$

In discrete time t the final product Y_t is distributed to gross investments I_t and consumption C_t:

$$Y_t = I_t + C_t. \tag{9.28}$$

It is assumed that the part of the final product that falls on investments is set in the form of the rate of accumulation ρ:

$$I_t = \rho Y_t. \tag{9.29}$$

Obviously, investments contribute to the growth of the fixed production assets of the coming year K_{t+1} and are expressed via the obsolete funds K_t taking into account the share of the depleted fixed production assets μ:

$$K_{t+1} = (1 - \mu)K_t + I_t. \tag{9.30}$$

The number of those employed in the economy L_{t+1} is determined via the number of employed in the current year L_t taking into account the annual growth rate of the number of employees v:

$$L_{t+1} = (1 + v)L_t. \tag{9.31}$$

The interrelated (9.27), (9.28), (9.29), (9.30), and (9.31) represent Solow's model, and they can be used to model economic growth of some idealized system.

Instead of the Cobbs-Douglas production function, let us use in the economic dynamics model the complex-valued power production function with real coefficients that was considered in detail in Chap. 6. Then the economic dynamics model will have the following form:

$$I_t + iC_t = a(K_t + iL_t)^b, \tag{9.32}$$

$$K_{t+1} = (1 - \mu)K_t + I_t, \tag{9.33}$$

$$L_{t+1} = (1 + v)L_t. \tag{9.34}$$

We see that the model obtained involves only three equalities since the production function immediately calculates accumulation I_t, which is to be used later on, and consumption C_t. We see that the rate of accumulation is not specified. If necessary, it can be calculated separately:

$$\rho = \frac{I_t}{I_t + C_t}.$$

It is obvious that the exponent of model (9.32) fulfills this role because

$$\rho = f(b).$$

If we use the other complex-valued production functions studied in Chap. 6 instead of the complex-valued power production function, we obtain completely different models of economic dynamics and other development trajectories. The model can be expanded by replacing other equations with complex-variable models, thereby developing the instrumental base of economists who model economic dynamics.

Ilyas Abdullayev, a lecturer at Khorezm National University, used the basic conditions of a complex-valued economy to build a system of models of economic dynamics of particular branches of the Republic of Uzbekistan and Khorezm Region. Let us use only one result to demonstrate how to apply complex-valued economics to the modeling of real economic dynamics.

For the manufacturing dynamics of the Republic of Uzbekistan from 1995 to 2008 using Solow's economic dynamics model, Abdullayev set out to build models with the following production functions:

Cobb-Douglas production function,
Neoclassical production function,
Power production function and
Complex-valued production function

The Cobb-Douglas function is not defined on this set – the exponent with labor resources is negative; in addition, the "neoclassical" production function is undefined as well. LSM makes it possible to find on this set only the power production function, which has the form

$$Q = 0.13K^{1.2}L^{5.8}. \tag{9.35}$$

This model poorly approximates the original data. The mean absolute approximation error is 132 %. Therefore, this function should not be used in economic dynamics models.

Using the method of synthesizing monofactor complex-valued models into one multifactor complex-valued model described in Sect. 6.7 Abdullayev obtained the following production function model of Uzbekistan manufacturing:

$$Q = 0.832e^{(-0.53-i2.7)}K^{(0.37+i0.121)}(L_{nnt} + iL_{it})^{3.03}. \tag{9.36}$$

Here, L_{nnt} is the number of manufacturing and production workers, and L_{it} is the number of nonproduction workers in Uzbekistan manufacturing (in dimensionless relative values).

Model (9.36) describes the dynamics of Uzbekistan's manufacturing with an average absolute approximation error of 32.8 %, which is almost five times more accurate than real-variable models.

9.4 Modeling Regional Socioeconomic Development

The socioeconomic development of a region is a very complex subject in scientific research because the level of this development results from many factors and conditions.[1] Any region of a given country represents a complex system of interrelating economic, social, political, historic, cultural, and other subsystems. Uneven regional socioeconomic status, the objective result of a great variety of development conditions, leads to the need for redistributing income at the state level. The general scheme of this redistribution is similar all over the world – each region pays a legally stipulated part of the collected taxes to the state treasury, leaving the other part at the disposal of the regional budgets.

The regions, called "donors," collect enough taxes so that the remaining funds exceed the sum necessary to support subsistence living. This surplus allows regional authorities to pass more active social and economic policies in the region. As a rule, they receive no return on the funds they transfer to the state budget (except for some state target programs).

Economically underdeveloped regions transfer to the treasury part of the collected taxes in the same proportion to the funds retained for their use as donor regions. But the funds remaining in the regional budget do not allows the region to work at a breakeven level: the authorities cannot "feed" themselves, much less implement planned social programs. For these regions subsidies come from the state budget at the expense of the funds coming from donor regions. These subsidies make it possible to solve the set social problems and to a certain extent level out social development of the regions.

In this situation where the level of social development is subject to thorough regulation by state authorities, and it is quite similar in state regions but the level of economic development is different, it is not very correct to speak about socioeconomic development of regions as a single parameter of regional authorities' activity. It makes sense to talk about two components – the social development of a region and its economic development. Modern economics does not do this – all approaches to studying regional development are based on the principle of combined estimation of regional social and economic development. It is for this reason that the estimation process involves numerous development indexes representing a convolution of an aggregate of various parameters – economic and social – in one

[1] This work performed under the International grant Russian Foundation for Humanities – The National Academy of Sciences of Ukraine № 10-02-00716/U "Valuation models are uneven and cyclical dynamics of the socio-economic Development of Regions of Ukraine and Russia".

index, though they are measured on various scales. Constructive critical analysis of these indexes shows that they differ in either the convolution method or the convolution filling. In any case these indices are obtained as the results of the summation of two different parts.

As it is not allowed to add the demand price and the volume of demand to estimate the demand, so similarly, it is senseless to combine the characteristics of the social and economic development of a region into one parameter to estimate the level of its development. It was shown earlier that the former are the result of state policy, the latter of regional policy.

This is why a complex-valued indicator of socioeconomic development was proposed that was free of this defect [3]. This complex-valued parameter *Z* includes a real part as a ratio of the average per capita income *C* to the subsistence minimum *LV*, which can be called "wealth level" *d*, and an imaginary part as the ratio of paid services to the population *PS* to the amount of the total commodity turnover of the region *CC*, which can be called "social satisfaction level" *s*:

$$Z = \frac{C}{LV} + i\frac{PS}{CC} = d + is. \tag{9.37}$$

The real part of this complex-valued parameter characterizes the income level of the residents of this region and therefore is a generalizing indicator of the level of its economic development. Indeed, the average income per capita *C* characterizes the entire aggregate of cash funds received by the average resident of this region from various sources, primarily from the earnings at production enterprises in the region. It is one of the rather accurate and broad indicators of the level of economic development of a region. The subsistence minimum *LV* characterizes the development of market relations in the region – the more competition there is in the regional market, the more participants compete in the market, the lower the prices for the goods included in the subsistence minimum, the lower the subsistence minimum. Then it is obvious that the ratio of the average per capita income to the subsistence minimum, called the wealth level, shows the average economic well-being of the region. Table 9.1 shows data on the change in the real part of the complex-valued parameter of regional development for Northwest Russia.

One can see from the table that with respect to the dynamics of economic development, the leader in 2007 was the Nenets autonomous district, followed by Saint-Petersburg. It is interesting that the gross regional product per capita was 1.156 million rubles per person in the Nenets autonomous district in 2007 and 0.243 million rubles per person in Saint-Petersburg.

The Republic of Karelia, Archangelsk region, Vologda region, Kaliningrad region, Leningrad region, Murmansk region, Novgorod region and Pskov region are significantly behind the leaders – the average wealth level for them in 2007 was 2.57, which is 2.3 times less than the leading region.

The dynamics of the imaginary part of this parameter – the level of social satisfaction of residents in the region –is also interesting. It is given in Table 9.2.

Table 9.1 Dynamics of wealth level in certain regions of Russia

Number	Region	2001	2002	2003	2004	2005	2006	2007
1	Republic of Karelia	–	2.11	2.09	2.15	2.25	2.43	2.29
2	Komi Republic	2.53	2.71	2.80	2.97	3.10	3.13	3.26
3	Archangelsk region	2.33	1.86	2.01	2.23	2.39	2.40	2.52
4	*Nenets autonomous district*	*2.27*	*2.65*	*3.62*	*4.52*	*4.27*	*4.70*	*5.97*
5	Vologda region	–	1.91	2.18	2.33	2.27	2.51	2.70
6	Kaliningrad region	1.85	1.41	1.69	1.78	2.05	2.50	2.89
7	Leningrad region	–	1.27	1.43	1.86	2.10	2.55	2.87
8	Murmansk region	2.13	2.14	2.19	2.27	2.27	2.39	2.68
9	Novgorod region	–	1.72	1.83	1.86	2.00	2.29	2.26
10	Pskov region	1.43	1.78	2.02	2.16	2.09	2.21	2.36
11	*Saint-Petersburg*	–	*2.03*	*2.67*	*3.25*	*3.92*	*4.08*	*4.31*

Table 9.2 Dynamics of rate of social satisfaction in certain regions of Russia

Number	Region	2000	2001	2002	2003	2004	2005	2006	2007
1	Republic of Karelia	0.22	0.22	0.23	0.28	0.30	0.31	0.31	0.32
2	Komi Republic	0.19	0.19	0.20	0.23	0.24	0.24	0.25	0.25
3	Archangelsk region	0.24	0.22	0.26	0.26	0.31	0.34	0.37	0.42
4	*Nenets autonomous district*	*0.16*	*0.13*	*0.14*	*0.16*	*0.17*	*0.15*	*0.21*	*0.24*
5	Vologda region	0.27	0.29	0.32	0.33	0.37	0.45	0.46	0.43
6	Kaliningrad region	0.23	0.25	0.30	0.32	0.32	0.33	0.35	0.34
7	*Leningrad region*	*0.20*	*0.21*	*0.25*	*0.28*	*0.23*	*0.24*	*0.23*	*0.21*
8	Murmansk region	0.21	0.25	0.31	0.36	0.41	0.50	0.52	0.48
9	Novgorod region	0.22	0.23	0.26	0.31	0.35	0.36	0.36	0.34
10	Pskov region	0.19	0.20	0.21	0.23	0.24	0.25	0.27	0.27
11	Saint-Petersburg	0.36	0.38	0.41	0.47	0.47	0.45	0.42	0.42

An interesting fact is that practically all the regions, districts, and the Komi Republic are characterized by an increase in the level of social satisfaction, except for the Leningrad region.

It is surprising that the Nenets autonomous district, which is the leader with respect to wealth level (Table 9.1), is an outsider, along with the Leningrad region, with respect to the level of social satisfaction (Table 9.2).

This can be explained by the fact that many workers in the region are not permanent residents. These "shift workers" agree to refuse social services in exchange for higher pay and subsequent satisfaction of their needs and demands in other regions of the country.

The low position of the Leningrad region in the social satisfaction rating is quite clear – some residents in the Leningrad region live in the suburbs of Saint-Petersburg, where they work and satisfy their social needs (salons, clinics, clubs, theaters, museums, etc.).

It is obvious that consideration of the real and imaginary parts of the complex parameter of socioeconomic development provides the researcher with important information that makes it possible to estimate the level of the results obtained.

However, it is not the two real variables that are of interest, but one complex one. Simple consideration of these complex-valued variables does not tell the researcher much (Table 9.3).

Since a complex variable can be written not only in arithmetic but also in exponential and trigonometric form

$$Z = \frac{C}{LV} + i\frac{PS}{CC} = d + is = Re^{i\theta} = R\cos\theta + iR\sin\theta, \tag{9.38}$$

its additional characteristics are the modulus of the complex variable

$$R = \sqrt{d^2 + s^2} \tag{9.39}$$

and its polar angle

$$\theta = arctg\frac{s}{d}. \tag{9.40}$$

It is easy to see that a decrease in the modulus values (9.39) can occur only in the event of a decrease in one of its constituents characterizing either economic or social development. The constant nature of the modulus reflects both the growth in one parameter and a decrease in the other. If the polar angle decreases, the numerator of fraction (9.40) decreases compared to the denominator, i.e., the level of social development is lower than that of economic development.

Table 9.4 shows the dynamics of the modulus of the complex-valued indicator of socioeconomic development of the same regions of Northwest Russia. Analysis of this parameter makes it possible to draw some conclusions. In the last year under observation socioeconomic development of such regions as the Republic of Karelia and the Vologda and Novgorod regions has slowed down.

Modules of complex-valued parameters of other regions have positive dynamics, which reflects the growth in the level of socioeconomic development of each region as the whole. Especially high rates are characteristic for the Nenets autonomous district. Previously, we looked at the fact that this region is characterized by a high rate of economic development and a low rate of the social component. Therefore, additional information on the correlation of the economic and social constituents and their dynamics can be obtained from an analysis of changes in the polar angle of the complex-valued parameter. Table 9.5 gives this dynamic for the regions under consideration.

It follows from (9.40) that the polar angle will grow if growth of the social component exceeds the growth of the economic one, and the polar angle will decrease if there is a more intensive growth of the economic component and less noticeable growth of the social one.

The polar angle may remain unchanged when both the economic and social components characterizing the region's condition have similar growth rates.

Table 9.3 Dynamics of complex-valued parameter of socioeconomic development in certain regions of Russia

Number	Region	2001	2002	2003	2004	2005	2006	2007
1	Republic of Karelia	–	2.11 + 0.23i	2.09 + 0.28i	2.15 + 0.3i	2.25 + 0.31i	2.43 + 0.31i	2.29 + 0.32i
2	Komi Republic	2.53 + 0.19i	2.71 + 0.2i	2.8 + 0.23i	2.97 + 0.24i	3.1 + 0.24i	3.13 + 0.25i	3.26 + 0.25i
3	Archangelsk region	2.33 + 0.22i	1.86 + 0.26i	2.01 + 0.26i	2.23 + 0.31i	2.39 + 0.34i	2.4 + 0.37i	2.52 + 0.42i
4	Nenets autonomous district	2.27 + 0.13i	2.65 + 0.14i	3.62 + 0.16i	4.52 + 0.17i	4.27 + 0.15i	4.7 + 0.21i	5.97 + 0.24i
5	Vologda region	–	1.91 + 0.32i	2.18 + 0.33i	2.33 + 0.37i	2.27 + 0.45i	2.51 + 0.46i	2.7 + 0.43i
6	Kaliningrad region	1.85 + 0.25i	1.41 + 0.3i	1.69 + 0.32i	1.78 + 0.32i	2.05 + 0.33i	2.5 + 0.35i	2.89 + 0.34i
7	Leningrad region	–	1.27 + 0.25i	1.43 + 0.28i	1.86 + 0.23i	2.1 + 0.24i	2.55 + 0.23i	2.87 + 0.21i
8	Murmansk region	2.13 + 0.25i	2.14 + 0.31i	2.19 + 0.36i	2.27 + 0.41i	2.27 + 0.5i	2.39 + 0.52i	2.68 + 0.48i
9	Novgorod region	–	1.72 + 0.26i	1.83 + 0.31i	1.86 + 0.35i	2 + 0.36i	2.29 + 0.36i	2.26 + 0.34i
10	Pskov region	1.43 + 0.2i	1.78 + 0.21i	2.02 + 0.23i	2.16 + 0.24i	2.09 + 0.25i	2.21 + 0.27i	2.36 + 0.27i
11	Saint-Petersburg	–	2.03 + 0.41i	2.67 + 0.47i	3.25 + 0.47i	3.92 + 0.45i	4.08 + 0.42i	4.31 + 0.42i

Table 9.4 Dynamics of modulus of complex-valued indicator of socioeconomic development of some Northwest regions of Russia

Number	Region	2001	2002	2003	2004	2005	2006	2007
1	Republic of Karelia	–	3.027	3.320	3.420	3.576	3.692	3.633
2	Komi Republic	3.152	3.368	3.542	3.763	3.848	3.902	4.007
3	Archangelsk region	3.143	3.013	3.138	3.541	3.773	3.936	4.240
4	Nenets autonomous district	2.606	2.992	3.962	4.822	4.516	5.066	6.329
5	Vologda region	–	3.406	3.596	3.878	4.247	4.397	4.389
6	Kaliningrad region	2.983	3.037	3.288	3.335	3.543	3.913	4.115
7	Leningrad region	–	2.680	2.897	2.866	3.041	3.348	3.510
8	Murmansk region	3.164	3.461	3.755	4.072	4.487	4.625	4.634
9	Novgorod region	–	2.961	3.320	3.564	3.681	3.806	3.708
10	Pskov region	2.459	2.685	2.979	3.141	3.152	3.329	3.423
11	Saint-Petersburg	–	3.951	4.587	4.925	5.248	5.330	5.496

Table 9.5 Dynamics of polar angle (in radians) of complex-valued rate of socioeconomic development in certain regions of Russia

Number	Region	2001	2002	2003	2004	2005	2006	2007
1	Republic of Karelia	–	0.799	0.889	0.891	0.891	0.852	0.889
2	Komi Republic	0.639	0.636	0.659	0.661	0.634	0.639	0.621
3	Archangelsk region	0.736	0.905	0.876	0.889	0.885	0.915	0.934
4	Nenets autonomous district	0.513	0.483	0.418	0.356	0.332	0.382	0.338
5	Vologda region	–	0.975	0.919	0.926	1.006	0.963	0.908
6	Kaliningrad region	0.902	1.088	1.031	1.008	0.954	0.878	0.792
7	Leningrad region	–	1.077	1.055	0.864	0.809	0.705	0.613
8	Murmansk region	0.832	0.904	0.948	0.979	1.04	1.028	0.954
9	Novgorod region	–	0.951	0.987	1.022	0.996	0.925	0.915
10	Pskov region	0.950	0.846	0.826	0.812	0.846	0.845	0.810
11	Saint-Petersburg	–	1.031	0.949	0.850	0.727	0.699	0.669

Table 9.5 shows that the polar angle is growing only for the Archangelsk region. Earlier, separate analysis of the real and imaginary parts of the complex-valued parameter (9.37) and analysis of the modulus dynamics did not distinguish the Archangelsk region in any way. Now we have discovered that it actively implements social programs and does everything for the residents to live more comfortably in their region.

Regions where the dynamics of the polar angle are approximately constant are the Republic of Karelia and Pskov region. The low modulus shows that no active state programs – either economic or social – are implemented there. These regions are developed independently, without active state support.

All the other regions of the Northwest are characterized by nonlinear dynamics with a decrease in the polar angle with time. This means that improvement of the economic conditions of these regions exceeds the satisfaction of the social needs and demands of their residents.

Since the complex-valued indicator of the regions' condition made it possible to reveal trends of socioeconomic development of Northwest regions of Russia, a natural question arises as to whether these trends can be described by some law. This is crucial because when analyzing the socioeconomic development of a region, one should judge the trends in general but not separate their components, i.e., it is necessary to aggregate the information.

There is a simple way to do this –build a trend of development of the economic component d of the complex parameter (9.37) showing the socioeconomic development of a region:

$$d = f_d(t) \tag{9.41}$$

and a trend of its social constituent:

$$s = f_s(t). \tag{9.42}$$

After that it is rather easy to make a forecast for each of the components for a certain period. However, this approach does not take into account the interrelation between the economic and social components of the common complex-valued parameter, which definitely exists. Therefore, it is more correct to build a complex-valued trend where the complex parameter (9.37) is represented as a certain complex-valued function of time that is represented as a real discrete variable:

$$d = is = f(t). \tag{9.43}$$

Various functions that may be used as these models are considered in detail in the Chap. 2. The power function is the most suitable for our problem. With reference to the class of models under consideration it will have the form

$$y_{rt} + iy_{it} = (a_0 + ia_1)t^{(b_0+ib_1)}. \tag{9.44}$$

The complex proportionality coefficient can be represented in exponential form:

$$a_0 + ia_1 = R_a e^{\alpha}.$$

Then (9.44) can be represented in a form that is convenient for analyzing the model properties:

$$y_{rt} + iy_{\text{it}} = R_a e^{\alpha} t^{b_0} e^{ib_1 \ln t} = R_a t^{b_0} e^{\alpha + ib_1 \ln t}. \tag{9.45}$$

It is obvious from the formula that the modulus of the complex result is

$$R_{yt} = R_a t^{b_0} \tag{9.46}$$

and the modeled polar angle is

$$\theta = \alpha + b_1 \ln t. \tag{9.47}$$

It is easy to see that each constituent of the complex exponent influences the complex-valued result modeled by the trend. With a positive real component of the complex exponent b_0 the trend will be characterized by a growing modulus of the complex result, and with negative values it will be characterized by a decreasing modulus and all its component (economic and social). If this coefficient is equal to zero, then the modulus of the complex result does not change but the proportions between the economic and social result (on the circle) do. This is certainly an incredible case.

The imaginary part of the complex exponent has another influence on the result since it affects the polar angle of the modeled result. If coefficient b_1 is positive, then the polar angle increases, which is possible when the growing trend of the social component of the region development dominates over the growing trend of the economic one. If the imaginary component is less than zero, then the modeled trend represents a sharper increase in the economic component over the social one. If this coefficient is close to zero, the polar angle of the modeled parameter is constant and equal to α. This is possible when the proportions between the rates of social and economic development of a region do not vary with time.

It is evident now that the trend's complex parameter (9.44) operates as a certain diagnosing coefficient, where the real part characterizes rates of the region's development and the imaginary one shows which part is developing more actively – the economic or the social one. Knowing these specifics, one can assume that for regions developing in postindustrial society the real and imaginary parts of the complex exponent will be positive.

In order to use trend (9.44) for purposes of forecasting socioeconomic development of regions, it is necessary to estimate the coefficients of the trend models using statistical data. The problem of estimating complex-valued model coefficients by LSM has already been solved; thus, let us show how to do it with respect to the trend under consideration.

Since the trend is nonlinear, it should be linearized by taking logarithms of the left- and right-hand side of equality (9.44):

$$\ln(y_{rt} + iy_{it}) = \ln(a_0 + ia_1) + (b_0 + ib_1)\ln t. \tag{9.48}$$

To avoid solving huge systems of normal equations to estimate four real coefficients, let us center the variables of the linear model around their averages:

$$Y_{rt} + iY_{it} = \ln(y_{rt} + iy_{it}) - \bar{y}, \quad T = \ln t - \bar{\tau}, \tag{9.49}$$

where

$$\bar{y} = \sum \frac{\ln(y_{rt} + iy_{it})}{T}, \quad \bar{\tau} = \sum \frac{\ln t}{T}.$$

Table 9.6 Trends in socioeconomic development in certain regions of Russia

Number	Region	Trend in socioeconomic development
1	Republic of Karelia	$y_{rt} + iy_{it} = (2.055 + i0.240)t^{0.071+i0.013}$
2	Komi Republic	$y_{rt} + iy_{it} = (2.492 + i0.188)t^{0.130+i0.002}$
3	Archangelsk region	$y_{rt} + iy_{it} = (2.041 + i0.206)t^{0.079+i0.030}$
4	Nenets autonomous district	$y_{rt} + iy_{it} = (2.129 + i0.120)t^{0.479-i0.010}$
5	Vologda region	$y_{rt} + iy_{it} = (1.907 + i0.307)t^{0.173+i0.008}$
6	Kaliningrad region	$y_{rt} + iy_{it} = (1.467 + i0.263)t^{0.241-i0.014}$
7	Leningrad region	$y_{rt} + iy_{it} = (1.157 + i0.249)t^{0.456-i0.073}$
8	Murmansk region	$y_{rt} + iy_{it} = (2.042 + i0.240)t^{0.101+i0.046}$
9	Novgorod region	$y_{rt} + iy_{it} = (1.660 + i0.273)t^{0.162+i0.002}$
10	Pskov region	$y_{rt} + iy_{it} = (1.481 + i0.195)t^{0.239-i0.009}$
11	Saint-Petersburg	$y_{rt} + iy_{it} = (2.018 + i0.422)t^{0.429-i0.062}$

Then trend (9.44) with reference to the case under consideration will look like a simple linear complex-valued function of a real argument:

$$Y_{rt} + iY_{it} = (b_0 + ib_1)\, T. \tag{9.50}$$

Now, one can use LSM to find the complex exponent values:

$$b_0 + ib_1 = \frac{\sum (Y_{rt} + iY_{it})T}{\sum T^2}. \tag{9.51}$$

We used (9.51) and the data of Table 9.3 to build regression models of trends of each of the regions under consideration (Table 9.6):
lHere in this section we are not looking to determine confidence limits of LSM estimations of trend coefficients; for our purposes this problem is not crucial. Our target is values of the complex exponent of each trend.

The maximum value of the real part of this complex exponent falls on the Nenets autonomous district (0.479), Leningrad region (0.456) and Saint-Petersburg (0.429). This means that during the period under consideration, these three regions were developing most dynamically. The lowest dynamic of socioeconomic development belongs to Archangelsk region (0.079) and the Republic of Karelia (0.071).

According to the imaginary part of the complex-valued trend, one can judge the type of regional development – social, economic, or socioeconomic.

The imaginary components of the complex exponent are close to zero in the Nenets autonomous district (−0.010), Kaliningrad region (−0.014), Pskov region (−0.009), Republic of Karelia (0.013), Komi Republic (0.002), Vologda region (0.008), and Novgorod region (0.002). This means that the proportions of development of the economic and social components in these regions remain stable and one can speak of balanced socioeconomic development.

Negative values of the imaginary constituent of the complex exponent are assigned to the Leningrad region (−0.073) and Saint-Petersburg (−0.062). This means that in these regions the economic component is more active than the social one, Saint-Petersburg being the leader with its rather huge real component.

Regions with a social type of development for the period under consideration are those where the imaginary component of the complex exponent is much greater than zero. These are two northern regions – Archangelsk region (0.030) and Murmansk region (0.046).

It should be noted that instead of numerous and various parameters of socioeconomic development of a region (like per capita Gross Regional Product, per capita investments in fixed capital, per capita volume of foreign trade turnover, monthly average income, per capita number of investments, local budget deficit, number of beds in hospitals, etc.) we used only four – average per capita income, subsistence level, social services, and total turnover in region. These four parameters help to make a comparative analysis of the socioeconomic development of the regions and draw certain conclusions. Certainly, studying the rates of regional development in full detail requires additional analysis of numerous parameters characterizing the particular features of each region. Nevertheless, to form a general picture of the regional development in the country the proposed approach seems quite suitable.

However, research is aimed not so much at estimating the state of an entity as at the possibility of governing this entity to reach some optimum state. With reference to the problem of regional management this goal implies finding cause-effect relationships between the parameters showing the level of socioeconomic development and factors predetermining the regional development.

It is evident that there are many factors influencing the level of regional development. This influence can be direct or indirect, immediate or delayed, strong or weak, etc. It is impossible for a researcher to take into account all these factors, so they restrict themselves to the most important ones. Formally, correlation analysis handles this with a certain degree of success. With reference to the complex-valued parameter (9.37), correlation analysis is inapplicable as it was developed to determine the degree of approximation of the relationship between random factors to the linear one, provided the factors themselves are real variables. In Chap. 4 we proposed a specially developed apparatus of complex-valued econometrics, one part of which is the correlation analysis of random complex variables. In that chapter we obtained a formula for the pair correlation coefficient between two random complex variables:

$$r_{XY} = \frac{\sum (y_{rt} + iy_{it})(x_{rt} + ix_{it})}{\sqrt{\sum (x_{rt} + ix_{it})^2 \sum (y_{rt} + iy_{it})^2}}. \tag{9.52}$$

This coefficient is complex. Its real part, as in the case of the pair correlation coefficient of real random variables, characterizes the degree of approximation between two random complex variables to the linear form and the imaginary part is the degree of scattering of actual points with respect to linear regression dependence.

Table 9.7 Coefficients of complex correlation between socioeconomic development in regions of Russia and first complex factor

Northwest region of Russia	Per capita monthly average cash income (rubls) Per capita rubles/monthly average cash expenses (rubles)
Republic of Karelia	0.84041 + i0.02443
Komi Republic	0.98036 – i0.00815
Archangelsk region	0.96443 + i1.49141
Nenets autonomous district	0.98504 + i0.00970
Vologda region	1.02798 + i0.04992
Kaliningrad region	1.00522 – i0.00513
Leningrad region	0.99044 + i0.00703
Murmansk region	1.04353 + i0.07393
Novgorod region	0.96741 – 0.02843
Pskov region	0.96027 + i0.03651
Saint-Petersburg	0.97921 – i0.02597

While creating complex variables to describe a certain economic process it should be noted that complex variables represent a convenient form of recording two real variables. This is why one can combine in a complex variable such pairs of socioeconomic parameters that above all reflect various sides of the same phenomenon or entity but have similar scales and dimensions.

With reference to the problem of finding factors determining socioeconomic dynamics of regional development, we created several complex variables that, according to economic analysis, imply an influence on regional development:

1. Per capita monthly average cash income of the population x_{1r} and per capita monthly average cash income x_{1i} of the population $x_{1r} + ix_{1i}$;
2. Gross regional product x_{2r} and agricultural produce x_{2i} of the region $x_{2r} + ix_{2i}$;
3. Crops x_{3r} and livestock x_{3i} of the region $x_{3r} + ix_{3i}$.

The complex pair correlation coefficients between the complex parameter of socioeconomic development of each of the regions of the Northwest of Russia (Table 9.3) and the first of the three factors are given in Table 9.7.

As follows from the analysis of values of the calculated complex pair correlation coefficients, for all the regions of the Northwest of Russia, one can use the following linear complex-valued model:

$$d_t + is_t = (a_{01} + ia_{11}) + (b_{01} + ib_{11})(x_{1rt} + ix_{2it}). \qquad (9.53)$$

The Republic of Karelia is an exception. Its real part of the complex pair correlation coefficient is relatively far from one; therefore in this case it is recommended to use a nonlinear function.

The degree of influence of the second complex variable – gross regional product x_{2r} and agricultural produce x_{2i} of the region – on the complex parameter of socioeconomic development Z_t is shown in Table 9.8, which contains the results of the complex-valued correlation analysis.

Table 9.8 Coefficients of complex correlation between indicator of socioeconomic development in regions of Russia and second complex factor

Northwest region of Russia	Gross regional product, millions of rubles/agric. produce – total, millions of rubles
Republic of Karelia	0.874985 + i0.05414
Komi Republic	0.94775 – i0.00561
Archangelsk region	0.96857 + i0.017192
Nenets autonomous district	0.90402 + i0.00096
Vologda region	0.91992 + i0.00571
Kaliningrad region	0.99416 + i0.00020
Leningrad region	0.99332 + i0/00012
Murmansk region	0.96576 + i0.11550
Novgorod region	0.96684 – i0.00482
Pskov region	0.91702 + i0.02245

Some commentary on this table is warranted. First, it should be mentioned that the line for the Saint-Petersburg complex correlation coefficient is not filled because there are no livestock or crops as a branch of the regional economy.

Practically for all the regions except the Republic of Karelia one can use the linear complex-valued model:

$$d_t + is_t = (a_{02} + ia_{12}) + (b_{02} + ib_{12})(x_{2rt} + ix_{2it}). \quad (9.54)$$

However, for the Republic of Karelia, the real part of the complex pair correlation coefficient is close to one, but it is possible that the model of the complex-valued power function will be more accurate for approximation and for forecasting dynamics.

Small values of the imaginary part of the complex correlation coefficient for practically all regions reflect the small variance of the linear models. The Murmansk region is an exception. Here, the imaginary part of the complex pair correlation coefficient is 0.11550, which is much greater than in other regions. This means that the linear regression for this region will be accompanied by a higher variance in the actual values of the complex socioeconomic parameter compared to the calculated values in other regions. The multifactor relationship in this case is preferable.

The role and influence of the last of the considered complex factors on the rate of socioeconomic development of regions, i.e., crops x_{3r} and livestock x_{3i}, can be determined from Table 9.9. This factor also excludes Saint-Petersburg from consideration as a city without such a branch.

It should also be noted that for the Republic of Karelia the real part of the complex correlation coefficient is not that close to one. The relatively high value of the imaginary part of this coefficient must also mean that the influence of local agricultural produce in the republic does not prevail and this complex factor can be neglected for the purpose of building regional development models.

Table 9.9 Coefficients of complex correlation between indicator of socioeconomic development in regions of Russia and third complex factor

Northwest region of Russia	Agricultural produce (crops), millions of rubles/agric. produce (stock breeding), millions of rubles
Republic of Karelia	0.85823 − i0.10777
Komi Republic	0.99021 + i0.00773
Archangelsk region	1.05721 + i0.15713
Nenets autonomous district	0.81765 + i0.00932
Vologda region	0.93737 + i0.07753
Kaliningrad region	0.99772 + i0.01498
Leningrad region	0.99477 + i0.01988
Murmansk region	1.00994 + i0.00782
Novgorod region	0.98000 − i0.01741
Pskov region	0.95432 − i0.00634

A similar situation is observed in the Nenets autonomous district. This region gets its supply of agricultural products from other territories.

For other territories agriculture is an important part of the regional socioeconomic system; therefore, the level of their socioeconomic development can be described by a linear complex-valued model:

$$d_t + is_t = (a_{03} + ia_{13}) + (b_{03} + ib_{13})(x_{3rt} + ix_{3it}). \tag{9.55}$$

Thus, modeling the regional development of practically all the territories of Northwest Russia can involve linear complex-valued models and with all three complex factors. Since the three considered factors influence one complex socioeconomic parameter, it is advisable to use a multifactor linear model of the following type:

$$\begin{aligned} d_t + is_t = {} & (a_0 + ia_1) + (b_{01} + ib_{11})(x_{1rt} + ix_{1it}) + (b_{02} + ib_{12})(x_{2rt} + ix_{2it}) \\ & + (b_{03} + ib_{13})(x_{3rt} + ix_{3it}). \end{aligned} \tag{9.56}$$

We see that the apparatus of a complex-valued economy can indeed be used for modeling regional development.

9.5 Conclusion

In December 2004 when it was first proposed to use complex numbers as a form of representing socioeconomic variables, it was clear how to develop this proposal but difficult to predict that complex variable models would provide researchers with not so much alternative results as new ones.

Six years of hard and intense work by a team of authors made it possible to formulate the principles of the new scientific direction called "complex-valued economics." The neglect of education and science by Russian state authorities that prevailed in the country from the 1990s made it impossible for us us to focus on this problem – all those involved in complex-valued economics had to think first about how to earn a living – and only then turn to scientific research.

In these circumstances the support rendered to us by the Russian Foundation of Fundamental Research turned out to be invaluable. Under the aegis of this foundation we conducted work within the framework of Grant No. 07-06-00151, "Development of principles of economic and mathematical modeling using complex variables," which lasted for 3 years from 2007 till 2009. In addition, with the material support of the RFFR a study, "Production functions of complex variables" (Grant No. 07-06-07030-d), was issued; it was the first serious work on complex-valued economics presented to a broad scientific society.

The present study shows that it represents a completed work that provides economists with the principles of complex-valued economics – from the general idea of complex-valued economic and mathematical models to particular methods of application of complex-valued models. On the basis of the results obtained one can further develop this scientific direction by uncovering an increasing number of practical economic problems where complex-valued economics would help to deepen one's understanding of the processes and complement the arsenal of economic and mathematical methods and models with a new more "fine-tuned" research instrument.

Due to a lack of time and strength, many provisions of the study are stated briefly without delving deeper into the given problem.

Econometrics of complex variables serves as a good example. It comprises a thorough description of the application of LSM and shows that assumptions adopted in modern mathematical statistical theory about random complex variables are not true – assumptions about how fundamental characteristics such as variance, correlation moment, and covariance should be real since they characterize the measure of variability further lead to a dead end. Elimination of this assumption and laying out of a hypothesis on the possibility of the existence of complex variance and correlation moment immediately eliminate the imbalance of further econometric structures and make it possible to obtain the required statistical estimations.

Since the concept of "function minimum" does not exist in the TFCV, a certain criterion was stated when LSM estimations were obtained that made it possible to determine the required model coefficients. The first structures of this type were estimated by means of a Hessian matrix for the correspondence of the estimations to the minimum of the formed criterion. However, we would like to obtain such a conclusion in a universal form – to prove that in all cases these estimations correspond to this minimum. A lack of time and strength made it impossible for us to solve this problem completely.

We were able to state the approach and method of interval estimation of coefficients of an econometric linear complex-valued model, but we would like to extrapolate it to other statistical characteristics, for example, to the value of a complex pair correlation coefficient.

We studied quite fully the properties of complex-valued production functions – complex argument functions, complex variable functions, and even functions of several complex arguments; however, so far we have not had sufficient time or strength to include such an important factor as scientific–technical progress (innovations) for a consideration of this part of the production function theory.

New results were obtained in stock market theory; however, they also require a more thorough development and explanation. The cyclic character of the market as an objective property can be best shown using complex-variable models; however, this direction is also spotlighted quite briefly.

A new direction of scientific research was set out in the part regarding the building of two-factor models under conditions of multicollinearity; however, so far, it has not been possible to extrapolate this approach to the problem of building models with a large number of factors – the hypotheses based on this model need verification and substantiation.

This is why the materials stated in this study represent only the foundation for the models and methods of complex-valued economics – the formation of complex-valued economics as a complete and balanced subdivision of economics is yet to come.

References

1. Svetunkov IS (2011) Short-term forecasting of socio-economic processes with the use of model with correction. Bus Inf 1(5):109–112
2. Svetunkov SG, Butukhanov AV, Svetunkov IS (2006) Zapredelnye sluchai metoda Brauna v ekonomicheskom prognozirovanii. SPbGUEF Publishers, St. Petersburg (in Russian)
3. Svetunkov SG, Svetunkov IS (2011) Power production functions of complex variables. Econ Math Methods 48:67–79

Printed by Printforce, the Netherlands